# Lecture Notes on Anatomy

# Lecture Notes on Anatomy

**D.B. MOFFAT** VRD, MD, FRCS
*Emeritus Professor of Anatomy*
*University of Wales College of Cardiff*

SECOND EDITION

OXFORD

BLACKWELL SCIENTIFIC PUBLICATIONS

LONDON EDINBURGH BOSTON

MELBOURNE PARIS BERLIN VIENNA

© 1987, 1993 by
Blackwell Scientific Publications
Editorial Offices:
Osney Mead, Oxford OX2 0EL
25 John Street, London WC1N 2BL
23 Ainslie Place, Edinburgh EH3 6AJ
238 Main Street, Cambridge
  Massachusetts 02142, USA
54 University Street, Carlton
  Victoria 3053, Australia

Other Editorial Offices:
Librairie Arnette SA
2, rue Casimir-Delavigne
75006 Paris
France

Blackwell Wissenschafts-Verlag GmbH
Düsseldorfer Str. 38
D-10707 Berlin
Germany

Blackwell MZV
Feldgasse 13
A-1238 Wien
Austria

First published 1987
Second edition 1993
Four Dragons edition 1993

Set by Semantic Graphics, Singapore
Printed and bound in Great Britain
at the University Press, Cambridge

DISTRIBUTORS

  Marston Book Services Ltd
  PO Box 87
  Oxford OX2 0DT
  (*Orders*: Tel: 0865 791155
          Fax: 0865 791927
          Telex: 837515)

USA
  Blackwell Scientific Publications, Inc.
  238 Main Street
  Cambridge, MA 02142
  (*Orders*: Tel: 800 759–6102
          617 876–7000)

Canada
  Times Mirror Professional Publishing Ltd
  130 Flaska Drive
  Markham, Ontario L6G 1B8
  (*Orders*: Tel: 800 268–4178
          416 470–6739)

Australia
  Blackwell Scientific Publications Pty Ltd
  54 University Street
  Carlton, Victoria 3053
  (*Orders*: Tel: 03 347–5552)


A catalogue record for this title
is available from the British Library

ISBN 0–632–03696–6 (BSP)
ISBN 0–632–03768–7 (Four Dragons)

Library of Congress
Cataloging-in-Publication Data

Moffat, D.B. (David Burns)
    Lecture notes on anatomy/D.B. Moffat.—
    2nd ed.
       p.      cm.
    Includes bibliographical references
    and index.
    ISBN 0–632–03696–6
    1. Human anatomy.  I. Title.
    [DNLM: 1.  Anatomy.
    QS 4 M695L   1993]
    QM23.2.M64   1993
    611—dc20
    DNLM/DLC
    for Library of Congress

# Contents

# Preface to the Second Edition

For this new edition extensive revision of both text and illustrations has been undertaken and Chapter 1 has been largely rewritten to bring it up to date and to eliminate redundant material. Many of the old illustrations have been improved and 15 new figures added. In particular, more basic embryology has been included where necessary to explain the more important anomalies of development and a good deal more applied anatomy has been added. These changes should make the book of particular value to candidates for Part 1 of the FRCS examination as well as for the Second MB or its equivalent. As a past examiner for these examinations, I am well aware of the most common errors perpetrated by such candidates and these are pointed out and explained where necessary.

As in the first edition, I have excluded abstruse details of academic anatomy, such as examination candidates were once expected to memorize, and this accounts for the absence of lists of muscles' attachments and branches of this and that to aid such memorization. Keep it visual!

I hope that this new edition will continue to help students to understand, and thus remember, the structure of the human body and to realize that, while there is a good deal to learn, anatomy can be fun!

D.B. Moffat
*Cardiff*

# Preface to the First Edition

This is not a textbook of anatomy. As its name implies, it is meant to be a series of lecture notes from lectures of the proper type, i.e. they are not intended as a complete account of 'all you need to know about . . .' but rather offer explanations of those parts of human anatomy which students most often find difficult, and which therefore tend to be neglected, along with a variety of anecdotes, mnemonics and other aids to remembering the large amount of topographical detail which is needed by the student of medicine and associated subjects.

Anatomy is a visual subject. If you are explaining to a stranger how to get to the station you don't have to remember that the road is the third on the left from the church. In imagination you move along the road from the church, visualizing and counting the side turnings as you go until you reach the road to the station. Similarly in anatomy, you should never try to memorize the branches of the external carotid or the 15 origins of psoas major. This could only be done by a genius. Instead, learn to draw simple diagrams and to memorize these so that you can visualize the structures you wish to describe. This is helped, of course, by recourse to dissections and museum specimens but your own rough diagrams are better in the long run unless the dissections are of outstanding quality.

Don't try to blind your colleagues or your examiners with science. Every examiner is familiar with the type of candidate who volunteers that there is a pseudoganglion on the branch of the axillary nerve to teres minor but who forgets about the branch to deltoid. Similarly, when answering a question, always begin with the important things and leave the unimportant things to the end (or forget them). The foramen spinosum transmits the middle meningeal artery; forget the nervus spinosus, it won't be mentioned again in this book.

You will appreciate by now that these *Lecture Notes* are not intended for medal and prize winners, who will have no difficulty in coping with the more ponderous tomes and who may even know all about the nervus spinosus! It is to help the average candidate to acquire a knowledge of the basic essentials of anatomy as required by medical students and, while it possibly does not present all you need to know to pass the Second MB, or the first part of the examination for the Fellowship, it is fairly certain that it contains nothing that you do not need.

Some stress is laid on common sources of error such as confusion between the interosseous and lumbrical muscles and some repetition is used. This may annoy you, but you will remember the facts. Neuro-anatomy has been omitted, since there are numerous excellent short textbooks which include some physiology as well as anatomy but the relationships between the bones of the skull and the underlying brain are discussed.

Terminology presents a difficulty. Occasionally, at international congresses, a committee meets and makes various changes in anatomical terminology, for example the 'greater tuberosity' of the humerus has now been changed to 'greater tubercle'. When you start clinical work, however, you will find that clinicians (and indeed some senior anatomists) stick to the terminology with which they are most familiar. Clinically, too, eponymous terminology is often used so that you will hear surgeons speak of the foramen of Winslow instead of the epiploic foramen. In this book, therefore, the older terminology is sometimes given, as well as the most recent.

Diagrams, of course, play an important part in visualizing, and therefore in remembering, anatomical relationships and many of these have been taken from two other books with which I have been involved. These are *Anatomy and Physiology for Physiotherapists* and *Human Embryology*. I should like to thank Dr R.F. Mottram for permission to steal illustrations from the former and Professors F. Beck and D.P. Davies from the latter. The majority of the illustrations, however, are new and it is a pleasure to thank Dave Gardner and Clare Williams of Oxford Illustrators Ltd for the immense care and skill they have employed in preparing the diagrams from my rough sketches. I should also (publicly) like to thank my wife for typing the manuscript, no mean feat in view of my handwriting.

Finally, I must thank my friends and colleagues at Blackwell Scientific Publications, particularly Mr Per Saugman and Mr John Robson, for their part in bringing this book to fruition.

D.B. Moffat
*Cardiff*

# Chapter 1
# General Anatomy

While the greater part of this book is concerned with details of the topographical anatomy of various regions of the body, there are many general principles that can be applied to all regions. A knowledge of these is essential and this is the subject of this chapter.

## CARTILAGE

This is a connective tissue composed of *chondroblasts* and *chondrocytes* embedded in a matrix of ground substance and fibres. In hyaline cartilage the fibres are fine collagen, in fibrocartilage they are large bundles of collagen, and in elastic cartilage they are coarse elastic fibres. Cartilage is an avascular and alymphatic tissue and so, like the cornea of the eye, can be transplanted easily. With increasing age it tends to become calcified (*not* ossified, as no osteocytes or Haversian systems are present) and is then visible in X-rays. *Hyaline cartilage* is found, for example, in epiphysial plates, as costal cartilages and as articular cartilage in synovial joints. *Fibrocartilage* is tougher and is found in secondary cartilaginous joints such as the joints between vertebral bodies. The largest piece of *elastic cartilage* in the body forms the 'skeleton' of the pinna of the ear where its consistency can be appreciated. It also forms the 'skeleton' of the epiglottis.

## BONE

Bone is one of the connective tissues and has two components, *organic* and *inorganic*. The former is the connective tissue proper and comprises the bone cells, *osteocytes*, and the interstitial matrix, which itself consists of ground substance and fibres. The inorganic component consists of the 'bone salts', principally calcium, magnesium, phosphate and carbonate. Either of the two components may be removed artificially, the organic component by maceration and drying, and the inorganic by decalcification with a mineral acid or a chelating agent. The 'bones' which are used for the study of osteology are thus merely skeletons of bones, consisting only of the inorganic component. It cannot be emphasized too strongly that, *in vivo*, bone is an active, living tissue, needing an adequate blood supply and capable of changing its form and structure very rapidly in response to changes in stress and in the chemical composition of the blood.

The uptake of inorganic components into bone can be studied by a bone scan. $^{99}$Technetium ethylene diphosphonate is given intravenously and its incorporation into bone in a few hours can be studied by using a gamma camera or a whole body scanner. Local increased uptake indicates increased osteogenic activity. In various diseases, and with increasing age, bone loses some of its inorganic components and becomes less dense to X-rays. This condition of *osteoporosis* is particularly liable to occur in postmenopausal women. Pressure, for example from an adjacent tumour, can also cause loss of bone substance.

Bone, as a tissue, is found in two main forms, namely *compact* and *cancellous* bone. The former is a dense, ivory-like tissue and is found, for example, forming the shafts of the long bones. Cancellous, or spongy, bone consists of interlacing trabeculae and is found principally at the ends of long bones, in the bodies of the vertebrae and within various other bones. A third form of bone, *woven bone*, contains dense, interlacing bundles of collagen fibres and is found in embryonic bones and in healing fractures. Histologically, bone is a highly organized tissue with the osteocytes embedded in the calcified matrix forming *Haversian systems* but, in spite of their complicated appearance, these too are labile and may be broken down and re-formed quite quickly.

## The structure of long bones

A major part of the orthopaedic surgeon's work is concerned with the long bones and a knowledge of their structure, blood supply and growth is therefore of great importance. ('Long' bones are not necessarily long but are long in relation to their diameter, so the term includes the phalanges as well as such bones as the femur.) The shaft (*diaphysis*) of a long bone consists of a thick-walled tube of compact bone which thins out at the expanded ends to form a shell around a mass of cancellous bone. The arrangement of the trabeculae is the result of the patterns of stress, and their disposition may be altered when these patterns are changed. Running through the cancellous bone near the junction of the expanded end with the shaft is a thin plate of dense bone called the *epiphysial scar*. This marks the site of the original epiphysial plate of cartilage and is visible as a thin line in X-rays. It should not be confused with lines of arrested growth (Harris' lines). These are similar in appearance but are due to a period of illness during growth.

The *medullary cavity*, in the centre of the shaft, is lined by *endosteum* that can lay down new bone when necessary, by virtue of its contained osteoblasts. It contains *bone marrow*, as do the interstices of the cancellous bone at the ends. There are two types of marrow — *red* and *yellow*. The red marrow is haemopoietic and produces both red and white blood cells (*not* red cells alone — a popular error). The yellow marrow is simply fat. At birth, red marrow occupies the whole interior of the bones but after about five years most of it is gradually replaced by yellow marrow, although this can revert to red marrow again in certain diseases. In the adult, the haemopoietic marrow is only found in the vertebrae, ribs, sternum and ilium and the diploë of the skull bones and, perhaps, in the proximal ends of the femur and humerus. The sternum is a useful site for sampling the red marrow by the minor operation of *sternal puncture*.

In life, the surface of the shaft and part of the ends are covered by *periosteum*. This consists of two layers: an outer fibrous layer which acts as a limiting membrane, and a deeper vascular layer containing blood vessels and cells which can, when necessary, lay down new bone on the surface of the shaft. The fibrous layer sends bundles of collagen fibres — *Sharpey's fibres* — into the bone to anchor the periosteum and, as you would expect, these are particularly strong at the attachments of tendons or ligaments. Towards the ends of the bones the periosteum blends with the fibrous capsule of the joints in which the bone is involved. The ends of long bones are covered with the articular cartilage of the joints, although, in the case of the terminal phalanges, only one end is cartilage-covered.

One of the principal functions of long bones is to provide attachment for muscles, tendons and ligaments and an inspection of a dried bone will often give a good deal of information about these attachments. Muscles are attached to the periosteum by means of their contained connective tissue and they need a large area for their attachment since this is relatively weak. Such attachments, for example that of brachialis to the humerus, leave no mark on the bone. Tendons, ligaments and aponeuroses, on the other hand, are immensely strong and need only a small 'concentrated' area of attachment. They leave a mark on the (adult) bone in the form of a roughened tubercle or ridge (or occasionally a depression such as that formed by the femoral attachment of popliteus). Thus the very strong *iliofemoral ligament* leaves a roughened line, called the *intertrochanteric line*, on the femur while the massive tendon of *deltoid* leaves a prominent *deltoid tubercle* on the lateral side of the shaft of the humerus. On the back of the femur, where numerous muscle origins and insertions are packed closely together, the attachments are necessarily by means of fibrous tissue and the result is the *linea aspera*. On young bones, however, there are no such markings and the surface of the bone is relatively smooth. When tendons (or ligaments) are attached to the epiphysis of a bone there is a zone of fibrocartilage in the tendon at its insertion and, in this case too, the bone is smooth.

## The blood supply of bone

The blood supply of long bones (Fig. 1.1) is derived from three principal sources: the main nutrient artery, the anastomosing vessels in the vicinity of the ends of the bone and the small vessels in the vascular layer of the periosteum. The *nutrient artery*, which is accompanied by a vein, enters a *nutrient foramen* near the middle of the shaft, usually at an oblique angle directed away from the growing end (see below). Within the medullary cavity it divides and gives medullary branches to supply the bone marrow. These vessels drain into a central venous sinus. The main divisions pass towards each end of the bone, giving numerous branches which enter the deep surface of the compact cortical bone. These branches enter the Haversian canals and pass through the bone, mainly running longitudinally. They finally reach the surface, where they anastomose with the periosteal vessels. In addition to the main nutrient vessels, at each end of the bone some arteries and numerous veins enter and leave the bone through foramina of various sizes. The arteries are branches of local vessels and also are derived from the arterial anastomosis around the corresponding joint. They enter the bone on both sides of the epiphysial scar and are therefore often classified as *metaphysial* and *epiphysial* arteries. They anastomose within the bone across the epiphysial scar although in the young bone, when an epiphysial cartilage is present, they form separate systems. The epiphysial arteries end up as a series of fine loops lying just deep to the articular cartilage, some of them, in fact, entering the deepest (calcified) layer of cartilage.

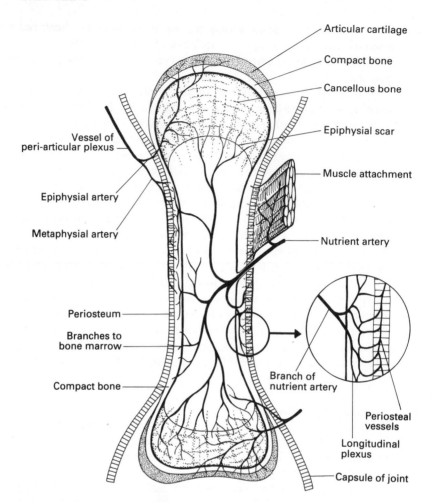

**Fig. 1.1** Blood supply of a typical long bone. See text for description.

The direction of blood flow is said to be largely centrifugal, i.e. from the branches of the nutrient artery outwards to the anastomoses with the periosteal vessels, but it is probable that the latter vessels are important in the nutrition of the outer layers of bone. Certainly in osteomyelitis, if the periosteum is stripped off the bone by subperiosteal pus and the medullary vessels are thrombosed, the area of bone affected will die and form a *sequestrum*. The main supply to the periosteal vascular plexus is by means of vessels from the locally attached muscles; for this reason, large bare areas of bone have a poor blood supply. Thus fractures of the lower third of the tibia, where no muscles are attached, often suffer from delayed union.

In bones other than long bones there is no set pattern for blood vessels, so that nutrient arteries may enter the bone at various points. The distribution of blood vessels often has a bearing on the healing of fractures. In fractures of the scaphoid, for example, delayed union is liable to occur if one of the fragments has a poor blood supply.

## Development of bone

There are two varieties of ossification, *intramembranous* and *intracartilaginous* (*enchondral*). The former is a process in which ossification occurs directly in embryonic mesenchyme and is found principally in the flat bones of the skull, the face (including the mandible) and the clavicle. The mesenchymal cells become transformed into *osteoblasts* which begin to lay down bone directly, becoming incorporated into the bone, as it develops, to form *osteocytes*. The process spreads from the primary centres of ossification until the whole bone has formed.

Enchondral ossification is more complicated and is the more important process clinically since all the limb bones (and many others) develop in this way. The process for a long bone is shown diagrammatically in Fig. 1.2. The bone is first laid down as a cartilaginous model which is approximately similar in shape to the final product. A *primary centre* of ossification then appears near the centre of the shaft and this spreads towards the ends. At the same time the periosteum lays down a collar of bone around the circumference of the shaft, this being carried out, in effect, by a process of intramembranous ossification. The primary centres of ossification begin to appear during intrauterine life, mostly during the 6th to 8th weeks. Much later, one or more secondary centres of ossification develop at each end of the cartilaginous model and these increase in size and (if more than one is present at each end) fuse with each other until a bony *epiphysis* can be recognized at each end. This is separated from the shaft, or *diaphysis*, by an *epiphysial plate*

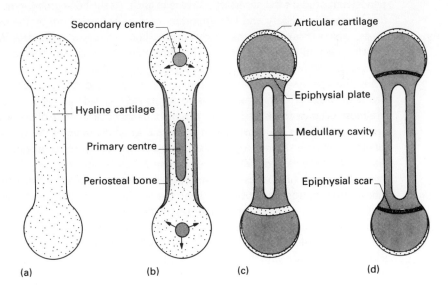

**Fig. 1.2** Ossification of a long bone. Bone is represented by the grey areas (see text for description).

of cartilage which, together with the articular cartilage at each end of the bone, is all that now remains of the original cartilaginous model. *The secondary centres of ossification all appear after birth except for that at the lower end of the femur (in the 9th intrauterine month) and possibly the upper end of the tibia.*

Before the mechanism of growth at the epiphysial cartilage can be described it is necessary to explain the process of ossification in more detail. In the first stage of ossification the cartilage cells become flattened and arranged in columns, and the matrix then becomes calcified. The cartilage cells swell up and die and this dead calcified cartilage is eroded and then removed as blood vessels grow in. *Osteoblasts* and *osteoclasts* are brought in with the vessels and the former begin to lay down true bone. The process of ossification spreads from the primary centre towards each end of the shaft and in a section through a developing bone all the stages of ossification can be seen simultaneously in the zone where the process is taking place. After ossification has been going on for some time, so that a definite long shaft can be recognized, the osteoclasts begin to remove the new bone in the centre of the shaft so that a *medullary cavity* is produced. Ossification at the secondary centres follows a similar pattern except that a true medullary cavity does not form although the interstices between the trabeculae of the cancellous bone make their appearance. When the bone is completely formed but still increasing in its dimensions, the various stages in the ossification process are found in a zone just on the diaphysial side of the epiphysial cartilage. This very vascular zone is the *metaphysis* and is particularly susceptible to infection in the immature bone.

The continuing process of ossification in the metaphysial region is accompanied by the continuing production of new cartilage in the epiphysial plate and in this way the bone is enabled to grow in length. Increasing deposits of bone by the periosteum produce the necessary increase in girth. As the bone grows, *remodelling* becomes necessary and this important process is carried out by the osteoclasts, which break down the newly formed bone wherever necessary. When growth in length is finally complete, the remaining epiphysial cartilage ceases to proliferate and becomes ossified, leaving only an epiphysial scar. In most long bones the epiphyses do not fuse simultaneously, one end fusing several years before the other. The end at which fusion occurs last is called the *growing end* and in almost all bones this is also the end at which the secondary centre of ossification appears first. Thus the growing end is the end at which *most* of the growth in length occurs, even though a great deal of growth also takes place at the other end. As has been mentioned already, the nutrient artery is directed away from the growing end and, since it passes towards the elbow and away from the knee (to the elbow, they go; from the knee, they flee), the growing end of the limb bones can easily be remembered. Disease or injury in the region of the epiphysial cartilage at the growing end is liable to produce shortening (i.e. failure of growth) of the bone. This is particularly important in the lower limb. When limbs are unequal in length, surgical interference with the epiphysis at the growing end of the longer limb may be used to remedy the inequality.

The epiphysial plate acts as a barrier to the spread of infection into the adjacent joint unless the metaphysis itself is intracapsular. This occurs at the head of the humerus, around the elbow joint, at the lower end of the ulna and at the upper and lower ends of the femur (see Fig. 1.4).

Once the epiphyses are fused, no further growth in length can occur although the periosteum can still lay down bone around the periphery of the shaft.

### Fracture healing

Immediately after a fracture there is local haemorrhage that forms a *haematoma* around the broken ends and this becomes invaded by cells derived from the blood and from the periosteum and endosteum. Osteoclasts and macrophages remove tissue debris and necrotic bone from the site and osteoblasts lay down a meshwork of collagen fibres in a glycoprotein matrix which becomes calcified to form *woven bone*. This unites the bone ends and extends under the periosteum to form an enlarged collar that encloses the fracture and the bone ends. This swelling, which is visible in X-rays, is known as *callus*. Three or four weeks later, osteoblasts in the callus and in the periosteum begin to lay down true bone that leads to bony union between the two ends and, finally, the excess of callus is removed in a process of remodelling.

## JOINTS

Joints are classified in various ways but the classification used here will be based on their structure. There are basically three types of joints, *fibrous, cartilaginous* and *synovial*. The synovial joints are by far the most important clinically and so will be described in more detail than the others.

### Fibrous joints

These are joints at which little or no movement is possible, the bones being held together by dense fibrous tissue. There are a number of different types of fibrous joints but only the two most important will be mentioned here. When there is sufficient fibrous tissue present to justify its being called an interosseous ligament, the joint is called a *syndesmosis*. An important example is the inferior tibiofibular joint. The other major type of fibrous joint is the *suture*. Sutures unite the flat bones of the skull, the fibrous pericranium being continuous with the endocranium (or fibrous layer of the dura) through the minute gap between the bones (Fig. 1.3).

### Cartilaginous joints

These are of two types, *primary* and *secondary*. In primary cartilaginous joints two or more bones are held together by a plate of hyaline cartilage (Fig. 1.3). This type of joint is also called a *synchondrosis* and, typically, with increasing age, the cartilage disappears and the two bones fuse to form a *synostosis*. Examples are the joint between an epiphysis and a diaphysis or between the occipital bone and the body of the sphenoid that is present before the age of 25.

In a secondary cartilaginous joint each bone is covered by a layer of hyaline cartilage and the two bones are united by a plate of fibrocartilage (Fig. 1.3). These joints are often called *symphyses* and by coincidence they all lie in the midline. Examples are the joints between the vertebral bodies, the manubriosternal joint and the pubic symphysis.

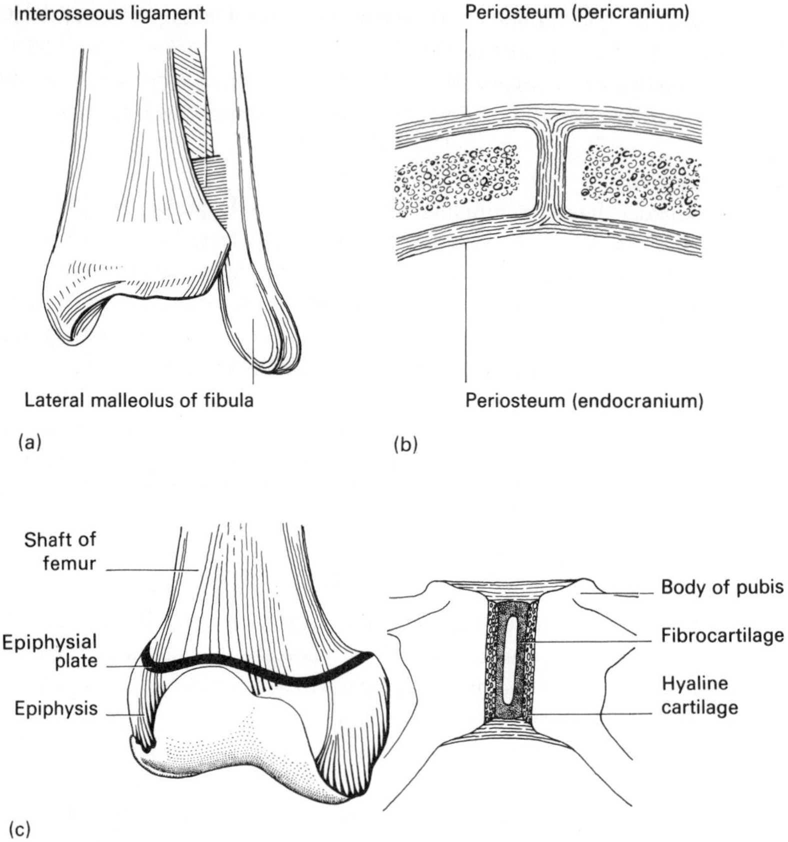

Interosseous ligament

Lateral malleolus of fibula

(a)

Periosteum (pericranium)

Periosteum (endocranium)

(b)

Shaft of femur

Epiphysial plate

Epiphysis

Body of pubis

Fibrocartilage

Hyaline cartilage

(c)

**Fig. 1.3** (a) A syndesmosis — the inferior tibiofibular joint. (b) A suture between bones of the skull. (c) Primary and secondary cartilaginous joints. The joint between the epiphysis and shaft of a young femur, and the pubic symphysis.

## Synovial joints

These are joints characterized by the presence of a synovial membrane. Most of them are freely movable although some, such as the sacro-iliac joint, display very little movement. The synovial joints are classified according to the movements that can occur at them. The most elaborate of the synovial joints are those in which movement can occur about three axes, i.e. they can produce flexion and extension, abduction and adduction and medial and lateral rotation. These are called *ball and socket joints* and they are exemplified by the hip and shoulder joints. Similar planes of movements are found in the first carpometacarpal joint, which is, however, classified as a *saddle-shaped joint*. In *ellipsoid joints*, such as the wrist and metacarpophalangeal joints, movement can occur in only two planes, viz. flexion and extension, abduction and adduction. *Hinge joints*, as their name implies, display movement in only one plane, i.e. flexion and extension. The elbow joint and the interphalangeal joints are hinge joints. Another simple type of movement is seen in *pivot joints* such as the radio-ulnar and atlanto-axial joints, in which the movement is rotational. The least complicated variety of synovial joint is the *plane* or *gliding joint*, in which only slight shifting can occur, as in the superior tibiofibular joint.

It must be mentioned that all the movements referred to above are active movements, i.e. performed by the patient. Most joints also show additional passive or *accessory movements*. These are movements which can be produced by the examiner, the patient's muscles being relaxed. For example, the wrist joint can be manipulated to display medial and lateral rotation, distraction and anteroposterior and sideways movements of the whole hand on the forearm.

### The features of a typical synovial joint

The principal components of a typical synovial joint are shown in diagrammatic fashion in Fig. 1.4 and they will now be described individually.

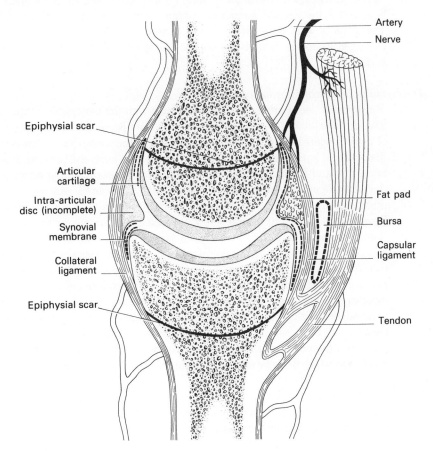

**Fig. 1.4** A diagrammatic section through a typical synovial joint. The epiphysial scar of the upper bone is intracapsular and that of the lower bone is extracapsular.

### The articular cartilage

This is a layer of hyaline cartilage (fibrocartilage in a few joints, such as the temporomandibular joint) that covers the end of each bone. It is thicker centrally on convex surfaces and thicker peripherally when the surface is concave. The articular cartilage adapts the shape of the bones more exactly to each other but the fit is still not perfect and the surfaces are very slightly non-congruent. In this way pockets of synovial fluid are pushed from one part of the joint to another and lubrication is much better than if the cartilage fitted perfectly over the entire contact area. The cartilage is white and smooth in youth but with increasing age becomes thinner and more brittle and has a more irregular surface.

The most superficial layer of the cartilage contains compressed and rather flattened cells and forms an extremely durable and relatively frictionless surface. The deepest layer is calcified and may be penetrated for a short distance by blood vessels from the underlying bone. Surprisingly, very few mitoses are seen in adult articular cartilage and it does not seem to renew itself to compensate for wear and tear. The cartilage is nourished by diffusion from blood vessels in the underlying bone and in the synovial membrane, which reaches to the edge of the cartilage. The synovial fluid itself also provides nutrition so that 'joint mice' or loose cartilaginous bodies in joints are not only living but are said actually to increase in size.

### The capsular ligament (fibrous capsule)

This forms a sleeve around the joint, blending with the periosteum of the bones that form the joint. It is thickened in places to form *ligaments* and the disposition of these is related to the movements that occur at the joint. Thus, in hinge joints, the principal ligaments are found at the sides of the joint, the capsular ligament being thin and flexible in front and behind (see, for example, the elbow joint). *Accessory ligaments* are found in certain joints. These are ligaments which are separate from the capsular ligament and they may be intra- or extracapsular. Examples of accessory ligaments are the cruciate ligaments of the knee joint (intracapsular) and the coracoclavicular ligament (extracapsular), which is an accessory ligament of the acromioclavicular joint. The function of the ligaments is to hold the bones together while allowing the appropriate movements to take place. They also help to limit movements to the normal range, the surrounding muscles also playing an important role in this activity. In most joints there is a *close-packed* position in which the ligaments are all taut and the joint surfaces are virtually fully congruent and have the maximum area of contact. The close-packed position is the position in which the joint is most firm and rigid but it is also the position in which damage to intra-articular structures is most likely to occur.

### The synovial membrane

The lining membrane of synovial joints forms from the innermost layer of cells in the developing joint but the cells never develop into synovial membrane over surfaces which are exposed to friction, so that in the fully developed joint the membrane lines the inside of the capsular ligament and covers the intra-articular parts of the bones but stops short at the edge of the articular cartilage. It also covers any intra-articular structures that are not exposed to wear and tear such as parts of intra-articular discs and also the fatty pads that fill up any dead space within a joint. In many joints the synovial membrane protrudes through a gap in the capsule and is continuous with the lining of one or more bursae around the joint. For example, the semimembranosus bursa may communicate with the knee joint and become palpable when there is an excess of synovial fluid in the joint cavity (*Baker's cyst*).

The functions of the synovial membrane are to produce synovial fluid and also to absorb the fluid when it is present in excess. The membrane contains two types of *synoviocytes*, phagocytic and secretory. They remove debris from the joint and transport it to the lymphatics and they synthesize glycosaminoglycans, particularly hyaluronate. Synovial fluid is a clear slippery fluid resembling egg-white, hence *syn-ovia*. It is a dialysate of the blood plasma and contains protein and the products of the synoviocytes. It is a highly efficient lubricant and also helps to nourish the articular cartilage.

Synovial membrane is also found in tendon sheaths and in bursae. Both are described in the next section.

### *Intra-articular discs*

Many joints contain intra-articular discs, partial or complete, which are composed of fibrocartilage or fibrous tissue. When complete, they subdivide the joint cavity into two separate joints at which different types of movement may occur, for example in the temporomandibular joint. The discs are covered by synovial membrane except over the regions that are exposed to friction.

### Blood and nerve supply of joints

The blood supply of joints (and of the bones that make up the joint) is derived from the peri-articular network of arteries that surrounds the joint. In the knee joint, for instance, the various genicular arteries, derived from major arteries in the vicinity, form a complex anastomosis. The fibrous tissue of the ligaments only requires a relatively poor supply but there is a rich plexus of vessels in the synovial membrane that also helps to supply the periphery of the articular cartilage.

*Hilton's law* states that the nerves that supply a joint are derived from nerves that supply (some of) the muscles that pass over the joint. The femoral nerve, for example, supplies quadriceps femoris. Each of the branches to the vasti (medialis, intermedius and lateralis) sends a branch to the knee joint while the branch to rectus femoris (the only member of quadriceps to cross the hip joint) sends a branch to that joint. The nerves end in various types of sensory nerve endings in the ligaments, including the capsular ligament, but the synovial membrane is poorly innervated and is relatively insensitive to pain.

## MUSCLE

There are three types of muscle, skeletal (voluntary or striped), cardiac and smooth. The latter two types are often classified as involuntary. In fact, the terms voluntary and involuntary are best avoided since some skeletal muscle is not under the control of the will (for example in the cremaster muscle and the upper part of the oesophagus) while some smooth muscle is voluntary (e.g. the ciliaris muscle in the eyeball, which is responsible for focusing the lens). The following account is restricted to skeletal muscle.

## Histology

The basic unit of skeletal muscle is the muscle fibre, which consists of only one cell but has a very large number of nuclei. Each fibre is between 10 and 100 µm in diameter but the length is very variable, some fibres being only a few millimetres long while others may be 30–40 cm in length in a long strap-shaped muscle like sartorius. Each muscle fibre is surrounded by a connective tissue sheath, the *endomysium*, while bundles, or fasciculi, of fibres are enclosed in a *perimysium*. The connective tissue which encloses the whole muscle is called the *epimysium* (compare these with the names of the connective tissue sheaths in nerves (p. 25). Finally, a muscle, with its sheath may be enclosed in its own compartment of deep fascia, alone or with other muscles with a similar function. The striations are caused by the overlapping of actin and myosin fibres but the functional histology of muscle need not concern us here. As might be expected, such a highly specialized tissue does not easily regenerate and a severely damaged muscle is repaired mainly by the formation of scar tissue.

## The structure of a typical muscle

A typical muscle is attached at each end to a bone but there are exceptions to this, for example in the facial muscles. The attachment may be directly to the periosteum and thence to the bone or by means of one or more tendons. The attachment to the bone that is usually moved by the muscle is traditionally called the *insertion* and that to the relatively fixed bone is the *origin*. These terms are misleading, however, as nearly all muscles can act in the opposite direction so that the fixed bone becomes movable and vice versa. Latissimus dorsi is a good example (p. 51). Many muscles are attached by *aponeuroses*, i.e. wide flattened tendons. This occurs particularly in the lower limb, where precise, accurate movements are not needed, whereas in the upper limb most muscles are inserted by tendons into fairly precise bony points.

### Tendons and tendon sheaths

Tendons, like ligaments, are composed of tightly packed collagen fibres with fibroblasts scattered throughout. They are extremely strong and, compared with muscle, require only a relatively limited blood supply. Therefore, wherever a muscle crosses a bone or is otherwise exposed to friction, it is replaced by a tendon or aponeurosis. Often, at these points, a bone is embedded in the tendon and such *sesamoid bones* may have one surface covered with cartilage, which articulates with a similar cartilage-covered surface on the other bone to form a synovial joint. The biggest sesamoid bone in the body is the patella. Even if no sesamoid bone is present, there is usually a *bursa* between the tendon and the bone. A bursa is a flattened sac of connective tissue lined by synovial membrane. Bursae often communicate with the cavity of neighbouring synovial joints. Bursae may also be found in the superficial tissues covering points of friction such as in the gluteal region. They are liable to become inflamed when subject to excessive pressure and to fill up with synovial fluid. Such *bursitis* is often occupational; if you work too hard you may suffer from *student's elbow* involving the bursa over the olecranon.

The blood supply of a tendon is derived mainly from the vessels of its own muscle at the proximal end. In addition the long flexor tendons of the hand and foot are supplied by vessels which travel in the *vinculae* (small folds of synovial membrane derived from the tendon sheath and containing blood vessels). The blood supply of tendons, although scanty compared with muscle, is extremely important and its maintenance is essential in all operations on tendons.

In many situations tendons are provided with *synovial sheaths* to enable them to move freely. Such a sheath consists of two layers, a visceral layer that covers the tendon itself and a parietal layer. The two layers are continuous at each end of the sheath and normally are separated only by a thin film of synovial fluid. Synovial sheaths should not be confused with *fibrous flexor sheaths*, which are the tough fibrous tunnels in the fingers and toes through which the tendons and their synovial sheaths pass. The synovial sheaths on the flexor aspects of the fingers are especially important since they are extensive, are particularly liable to trauma and infection and can give rise to severe disability as a sequel to an inadequately treated infection.

### The nerve supply of muscles

Nerves enter muscles at one or more *neurovascular hila*, which are fairly constant in position for each muscle. These hila are important to physiotherapists since they are the points at which faradic stimulation of the nerve is most effective. Although the nerve to a muscle is usually called a 'motor nerve', a large proportion of its fibres are, in fact, afferent, carrying impulses from *neuromuscular spindles* in the muscle and *Golgi nerve endings* in the tendon. The relative size of the nerve supplying a muscle depends upon the size of the *motor units* in that muscle. (A motor unit consists of an anterior horn cell, its axon and all the muscle fibres that the axon supplies.) Muscles that have precise and accurately controlled actions require correspondingly small motor units but muscles that have more generalized functions, such as those in the lower limb, have very large motor units. It is instructive to compare the relative sizes of the nerve and muscle in the lateral rectus muscle of the eye and the gluteus maximus.

### Muscle architecture

The arrangement of muscle fibres in individual muscles varies a great deal and is often dependent upon the function of the muscle concerned but here it will be sufficient to divide muscles into two types, those whose fibres run in the same direction as the line of pull and those whose fibres run more or less at an angle to the line of pull. The former group comprises mainly *fusiform triangular* or *strap-like muscles* such as sartorius, while the latter are the *pennate muscles* (Fig. 1.5). The pennate muscles are subdivided into *unipennate muscles*, in which the fibres are inserted into one side of a long tendon, *bipennate*, in which they are attached to both sides (resembling a feather, hence the name), and *multipennate*, which consist of a series of bipennate muscles fused together. Examples of these

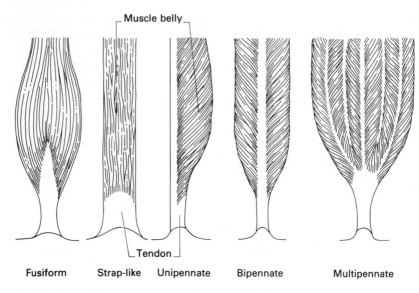

**Fig. 1.5** Some of the common arrangements of muscle fibres within a muscle.

three types of pennate muscles are, respectively, flexor policis longus, rectus femoris and the central portion of the deltoid muscle. As can be seen from Fig. 1.5, pennate muscles have relatively short fibres and hence have a relatively restricted range of movement compared with strap-like or fusiform muscles. Pennate muscles have, however, a very large number of muscle fibres inserted into their tendons and so are relatively strong. Soleus provides a good example: the most superficial part is fusiform since it is responsible for the extensive movement of plantar flexion; its deeper portion, however, is multipennate since it is a postural muscle that has to carry out the powerful but short-range adjustments necessary to keep the body upright.

## Group action of muscles

With rare exceptions, single muscles do not contract alone. Whole movements rather than individual muscles are represented on the cerebral cortex and each movement may require the contraction or relaxation of a whole group of muscles. Lifting the arm from the side, for example, requires the contraction of the abductors of the shoulder joint, simultaneous relaxation of the adductors, stabilization of the scapula by muscles passing from that bone to the trunk and a whole series of minor adjustments to posture to allow for the altered position of the centre of gravity. Muscles can therefore act in a number of different ways, not only to produce, but sometimes to prevent a movement.

### Prime movers

A muscle is said to act as a prime mover when it simply brings its origin nearer to its insertion. Thus, biceps brachii is acting as a prime mover when it produces flexion at the elbow joint.

## Antagonists

An antagonist is a muscle which is able to produce the opposite movement to a prime mover. The viscous and elastic properties of muscles acting as antagonists can also help to control the action of prime movers and, in the final stages of a movement, the contraction of the antagonist helps to bring the movement to a smooth stop. A prime mover and its antagonist may contract together to hold a joint in a fixed position and also when acting together as fixators (see next section).

## Fixators

A muscle is said to act as a fixator when it contracts to fix or stabilize a bone so that another muscle can act from that bone to produce a movement. Thus, in abduction of the little finger, flexor carpi ulnaris contracts in order to fix the pisiform so that abductor digiti mimimi can act from its origin and abduct the metacarpophalangeal joint. Much larger muscles are used to fix the scapula during movements of the upper limb and one of the best ways to demonstrate contraction of the main part of trapezius is for the patient to hold both arms out in front of him and to resist downward pressure on the arms.

## Synergists

A muscle acting as a synergist is used to prevent an unwanted movement at a joint. The long flexors of the fingers pass over the wrist joint and can flex it but this movement is disadvantageous if it is desired to clench the fist. The extensors of the wrist therefore contract as synergists to prevent wrist flexion. In radial nerve lesions, an attempt to make a fist leads to flexion of the wrist because of the loss of this synergistic action.

## Concentric and eccentric action

A muscle acting as a prime mover exhibits *concentric action* because its origin is being brought nearer to its insertion or vice versa. The antagonist to this movement will be 'paying out' and this is known as *eccentric action*, the distance between origin and insertion being increased. Many movements, however, can normally be carried out by gravity alone and in such cases a muscle that would normally be acting as antagonist to such a movement acts as an antagonist to gravity by its eccentric action. Thus, when the arm is lowered to the side, pectoralis major is relaxed because gravity is carrying out the movement but deltoid, an abductor, is contracting eccentrically.

## Insufficiency

A number of muscles that cross more than one joint are unable to function effectively in certain positions of the joints because they are too long or too short. In the former case the effect is known as *active insufficiency*. Thus flexor digitorum profundus is unable to flex fully all the joints over which it passes because it is unable to shorten sufficiently. In order to get a tight grip with the fingers, the wrist must therefore not be allowed to flex. If a joint is immobilized for a long time in such a position that a number of muscles are unduly stretched, they will suffer from active insufficiency when the limb returns to normal use.

*Passive insufficiency*, when a muscle is too short to act efficiently, is shown (in most people) by the hamstrings. These are extensors of the hip and flexors of the knee but, if the hip is fully flexed, tension in the hamstrings will prevent full extension of the knee.

### Trick movements

When certain groups of muscles are paralysed, their function can often be taken over, quite unexpectedly, by other muscles and such trick movements should be watched for during examination of neurological lesions. The tendons of flexor digitorum profundus and superficialis spread out to reach the four fingers after passing through the carpal tunnel. When the muscles contract, the fingers therefore tend to be adducted as well as flexed and, in the same way, extensor digitorum longus can act as an abductor during extension. In ulnar nerve lesions, these actions may mimic the actions of the paralysed interossei and it is therefore important that the hand should be kept flat when these muscles are being tested.

## SKIN, SUPERFICIAL FASCIA AND DEEP FASCIA

### Skin

The skin consists of two principal layers, the dermis and the epidermis.

### The dermis

The dermis is composed of dense connective tissue containing numerous nerves, blood vessels and lymphatics. It varies in thickness, being particularly thin in the eyelids and in the penis and scrotum. In places the dermis is bound down to the underlying fascia, such connections causing diminished skin mobility in these regions and also being responsible for the flexure lines of joints. Such lines do not necessarily indicate the position of the joint; for example the flexure line of the metacarpophalangeal joints is the distal palmar crease (the palmist's 'line of heart'). The creases at the bases of the fingers lie over the shafts of the phalanges.

### The epidermis

This is the non-vascular superficial layer of the skin and is a keratinized stratified squamous epithelium. The dermis projects into the epidermis in a series of roughly conical *dermal papilla* so that the dermo-epidermal junction is very irregular. The dermal papillae contain capillary loops and specialized nerve endings. The deepest layer of the epidermis is the *basal* or *Malpighian layer* (*stratum germinativum*), where the majority of mitoses are found. Superficial to this is the prickle cell layer (*stratum spinosum*), where the cells are more widely separated but linked by desmosomes. Outside the prickle cell layer is a thin layer of cells containing granules of keratohyalin (*stratum granulosum*). Beyond this again is the narrow and clear *stratum lucidum* and, finally, the *stratum corneum*, consisting of flattened dead cells. The stratum corneum varies greatly in thickness in different parts of the body, being thickest on the soles of the feet.

In the deepest layers of the epidermis are several other types of specialized cells including *melanoblasts*, which synthesize melanin. This pigment protects the deeper layers of cells, with their active mitoses, from the effects of ultraviolet light. *Langerhans* cells and other similar 'dendritic' cells play an important role in cell-mediated immunity.

## Skin appendages

Hairs, coarse or fine, are found in the skin of most parts of the body but are absent from the palms and soles, parts of the external genitalia, such as the glans penis, and the dorsal surface of the distal phalanges. Associated with the hairs (but also occurring independently in a few places) are the *sebaceous glands*; these secrete the greasy *sebum* and are particularly numerous on the scalp and face and around the main apertures of the body. Blockage of the duct may cause the gland to become distended with sebum to form a *sebaceous cyst*.

*Sweat glands* are of two types. The simple type consists of a coiled tube in the dermis which, by means of a single duct, opens on to the surface of the skin. These are found all over the body. The larger *apocrine glands* are found in the axilla, on the nipples and around the anus and external genitalia.

## Blood supply

The skin has a rich blood supply that reaches it via the superficial fascia. The vessels form a freely anastomosing plexus in the deepest part of the dermis. From this, vessels pass to sweat and sebaceous glands and the hair follicles and they also pass to the more superficial part of the dermis, where another plexus is formed. This gives rise to the capillary loops that lie in the dermal papillae. The plexuses are important in maintaining the circulation during undercutting of skin flaps but death of areas of skin may occur if the undercutting is too extensive or the flap is too thin.

Arteriovenous anastomoses are common in the dermis; in certain situations, particularly in the nail beds, they are of the coiled, richly innervated type known as *glomera* (see below).

## Nerve supply

The skin is richly supplied with nerves, with specialized types of nerve ending being found particularly in the dermal papillae. Hair follicles are surrounded by a basketwork of fine nerves so that hairy skin is much more sensitive than hairless skin, the hairs acting as levers when touched.

## Superficial fascia

The superficial fascia consists of a layer of loose connective tissue, of varying thickness, which is mostly heavily infiltrated with fat. It is the main factor responsible for the external contours of the female; the distribution of fat may be adequately studied in the paintings of Renoir.

The superficial fascia is extremely thin in the eyelids, the external ear and the penis and scrotum. In the palms and soles and in the scalp it is very dense and infiltrated with collagen fibres which bind it down to the underlying structures. The superficial fascia acts as a distributing layer in which blood vessels, lymphatics and nerves can travel before entering the dermis. It allows for mobility of the skin on the underlying structures. This is particularly important in the region of joints since loss of the superficial fascia, for example in severe burns, can lead to restriction of movement. In the palms and soles and over the ischial tuberosities, where it contains both fat and fibrous tissue in abundance, it forms an important cushioning layer since this type of tissue only requires a relatively small blood supply. It provides a depot for food storage and forms a useful insulating layer over the body; note the thickness of superficial fascia in Channel swimmers, who enhance its action by adding a thick surface layer of grease or lanolin.

## Deep fascia

The deep fascia varies in thickness, being thickest over the limbs, particularly in the lower limb. It is very thin in the thorax, and non-existent in the abdominal wall, where it is replaced by the *membranous layer of superficial fascia (Scarpa's fascia)*. In the thigh it is given the special name of *fascia lata*. Deep fascia is usually attached to any bone that it encounters. For example, the deep fascia of the

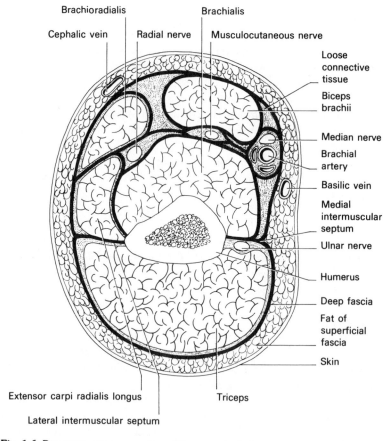

**Fig. 1.6** Diagrammatic cross-section of the upper limb above the elbow to show how the deep fascia divides the limb up into compartments.

posterior triangle of the neck and of the anterior wall of the axilla (*clavipectoral fascia*) are each attached to the clavicle, while the deep fascia of the gluteal region is attached to the iliac crest. In places, the deep fascia splits to enclose structures such as glands. The submandibular and parotid glands, for instance, are each enclosed in deep fascia which forms a rather inextensible compartment within which any acute swelling of the gland can cause a rapid increase in pressure and consequent pain.

Extensions from the investing layer of deep fascia pass deeply to form *intermuscular septa* and sheaths for neurovascular bundles such as the axillary sheath (Fig. 1.6). These septa enable groups of muscles to contract individually and to slide freely over adjacent muscle groups; hence the importance, when exploring bullet or stab wounds, of placing the patient in the position in which the injury occurred so that the sites of penetration of the various layers correspond. In the limbs, the investing layer of deep fascia separates the superficial and deep systems of veins and lymphatics, and communicating vessels pierce the deep fascia at intervals.

The layers of deep fascia are not well seen in embalmed cadavers in the dissecting room but they form very well-defined layers in life and, in particular, they have an important influence on the tracking of pus and of body fluids so that a deep-seated abscess may present at a position remote from the primary lesion. This is well seen in a psoas abscess, in which, although the psoas fascia is not particularly thick, it is nevertheless sufficiently substantial to channel the contained pus down to the inguinal region.

# BLOOD VESSELS AND LYMPHATICS

## Arteries

The arterial wall consists essentially of three basic components, namely *elastic tissue, smooth muscle* and *collagen fibres*, but the respective proportions of each may vary in different portions of the arterial tree. Each of the three components has its own particular function. The elastic tissue causes the artery to resume its original size after distension, the smooth muscle can contract to narrow or even to close the lumen and the collagen resists excessive distension.

### Elastic arteries

In the walls of the larger arteries nearest to the heart, there is a preponderance of elastic tissue, particularly in the media. During systole, part of the energy of myocardial contraction is expended in propelling the blood through the arterial tree, and part in distending the large arteries. During diastole, the elastic recoil of the arterial wall maintains the diastolic pressure.

### Muscular arteries

Smaller vessels, such as the radial or femoral arteries, are muscular arteries, in which the media consists almost entirely of smooth muscle with only a few scattered elastic and collagen fibres. The contraction of the circular, or rather, helical layers of smooth muscle fibres can be advantageous in preventing excessive haemorrhage when arteries are cleanly divided. Partially divided arteries may bleed severely, however, since contraction of the smooth muscle actually enlarges the gap in the arterial wall. Injuries to, or in the vicinity of, a muscular artery may cause severe spasm of the arterial wall so that circulatory problems may arise distal to the lesion.

### Anastomoses and end arteries

A characteristic feature of the arterial tree is the presence of frequent anastomoses between both large and small arteries. Examples are seen in the important arterial anastomosis round the scapula (p. 59), in the arcades formed between adjacent jejunal and ileal arteries in the mesentery of the small intestine and in the marginal artery around the colon. In certain situations, however, anastomoses are virtually absent so that the sudden occlusion of such an artery leads to loss of blood supply and death of the area of supply (*infarction*). Vessels of this type are known as *end arteries*. The central artery of the retina is such an end artery, as are the segmental branches of the pulmonary and renal arteries. In some other organs, such as the heart and the brain, arterial anastomoses are present but are insufficient to provide an adequate collateral circulation except when occlusion takes place slowly in a young and otherwise healthy person. Such vessels are therefore often known as *functional end arteries*.

### Arterioles

As the arterial tree is followed peripherally, the amount of collagen and elastic tissue in the walls diminishes until in the arterioles the wall consists only of one or two layers of smooth muscle cells with no elastic tissue and only very little collagen. The muscle in the arteriolar walls is important since it is here that the greatest drop in blood pressure occurs and the arterioles thus provide the major contribution to the peripheral resistance. The contraction or relaxation of the arteriolar muscle therefore plays a crucial role in the control of blood pressure.

Arterioles, like arteries, normally anastomose freely one with another. In many tissues they also form anastomoses with veins. These may be direct A–V shunts or they may take the form of more complex glomera. A *glomus* consists of a knot of arterioles whose walls contain modified smooth muscle cells (epithelioid cells). The arterioles open into one or more veins. The whole complex is surrounded by an extremely rich network of nerves. Glomus tumours, which occur particularly in the nail bed, are usually excruciatingly painful and, in some cases, blocking of both somatic and sympathetic nerves in the region may fail to control the pain.

### The capillary bed

The terminal arterioles open into the capillary bed, which is a complex network of vessels whose walls contain no smooth muscle at all. A capillary is in fact, little more than an endothelial tube so that it is ideally adapted for exchanges of fluid, electrolytes and gases between the blood and the tissue fluid. The flow of blood into the capillaries is probably controlled by *precapillary sphincters*, one or two layers of smooth muscle that surround the origin of capillaries from the arterioles, but it is also possible that the endothelial cells may play a part, as recent work has shown them to be more complicated than was once thought. The shape of the meshes of the capillary network depends on the disposition of the surrounding structures. Thus, in the renal medulla and in skeletal muscle, the network is very elongated, with most of the capillaries running longitudinally, while, in the exocrine glands, the capillaries form a basketwork around the acini.

*Sinusoids* are found in the liver lobules, spleen, bone marrow and endocrine glands. They are a little wider than capillaries, have a more irregular outline and do not have a continuous endothelial lining. They may be lined partially by phagocytic cells such as the *Kupffer cells* in the liver.

## Veins

### General features

Capillaries open into venules, which, in turn, unite to form veins. Veins have a similar structure to arteries but, since they only have to withstand a relatively low blood pressure, the walls are much thinner and only a relatively small amount of smooth muscle is present. In cases of large abnormal arteriovenous communications (*fistulae*), however, when blood enters the venous system at arterial pressure, a process of 'arterialization' soon increases the thickness of the venous wall to approximately that of arteries. Such fistulae are sometimes created deliberately in patients undergoing dialysis treatment in order to provide a large vessel for the repeated insertion of needles.

Individually, veins have a larger diameter than the corresponding artery and, in general, they have extensive anastomoses with neighbouring veins. In the kidney, for example, the segmental arteries are end arteries but the corresponding veins anastomose via the arcuate veins, a fact which was recognized by John Hunter. With certain exceptions, the larger arteries are each accompanied by a single vein with a similar name, axillary, femoral, etc., but the smaller arteries are usually accompanied by a pair of *venae comitantes* which have interconnecting cross channels at intervals. Some arteries are accompanied by a dense venous plexus, for example the vertebral and testicular arteries. The testicular venous plexus is said to form a countercurrent heat exchange system to keep the testis cool.

### Valves

The most characteristic feature of the venous system is the presence of valves that prevent backflow of blood. No valves are present in the intracranial veins, the venae cavae or the portal system but elsewhere they are a prominent feature. Their importance lies in the part they play in promoting venous return. Contraction of the surrounding muscles, enclosed in their sheaths of deep fascia, compresses the veins, whose valves ensure that the blood within them flows only towards the heart. This is particularly important in the lower limb, where the capacious venous plexus in the larger calf muscles is powerfully compressed with each step.

In the limbs there are both superficial and deep systems of veins, separated by the deep fascia. The deep fascia is, however, pierced by perforating veins ('perforators') that connect superficial and deep systems. The valves in the perforating veins are so arranged as to direct the blood from superficial to deep systems so that the muscle pump, while emptying the deep veins, also plays a part in draining blood from the superficial system.

### Portal vessels

A portal vessel (or vein) is one that has a capillary network at each end. Thus, the (hepatic) portal vein is formed by the junction of veins draining the capillaries of the intestine and the spleen. Upon reaching the liver, the portal vein branches out again, eventually to form the sinusoids of the liver. The hypophysial portal system consists of vessels that drain the hypothalamus, pass down the pituitary stalk and then break up again to supply the adenohypophysis. Other portal systems are found elsewhere in the body.

## The lymphatic system

The lymphatic system consists of a series of lymph nodes and lymphatic channels. They are found almost throughout the body but there are no lymphatics in the brain and spinal cord, in the bone marrow or in the alveoli of the lung. They are also absent from avascular tissues such as cartilage and the cornea of the eye. It is therefore possible to graft the cornea and pieces of cartilage without producing an immune response although, unfortunately, attempts to restore bone growth by transplanting epiphysial cartilage have so far been unsuccessful.

### The lymph nodes

Lymph nodes are scattered throughout the body in strategic situations, being found particularly at the roots of the limbs (axillary and inguinal regions), in the mediastinum, in the posterior abdominal wall and in the mesenteries. Each lymph node receives lymph from its own drainage area, removes particles, including bacteria, and sets up immune reactions. The nodes become enlarged in severe infections, with secondaries (*metastases*) from malignant tumours and in diseases of the lymphatic system itself. You will need to know the position of the groups of nodes and their areas of drainage for two reasons: if a patient has a severe infection or a malignant tumour, you need to know which groups of nodes to examine to check the spread of the disease, while, if a patient presents with enlarged nodes, you need to know which region to examine for the primary lesion.

### The lymph vessels

These begin as lymphatic capillaries in the tissues. Here they form a lymphatic network with blind-ended diverticula. The lymphatic capillaries drain tissue fluid and extravascular protein from the tissues and transport them to larger lymphatic vessels. In the intestine, the terminal lymphatic capillaries are the *lacteals*. These are blind-ended vessels in the villi whose contained lymph is rich in absorbed lipids.

The lymphatics, in general, follow the veins so that in the limbs they form superficial and deep systems, with some interconnections. The lymphatic vessels contain numerous valves, which have the same function as those of veins. Movement and muscle contraction are as important in promoting drainage of lymph from the tissues as they are of blood. The flow of lymph is also encouraged by massage. Like veins, lymphatics anastomose freely with neighbouring vessels and a collateral circulation can be developed after blockage of lymphatics by tumour cells. The spread of malignant tumours does not, therefore, always follow exactly the pathways of drainage described in anatomical textbooks.

## Major lymph channels

From the tissues, the lymph traverses one or more groups of peripheral lymph nodes, and eventually passes into one of the main lymph channels. In the case of the lower limb, the lymph finally drains into the *cisterna chyli*, along with lymph from the abdomen. From this thin-walled receptacle, which lies in front of the 1st and 2nd lumbar vertebrae, the *thoracic duct* passes through the aortic opening in the diaphragm and then up the anterior aspect of the thoracic spine slightly to the right of the midline. At about the level of the 4th thoracic vertebra it crosses over to the left, runs up the left border of the oesophagus and finally arches laterally across the vertebral vessels and behind the carotid sheath to open into the junction between the left internal jugular and subclavian veins. It is joined by lymphatics from the upper part of the abdomen, from the posterior mediastinum and left intercostal spaces, from the left side of the head and neck via the *jugular trunk* and from the left upper limb via the *subclavian trunk*. It may also be joined by the left *bronchomediastinal trunk*, which drains the left lung and part of the heart, but this often opens separately into the venous system.

On the right side, there may be a *right lymphatic duct*, which is formed by the junction of right jugular, subclavian and bronchomediastinal trunks, but more often these three vessels open independently into the junction between the right internal jugular and subclavian veins.

Thus the thoracic duct drains lymph from all parts of the body except the right side of the head and neck, the right upper limb and the right side of the thorax. However, the lowest part of the thoracic wall on the right side also drains indirectly into the thoracic duct.

## THE NERVOUS SYSTEM

The nervous system consists of the brain and spinal cord together with the peripheral nerves and their associated ganglia. It may be subdivided into two basic parts, the *cerebrospinal* and the *autonomic* systems, the latter being further subdivided into *sympathetic* and *parasympathetic* systems. It is sometimes also subdivided into the *central nervous system* (CNS) and the *peripheral nervous system*. The former consists of the brain and spinal cord and is entirely encased within the bony confines of the skull and vertebral column, while the latter comprises the cranial and spinal nerves and the autonomic nerves. The basic unit of the nervous system is the *neuron*, and, since the bizarre shape of some of these cells is seldom realized, the neurons will now be described in detail.

### The neuron

A neuron consists of a nerve cell together with its processes (Fig. 1.7). The cell body varies in size and may be very large. It has a large nucleus with a prominent nucleolus and various organelles, such as mitochondria, in its cytoplasm. In particular, the cell contains a great deal of granular endoplasmic reticulum which,

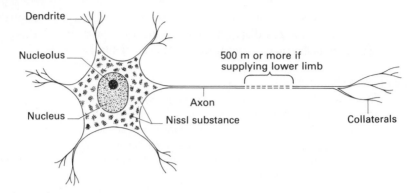

**Fig. 1.7** A nerve cell.

since it contains much RNA, stains with basic dyes such as thionin. The prominent blue granules that result from such staining are called *Nissl granule* (or Nissl substance) and their presence or absence is an important sign of the health or otherwise of the cell. The shorter processes, which convey afferent impulses (i.e. towards the cell), are called *dendrites* and the one long process, which carries efferent impulses, is the *axon*. It is important to realize that the axon may be very long indeed, much too long to be illustrated in this book. For example, the axons that pass from the lower end of the spinal cord to the muscles of the foot are over 1 m in length. Thus, in a scale diagram, if a cell body has a diameter of 100 μm, illustrated in this book with a diameter of 10 cm, the magnification would be × 1000 so that the axon would have to be drawn 1 km in length, reaching from the Royal College of Surgeons to the Old Bailey!

It is the axons, along with the various supporting cells, that form the *tracts* in the central nervous system and the *nerves* in the peripheral nervous system. They are sometimes loosely known as 'nerve fibres'. Most, but not all, axons are surrounded by a *myelin sheath*. This is a layer of lipid material that can clearly be seen by light microscopy when sections are treated with a stain for fats such as osmium tetroxide. These nerve fibres are called *myelinated fibres*, as opposed to *non-myelinated fibres* such as the grey rami communicantes in the sympathetic nervous system.

In the peripheral nervous system, the axons are embedded in a series of *Schwann cells*, which, in myelinated nerves, are wrapped around the axon in a complicated way to form the myelin sheath. The presence of Schwann cells is essential for the regeneration of a damaged nerve and, in their absence, regeneration is impossible. In the central nervous system the Schwann cells are replaced by *oligodendrocytes*, which do not have similar properties. Thus severe injury to the spinal cord causes permanent paraplegia or tetraplegia.

## The central nervous system

The central nervous system consists of the brain and spinal cord. It is made up of neurons and supporting cells; the latter are of various types and are known collectively as *neuroglia*. They correspond to connective tissue cells in the ordinary tissues. The brain develops from the cranial end of the neural tube and is subdivided embryologically into the *forebrain* (which becomes the cerebral hemispheres and the diencephalon), the *midbrain* and the *hindbrain* (which becomes the pons, the medulla and the cerebellum). The remainder of the neural tube becomes the spinal cord, which extends over the whole length of the spine in the embryo but ends at or near the lower border of the 1st lumbar vertebra in the adult. In the central nervous system, collections of nerve cells form the *grey matter*, while bundles of axons are the *white matter*. In the cerebrum and cerebellum, grey matter forms the outermost thin rind or *cortex*; isolated collections of grey matter within the brain are called *nuclei*.

## The peripheral nervous system

Each peripheral nerve consists essentially of a bundle of axons, together with their supporting tissues. The axons are mostly myelinated except at their terminations and except for most of the components of the peripheral autonomic system (see Chapter 15). Around each of the axons is a delicate sheath of connective tissue called the *endoneurium*. The axons are gathered together to form fasciculi and each fasciculus is surrounded by another connective tissue sheath, the *perineurium*. The nerve itself is composed of a number of fasciculi and is surrounded by the *epineurium*. Nerves have a good blood supply, being a metabolically active tissue, and in the larger nerves the vessels can be seen with the naked eye.

Excluding the autonomic system, the peripheral nervous system comprises the twelve cranial nerves together with eight cervical, twelve thoracic, five lumbar, five sacral and one coccygeal nerves. A little embryology helps to explain the distribution of the peripheral nervous system. Early on, the mesoderm lies between the ectoderm and the endoderm. The mesoderm then splits into two layers, one adhering to the ectoderm to form the *somatopleure* and the other adhering to the endoderm to form the *splanchnopleure*. When the embryo becomes folded, in the future trunk region the somatopleure forms the body wall and the splanchnopleure forms the gut wall, as well as all the derivatives of the gut such as the lungs, liver and pancreas. This is shown in Fig. 1.8, which also shows the neural tube dorsally and the dorsal aorta below it. Condensations of mesoderm — the mesodermal somites — lie on either side of the neural tube, while near the aorta lie the sympathetic trunks. Muscle from the somites migrates ventrally into the body wall to form the trunk musculature, bringing the spinal nerves with it. Smooth muscle in the gut, however, develops *in situ* and branches from the sympathetic trunk supply all such muscles. Thus the intercostals, abdominal muscles, skin, parietal peritoneum and pleura (which are part of the body wall) are all supplied by spinal nerves, while the muscles of the gut, biliary system, pancreatic ducts, etc., as well as the visceral pleura and peritoneum, are all supplied by the autonomic system.

## The spinal nerves

The 1st cervical nerve lies above the 1st cervical vertebra (atlas) but since there are eight cervical nerves but only seven cervical vertebrae, the 8th cervical nerve lies *below* the 7th cervical vertebra and, from here down, each nerve lies below the

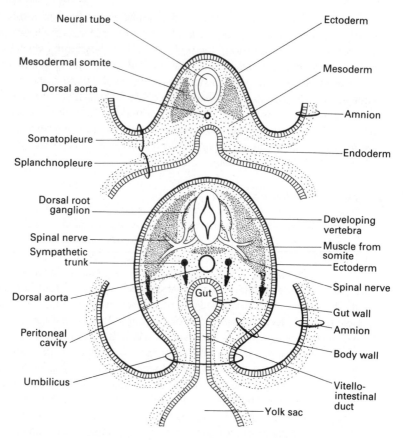

**Fig. 1.8** Transverse sections of embryos before and after folding to form the gut and body wall. See text for description.

vertebra with the corresponding number. Each spinal nerve arises from the spinal cord by two roots, *dorsal* and *ventral*. The dorsal roots are sensory and the cell bodies of these axons lie in the *dorsal root (spinal) ganglia*. Remember that a *ganglion* is a collection of cell bodies and is not, as is so often said, a 'relay station' or collection of synapses. Certainly synapses occur in the autonomic ganglia, but there are no synapses in the dorsal root ganglia, which contain *pseudo-unipolar nerve cells* whose processes pass both centrally and peripherally (Fig. 1.9). The anterior nerve roots are almost entirely motor and the cell bodies lie in the ventral column (anterior horn) of grey matter, except for the sympathetic efferent fibres, which lie in the lateral column. Spinal nerves are thus mixed nerves containing motor, sensory and sympathetic fibres.

Each spinal nerve gives rise to a *dorsal* and a *ventral ramus*. The dorsal ramus passes dorsally to supply the skin and muscles of the back. The ventral ramus communicates with the sympathetic trunk (see Chapter 15) and then, in the case of the thoracic and 1st lumbar nerves, passes ventrally into the chest wall or abdominal wall to supply the muscles (Fig. 1.9). Each nerve gives a *lateral*

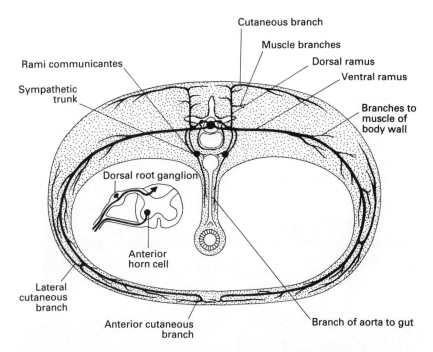

**Fig. 1.9** Schematic transverse section of the trunk to show the course of a typical spinal nerve. The diagram inserted into the peritoneal cavity is meant to show typical sensory and motor neurons.

*cutaneous* branch to supply the skin of the side of the trunk and an *anterior cutaneous* branch to supply the front. The sympathetic fibres supply the blood vessels and the sweat glands. The remaining spinal nerves are involved in supplying the head and neck and the limbs and will be dealt with in the appropriate chapters.

### The cranial nerves

The 12 pairs of cranial nerves all arise from the brain and they are, in general, very similar to the spinal nerves except for the nerves of the special senses (1, 2 and 8). A mixed cranial nerve, such as the trigeminal, has two roots, a motor and a sensory root. The motor root arises from a collection of nerve cells (a nucleus) within the brain so that this corresponds exactly to a collection of ventral horn cells. The sensory root corresponds to the dorsal root of a spinal nerve and therefore bears a ganglion. Some cranial nerves such as the hypoglossal are purely motor but any nerve that has a ganglion can be assumed to contain sensory fibres.

### The autonomic nervous system

The autonomic nervous system is so called because it is autonomous, i.e. self-regulating. It is therefore not, in general, under the control of the will. It is subdivided into *sympathetic* and *parasympathetic* systems. The detailed anatomy of these will be described in Chapter 15.

## Dermatomes

As mentioned above, the spinal nerves are mixed nerves containing motor, sensory and autonomic fibres. The thoracic and upper lumbar spinal nerves are arranged segmentally (with a good deal of overlap) and each nerve supplies sensory and motor fibres to its own segment (Fig. 1.10). Things are more complicated in the limbs, however, because the nerves that supply them form plexuses from which individual *named* nerves arise. Each of these named nerves may contain axons derived from a number of *spinal* nerves so that the area of skin supplied by one of the named nerves may be quite a different shape from the area supplied by one of the spinal nerves. This is explained in Fig. 1.11, which shows an imaginary limb innervated by three spinal nerves 1, 2 and 3. Suppose that

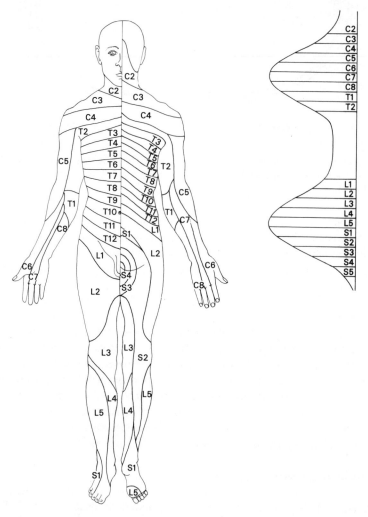

**Fig. 1.10** The dermatomes of the front and back of the body. The small diagram shows the regular arrangement of dermatomes in the upper and lower limbs of the embryo.

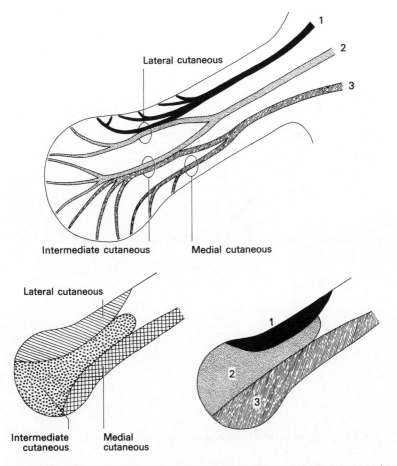

**Fig. 1.11** An imaginary limb innervated by three spinal nerves which give rise to three cutaneous nerves, two of which contain components of two spinal nerves. Note that the 'map' of the area supplied by each cutaneous nerve differs from the 'map' of the dermatomes.

branches of these spinal nerves form a plexus from which arise named nerves called the *lateral, intermediate* and *medial cutaneous nerves* of the limb, whose *root values* (i.e. the spinal nerves from which their fibres are derived) are, respectively, 1 and 2, 2 and 3, and 3. The map of the areas supplied by the three named nerves (on the left of the diagram) will be quite different from the map of the areas supplied by individual spinal nerves. The latter areas are known as *dermatomes* and it is essential for your clinical studies that you are familiar with the dermatome map shown in Fig. 1.10. It is easier to understand when you know how the rotation of the limbs that occurs during development distorts the simple segmental arrangement of the dermatomes. This is shown in Fig. 1.12 and described in Chapter 6. You will now understand why the dermatomes are arranged consecutively down the lateral side and up the medial side of the upper limb and down the medial side and up the lateral side and back of the lower limb.

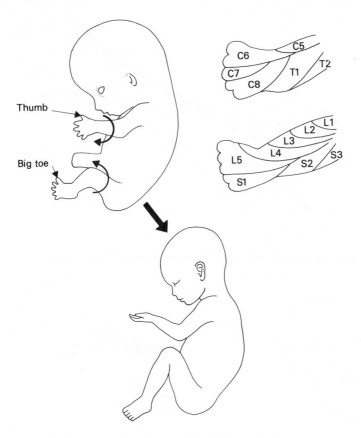

Thumb

Big toe

C5
C6
C7
C8
T1
T2

L1
L2
L3
L4
L5
S1
S2
S3

**Fig. 1.12** Diagram to show rotation of limbs during development.

The areas of skin supplied by both named nerves and spinal nerves vary slightly from one person to another and there is so much overlap that the area of skin affected by a lesion may be much less than you expect. For example, if one of the thoracic nerves is damaged there may be very little, if any, disturbance of sensation.

It must be added that there is also a segmental distribution of motor fibres to particular groups of muscles; for example, the 1st thoracic nerve supplies the small muscles of the hand. The more important of these *myotomes* will be described in the appropriate chapters.

## GLANDS

Glands are of two types, *endocrine* and *exocrine*. Endocrine glands pass their secretion directly into the bloodstream and are therefore without ducts. The secretions of many of them, such as the thyroid and the suprarenal cortex, are controlled by regulating hormones from the pituitary. The exocrine glands pass their secretions on to a surface or into a lumen and therefore have one or more ducts. Examples include the sweat glands, the salivary glands and the exocrine part of the pancreas. The embryology is easy — the gland develops as one or more outgrowths from the surface or cavity on to or into which they open. Thus the liver and gall bladder develop as two diverticula of an outgrowth from the embryonic duodenum. The pancreas often has two ducts and develops as two outgrowths from the duodenum. The prostate has 15–20 ducts and develops as 15–20 outgrowths from the future urethra. *But remember the exception to this rule —*

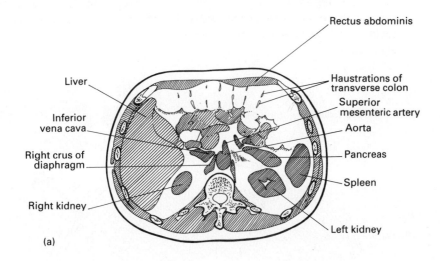

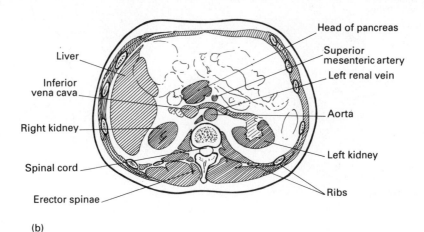

**Fig. 1.13** (a) CT scan of upper abdomen. The aorta is giving off the superior mesenteric artery behind the pancreas. (b) CT scan, two sections below. The left renal vein crosses the aorta to reach the inferior vena cava. It lies between the superior mesenteric artery and the aorta (see Fig. 10.12).

*the kidney and ureter develop as an outgrowth from the mesonephric duct, not from the bladder.* Knowledge of embryology will often help to explain anomalies of glands and their ducts, such as annular pancreas and duplex ureters.

## IMAGING METHODS

In addition to straightforward radiographs, with or without the use of contrast medium, such as are used to visualize fractured bones or the gastrointestinal tract, there are a number of methods of enhancing the image while, at the same time, reducing the dose of radiation received by the patient. In addition, in recent years, sophisticated methods have been developed to give a great deal of detailed information.

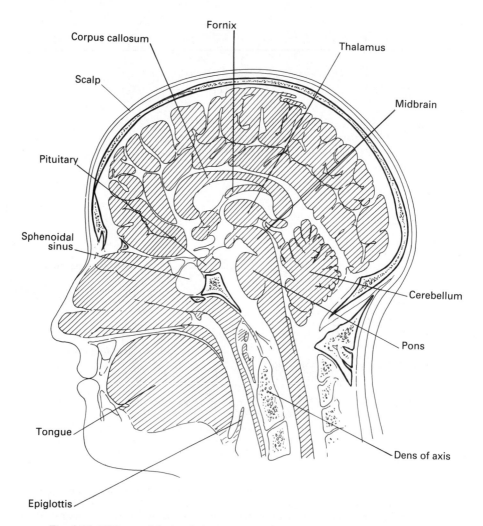

**Fig. 1.14** MRI scan of the head, shown in sagittal section.

## Computer-assisted tomography (CT scan)

In this technique, the X-ray source and the detector are opposite each other and are rotated together round the long axis of the patient's trunk (or limb). From the multiple images obtained, a computer assembles a cross-section of the body giving the appearance of a 5–10 mm slice. Additional information may be given by the use of contrast medium. When viewing such an image, remember that the slice is seen *from below* so that the liver, for example, is on the left of the picture (Fig. 1.13, p. 31).

## Magnetic resonance imaging (MRI)

The patient is placed in a magnetic field and radiofrequency waves are applied. The nuclei of atoms emit a radio signal that can be detected and assembled by a computer to form an image on a screen. The images are rather similar to those of a CT scan with the additional advantage that sagittal and coronal, as well as transverse, sections can be produced (Fig. 1.14). It has the disadvantage that it takes a long time (40–60 min), during which the patient is lying in a very claustrophobic situation.

## Ultrasonography

Ultrasonic energy is emitted from a transducer and is reflected to a varying extent by the body structures it encounters. The reflections are picked up by a detector and a two-dimensional image is formed (Fig. 1.15). The process is so rapid that multiple images can be produced in a *real-time scan* so that movements can be studied. As this is a non-invasive technique, it is used in obstetrics for the detection of fetal abnormalities such as spina bifida.

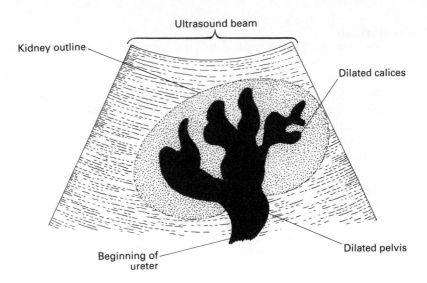

**Fig. 1.15** Ultrasound scan of the kidney. In this case there was some dilatation of the pelvis and calices due to obstruction.

# Chapter 2
# The Bones of the Upper Limb

The main function of the upper limb is to provide for manipulation of objects, both heavy, such as sledge-hammers, or light and delicate, such as pins and needles. The lower limb, on the other hand, is mainly for standing and walking. It is important to keep these separate functions in mind and to understand the corresponding differences in structure between the limbs. Thus the bones of the lower limb are large and strong, the lower limb girdle is massive and attached very firmly to the trunk, and the muscles, in spite of being powerful, are not necessarily attached to small areas of bone but are often inserted widely into extensive fascial areas. In the upper limb, mobility is of prime importance and the joints mostly have a wide range of movement. The upper limb girdle is extremely mobile and muscles tend to be inserted into rather precise bony points. It is often not realized that there is only one joint that unites the upper limb and its girdle with the trunk, i.e. the sternoclavicular joint. Thus wide movements of the upper limb are accompanied by movements of the scapula and clavicle.

## THE BONES OF THE UPPER LIMB

It is not proposed here to give formal and detailed descriptions of all the bones; this section is intended only to draw your attention to some of the salient points, although most of the names of the various parts of the bones are shown in the illustrations.

### The clavicle

Feel your own clavicle and note that it is concave forwards at its medial end and convex forwards at its lateral end. It is subcutaneous along its whole length, being crossed only by the platysma muscle and the medial, intermediate and lateral supraclavicular nerves. The under-surface is slightly grooved (Fig. 2.1) for the subclavius muscle, and the medial end is expanded to articulate (via an intra-articular disc) with the manubrium sterni. The lateral end, however, is flattened and not very strong. It is hardly visible on an X-ray. The acromioclavicular joint is thus a weak joint and the main band between the two bones is the *coracoclavicular ligament* (p. 47). It is important to understand the relationship between the clavicle and scapula. This is shown in Fig. 2.2. You will see that the outer border of the acromion forms the lateral border of this region, the end of the clavicle articulating with its *medial* surface. The conoid tubercle of the clavicle lies directly above the base of the coracoid process so that the conoid ligament is vertical.

A roughened area on the under-surface near the medial end is for the *costoclavicular* or rhomboid ligament. The clavicle almost conceals the 1st rib in

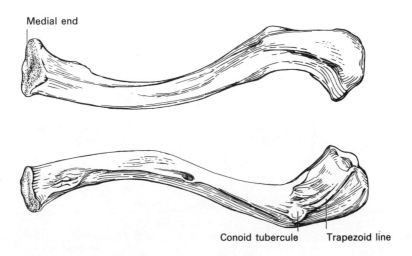

Medial end

Conoid tubercule    Trapezoid line

**Fig. 2.1** The upper and lower surfaces of the left clavicle.

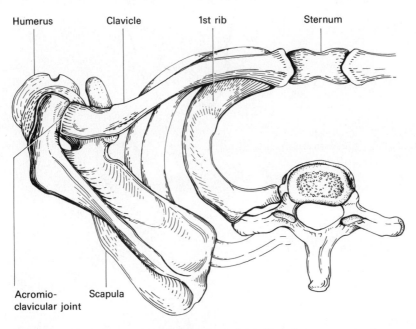

Humerus    Clavicle    1st rib    Sternum

Acromio-    Scapula
clavicular joint

**Fig. 2.2** A view of the upper limb skeleton from above. Note the narrow triangular space between clavicle, scapula and 1st rib through which nerves and vessels have to pass into the axilla.

front so that when using the ribs as landmarks it is easier to start counting from the 2nd rib, whose costal cartilage articulates with the sternum at the level of the manubriosternal joint (p. 166).

## The scapula

When the arm is by the side, this extends from the 2nd to the 7th ribs posteriorly. Most of it can only be felt indistinctly but the spine, acromion and coracoid process (Fig. 2.3) are fairly easy to feel. There are muscles (supraspinatus and infraspinatus) above and below the spine and two other

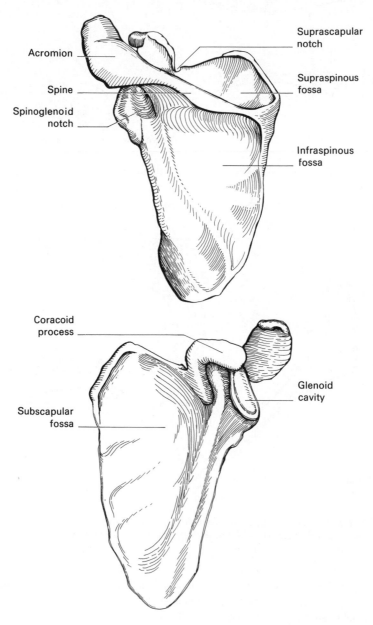

**Fig. 2.3** The back and the front of the left scapula.

muscles (trapezius and deltoid) attached to its surface so that it is not as prominent as might be expected. The coracoid process can be felt through the medial part of deltoid about one finger's-breadth below the clavicle.

The subscapular, or costal, surface (Fig. 2.3) is marked by some roughened ridges. These markings are not caused by the ribs (there are two muscles between the scapula and the ribs; see Fig. 3.12) but are the result of tendinous intersections in the subscapularis muscle, which is attached here. A smooth ridge runs down the lateral part of this surface from the glenoid fossa to the lower angle. This is a strengthening bar which serves as a lever when the strong lower part of the serratus anterior muscle rotates the scapula to turn the glenoid upwards.

### The humerus

The head and the greater and lesser tubercles (tuberosities) form a composite mass at the upper end of the humerus. Although each ossifies from a separate centre, they fuse together to form the upper epiphysis, which is separated from the diaphysis by an epiphysial cartilage up to the age of 20 years. Thus, on cursory inspection, this tripartite structure appears to be connected to the shaft by a slender neck and this is certainly a place where fractures may occur. This region is therefore called the *surgical neck*. The true *anatomical neck*, of course, is where the head joins the rest of the bone (Fig. 2.4) and is of academic interest only. The relatively very large size of the head, compared with the small and shallow glenoid fossa, is one reason for the great mobility of the shoulder joint. Another reason lies in the relatively weak and lax ligaments around the joint so that, unlike the upper end of the femur, there are no roughened ligamentous markings in this region.

The lower end of the humerus is for the attachment of muscles and for articulation with the radius and ulna. The trochlea (a 'pulley') carries the ulna and, owing to the shape of the olecranon, it extends on to the back of the bone. The radius articulates by a pivot joint with the capitulum and, since the elbow joint cannot be extended much beyond 180°, the capitulum does not extend on to the back. Note that the medial edge of the trochlea projects further than the lateral edge. This throws the ulna (and therefore the forearm) slightly laterally, thus producing the *carrying angle* between arm and forearm. Figure 2.5 explains the reason for this name and it is said to be more pronounced in the female than the male because the former has wider hips. An interesting but unlikely story! The radial, coronoid and olecranon fossae accommodate the corresponding bony prominences in full flexion or extension.

### The radius and ulna (Fig. 2.6)

These bones are considered together since they are complementary in the make-up of the forearm. The upper end of the ulna has a firm articulation with the humerus and the lower end of the radius has a firm articulation with the scaphoid and lunate in the carpus. The ligaments and interosseous membrane between the two bones therefore have to transmit the weight from radius to ulna. During pronation and supination the head of the radius rotates on the capitulum about the centre of its head but its lower end, carrying the hand with it, swings round the

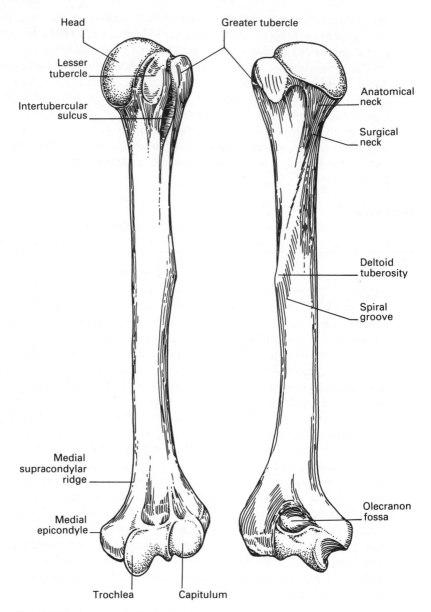

**Fig. 2.4** The front and the back of the humerus.

lower end of the ulna, its centre of rotation being the pit on the head of that bone. I have no idea why the head of the ulna is at the lower end.

### The carpus (Fig. 2.7)

It is customary to describe proximal and distal rows, each containing four bones, but this gives a false impression of the carpus. In the proximal row, the scaphoid

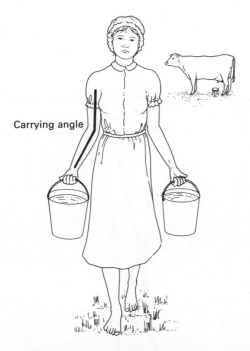

Carrying angle

**Fig. 2.5** The rural scene illustrates the carrying angle of the humerus. The flat-footed appearance is the result of not wearing shoes (p. 162). The cow is demonstrating one of the functions of serratus anterior (p. 53).

and the lunate are large and important, clinically as well as academically, the former frequently being fractured and the latter sometimes dislocated. The blood supply to the scaphoid mostly enters the distal part of the bone so that a proximal fracture may take up to 20 weeks to heal. The triquetral plays no part in the wrist joint and the pisiform is best regarded as a sesamoid bone in the tendon of flexor carpi ulnaris (p. 76). In the distal row, the bones are all important since they carry the metacarpals, the only point of possible confusion being between trapezium and trapezoid. Remember trape*zium* supports the *thumb*. There are two other important things to remember about the carpal bones. One is that they do not lie flat but are arranged to form a hollow on the palmar aspect (Fig. 2.7) so that only the pisiform, scaphoid, hook of hamate and possibly trapezium can be palpated on this surface. The other is that the bones do not lie in the wrist but in the hand, as may be appreciated if you palpate these bones in your own hand.

## The metacarpals and phalanges

The 1st metacarpal is the most important of these bones by virtue of the mobility of its carpometacarpal joint which is responsible for the movement of opposition of the thumb (p. 82). It is also an oddity since its epiphysis is at the proximal end, like the phalanges, whereas in the other metacarpals it is at the distal end. Some have, in fact, argued that it represents the proximal phalanx, the metacarpal having

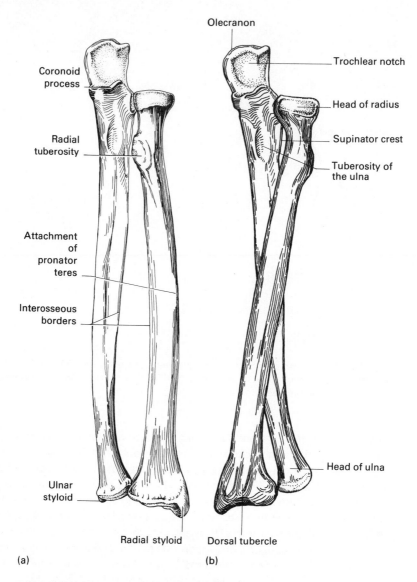

Olecranon

Coronoid process

Trochlear notch

Head of radius

Radial tuberosity

Supinator crest

Tuberosity of the ulna

Attachment of pronator teres

Interosseous borders

Head of ulna

Ulnar styloid

Radial styloid    Dorsal tubercle

(a)                                    (b)

**Fig. 2.6** The left radius and ulna seen from the front during (a) supination, and (b) pronation.

disappeared. Remembering which metacarpal bones articulate with which carpal bones (if you ever have to) is not difficult. The base of the 2nd metacarpal is large and takes on three carpal bones while the capitate is large and takes on three metacarpals (Fig. 2.8). The others fit naturally into place if you remember the uniqueness of the first carpometacarpal joint, which is quite separate from all the others.

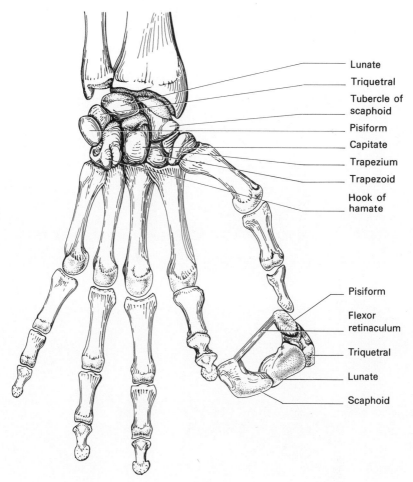

**Fig. 2.7** The skeleton of the left hand, holding a cross-section through the carpal tunnel.

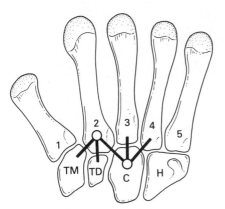

**Fig. 2.8** The 2nd metacarpal articulates with three carpal bones and the capitate with three metacarpals.

The only thing worth noting about the phalanges is the roughened ridge on each side which marks the attachment of the tough fibrous flexor sheath (p. 76). It fades out towards each end of the bone since the sheath gets thinner here to allow free movement at the joints.

# Chapter 3
# The Shoulder Region and Arm

## THE SHOULDER JOINT

This joint will be described first since it forms the 'foundation' of the region, many of the nearby muscles having the support of this otherwise weak joint as their principal function. It is, of course, a ball and socket joint, with a large ball and a small and very shallow socket. It is therefore very mobile but also rather easily dislocated. As in all synovial joints, the articular surfaces are covered with hyaline articular cartilage. The glenoid fossa is deepened slightly by a rim of fibrocartilage, the *glenoidal labrum*, but is still very shallow. The capsular ligament is fairly thin and is attached close to the margins of the articular surfaces but inferiorly its attachment moves down on to the surgical neck of the humerus so that the lax, inferior part of the capsular ligament allows free abduction at the joint (Fig. 3.1). The capsular ligament is strengthened by three not very strong *glenohumeral ligaments* anteriorly and a *coracohumeral ligament* above, and is weakened by two openings. Anterior to the joint is the tendon of *subscapularis* (Figs 3.2 & 3.6) and this is separated from the joint by the *subscapular bursa*. This communicates freely with the joint through a large defect in the capsular ligament. The other deficiency is smaller. The tendon of the long head of biceps arises from the supraglenoid tubercle of the scapula, passes through the joint (but outside the synovial membrane) and emerges at the upper end of the intertubercular sulcus, carrying with it a prolongation of the synovial membrane to form a bursa.

A much more important ligament is the *coraco-acromial ligament*, which links the acromion with the coracoid process (Fig. 3.2). The head of the humerus and the greater tubercle have to move under this ligament so, as you would expect, there is a large bursa, the *subacromial bursa* (Fig. 3.3), between them. This is sometimes known as the *secondary shoulder joint* because inflammation of this bursa can cause characteristic severe pain during part of the movement of abduction (*painful arc*) and it may also cause some restriction of movement. The reason inflammation may occur here is that in the floor of the bursa lies the tendon of supraspinatus (p. 47), which is itself liable to degenerative changes and these may involve the bursa.

The real strength of the shoulder joint lies not in the shape of the participating bones or in the ligaments, but in the four short muscles that surround the joint closely and are attached to the front, top and back of the upper end of the humerus. These muscles are subscapularis, supraspinatus, infraspinatus and teres minor (see Fig. 3.6) and they are known collectively as the *rotator cuff*. Each has its own action (p. 47) in producing movement but collectively they act as controllable ligaments, strengthening the shoulder joint when they are all contracted but allowing free movement when they are relaxed.

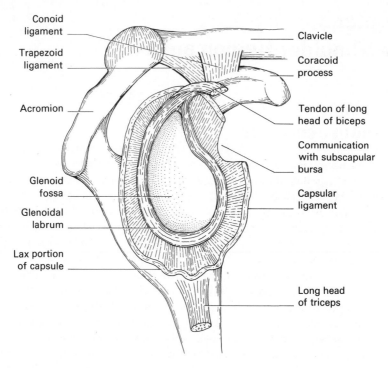

**Fig. 3.1** Lateral view of the shoulder joint after division of the capsular ligament and removal of the humerus. The coraco-acromial ligament has been omitted (see Fig. 3.2).

## Movements of the shoulder joint

The movements of the shoulder joint are those of any ball and socket joint, namely flexion and extension, abduction and adduction and medial and lateral rotation. The movements are not entirely straightforward and need a more detailed description.

### Flexion and extension

The scapula does not lie in the coronal plane but is orientated so that the glenoid fossa faces forwards as well as laterally. Therefore true flexion carries the arm forwards and across the body (Fig. 3.4) while extension is in the opposite direction.

### Abduction and adduction

The plane of the glenoid cavity indicates that true abduction carries the arm slightly forwards as well as from the midline. Abduction at the shoulder joint involves movement of the upper end of the humerus under the coraco-acromial ligament and during the movement the greater tubercle impinges upon the under-surface of the acromion and can only be released by lateral rotation of the arm (try it on an articulated skeleton). It is therefore necessary to rotate the arm

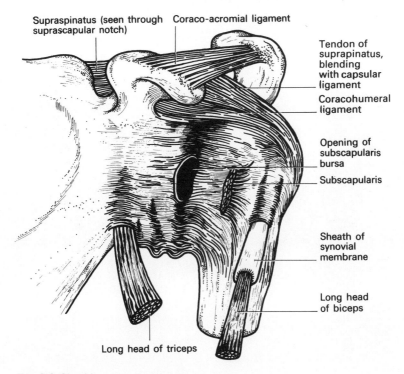

Supraspinatus (seen through suprascapular notch)

Coraco-acromial ligament

Tendon of suprapinatus, blending with capsular ligament

Coracohumeral ligament

Opening of subscapularis bursa

Subscapularis

Sheath of synovial membrane

Long head of biceps

Long head of triceps

**Fig. 3.2** Shoulder joint, anterior aspect.

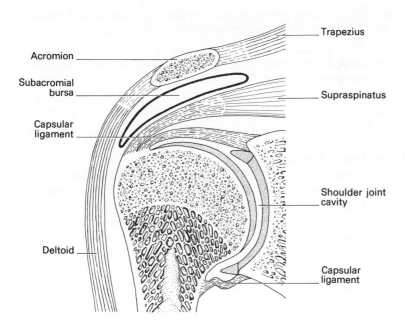

Trapezius

Acromion

Subacromial bursa

Capsular ligament

Supraspinatus

Shoulder joint cavity

Deltoid

Capsular ligament

**Fig. 3.3** A section through the shoulder joint to show the subacromial bursa.

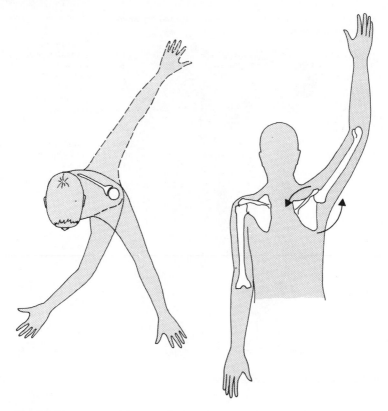

**Fig. 3.4** Movements of the upper limb as shown are due not only to movements at the glenohumeral joint but also to movements of the scapula.

laterally to attain full abduction and this must be remembered when carrying out the movement on an unconscious patient.

Full abduction of the upper limb is produced by rotation of the scapula as well as by movement at the shoulder joint, so it will be described in more detail later (p. 54).

### Medial and lateral rotation

If you attempt this with the arm straight while watching the plane of the palm of the hand you will get the impression that rotation is possible through 180°. Of course, this is because rotation at the shoulder joint is combined with pronation and supination. To estimate accurately the range of rotation you should first flex the elbow to a right angle and then imagine the hand moving round a protractor (Fig. 3.5).

## OTHER JOINTS IN THIS REGION

### The sternoclavicular joint

This is between the expanded medial end of the clavicle and the sternum. The ends

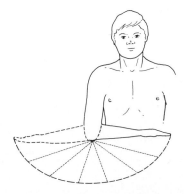

**Fig. 3.5** The correct way to estimate the amount of rotation at the shoulder joint.

of the bones are covered by fibrocartilage and there is an intra-articular disc. There is a capsular ligament with anterior and posterior thickenings. The *costoclavicular* (or rhomboid) ligament passes from the clavicle to the first costal cartilage. This forms a fulcrum so that when the outer end of the clavicle is raised, as in shrugging the shoulders, the medial end is depressed.

### The acromioclavicular joint
This joint is weak and the main bond between the two bones is the coracoclavicular ligament (see Fig. 3.1). This has two components, the *conoid* and *trapezoid* ligaments, which are attached to the conoid tubercle and trapezoid line of the clavicle respectively (see Fig. 2.1).

## THE ROTATOR CUFF MUSCLES (Fig. 3.6)
The main function of these muscles in helping to keep the head of the humerus in place has already been mentioned and it remains only to describe the attachments of the muscles and their roles in producing movements at the shoulder joint.

### Subscapularis
This is attached to most of the subscapular fossa of the scapula and, since it contains strong tendinous intersections, it leaves a number of ridges in the fossa. Note that the scapula is separated from the ribs by two muscles, *subscapularis* and *serratus anterior* (see Fig. 3.12). The thick tendon is an immediate anterior relation of the shoulder joint, being separated from it only by the large *subscapular bursa*. Its humeral attachment is to the lesser tubercle so that it is obviously a medial rotator (Fig. 3.7). The nerve supply is from upper and lower subscapular nerves.

### Supraspinatus
Arising from the supraspinous fossa, this muscle narrows down to a flattened tendon that lies in the floor of the subacromial bursa under the coraco-acromial ligament and is attached to the upper surface of the greater tubercle of the humerus, blending with the capsular ligament (Fig. 3.7). It is a weak abductor but

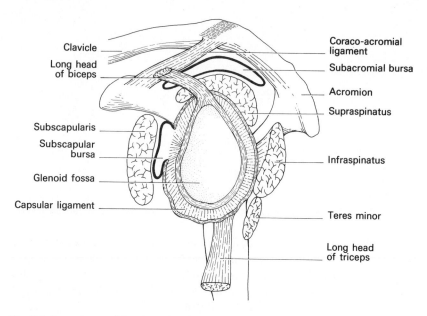

**Fig. 3.6** Lateral view of the glenoid and the muscles of the rotator cuff.

is important because it is the only muscle capable of starting the movement of abduction (p. 51). The nerve supply is from the suprascapular nerve.

### Infraspinatus

Arising from the infraspinous fossa, infraspinatus passes behind the shoulder joint to be attached to the back of the greater tubercle (Fig. 3.7). It is a lateral rotator. The nerve supply is from the suprascapular nerve after the nerve has passed through the spinoglenoid notch.

### Teres minor

Arising from the lateral margin of the scapula, the tendon of teres minor is attached to the back of the greater tubercle below infraspinatus (Fig. 3.7). It is another lateral rotator and a weak adductor. The nerve supply is from the axillary nerve by a branch which is less important than that to deltoid.

Remember that teres major passes to the *front* of the humerus but teres minor to the *back* so that the humerus is like a cigarette held between two fingers (but not the fingers of a medical student, most of whom hopefully do not smoke).

## DELTOID AND TRAPEZIUS

These two muscles are dealt with together although they have quite different actions. They may be regarded, however, as being one large continuous muscle sheet with the spine of the scapula, the acromion and the lateral third of the clavicle forming a bony outcrop between them (Figs 3.8 & 3.20). Thus, during abduction of the upper limb, trapezius elevates the tip of the shoulder, which

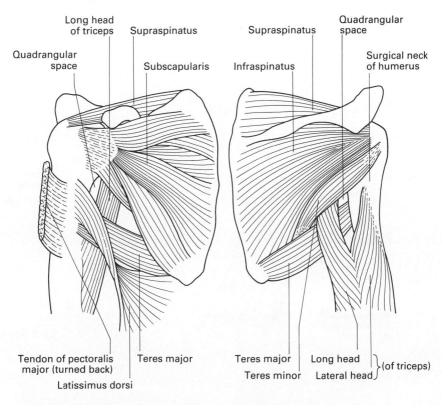

**Fig. 3.7** The quadrangular space, from in front and behind. In the posterior view, latissimus dorsi has been omitted to show the origin of teres major.

carries with it the deltoid muscle and the humerus, the latter being further elevated by deltoid.

Trapezius arises from the medial third of the superior nuchal line (directly adjacent to sternocleidomastoid), and from the spines of the cervical and thoracic vertebrae, the latter directly but the former via the ligamentum nuchae owing to the backward-facing concavity of the cervical vertebrae. It is inserted into the spine of the scapula, the lateral surface of the acromion and the lateral third of the clavicle, above the origin of deltoid (Fig. 3.8). Its nerve supply is the spinal accessory nerve. Owing to its widespread attachments, its actions are complicated. By means of its attachment to the skull, trapezius can extend the head as in looking upwards. Acting in the opposite direction, the muscle can shrug the shoulders (the standard clinical test for the integrity of the spinal accessory nerve), and the tone in the trapezius muscles is responsible for maintaining the position of the shoulders, especially when carrying heavy weights. Good tailors are aware of this and, if their client has poor muscle tone with sloping shoulders, can give the impression of health and virility by appropriate padding. Together, the two muscles (left and right) brace the shoulders back, and they are also important fixator muscles for the scapula when other muscles such as deltoid are acting on

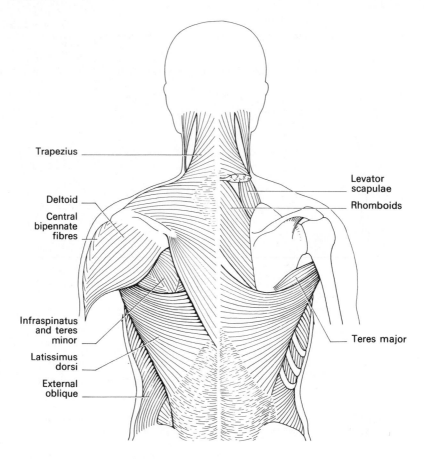

**Fig. 3.8** The muscles of the back.

the humerus. The muscle also helps in rotating the scapula during abduction of the upper limb. In Fig. 3.8 you will notice that the upper fibres will elevate the tip of the shoulder while the lower fibres depress the medial end of the scapular spine. The couple thus produced will rotate the scapula to turn the glenoid upwards (p. 54).

Deltoid arises from the scapula and clavicle immediately below the attachment of trapezius. It is inserted by means of a tough tendon into the deltoid tuberosity on the lateral surface of the humerus (Fig. 3.9). The central part of the muscle is multipennate. When viewed from the lateral side the muscle is thus triangular in shape, hence its name (Greek delta, $\Delta$). The nerve supply is from the axillary nerve which partly encircles the surgical neck of the humerus immediately deep to the muscle. The most important function of deltoid is abduction at the shoulder joint, this movement being carried out by the powerful multipennate part of the muscle that arises from the acromion. When the arm is thus abducted, the anterior and posterior fibres act as braces like the stays of a television mast. Deltoid cannot, however, initiate abduction since when the arm is by the side its central fibres are

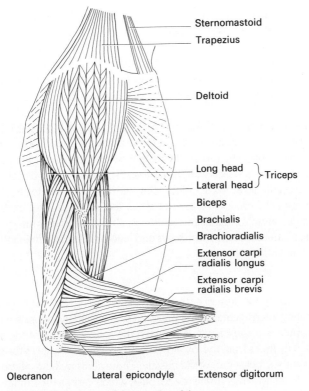

Sternomastoid
Trapezius
Deltoid
Long head ⎫
          ⎬ Triceps
Lateral head ⎭
Biceps
Brachialis
Brachioradialis
Extensor carpi
radialis longus
Extensor carpi
radialis brevis

Olecranon    Lateral epicondyle    Extensor digitorum

**Fig. 3.9** Lateral view of the muscles of the arm.

vertical so that the muscle will pull the humerus bodily upwards. The initial movement is therefore carried out by supraspinatus (p. 47). The anterior fibres acting on their own are flexors of the shoulder and posterior fibres are extensors. Deltoid acts eccentrically when lowering the arm to the side.

## LATISSIMUS DORSI

This is a very extensive flat muscle that connects the trunk to the humerus. Trapezius overlaps its origin by six spines, i.e. it arises from the lower six thoracic spines and continues downwards to be attached to the lumbar spines via the thoracolumbar fascia (p. 239). It continues even further by being attached to the medial part of the iliac crest. The fibres converge to form a very strong tendon that is inserted into the floor of the intertubercular sulcus. Its tendon is, at first, behind teres major but twists around that muscle to be inserted in front of it (Fig. 3.7). The upper border loops across the inferior angle of the scapula, holding it in a 'waistcoat pocket' (Fig. 3.8), while its lowermost fibres pass up the side of the abdomen and chest as a free border that can be nicely seen during a cough. The nerve supply is the thoracodorsal nerve (C 6, 7 and 8). Acting as a prime mover, latissimus dorsi is a powerful adductor and medial rotator of the humerus and it can extend the previously flexed arm. Working in the opposite direction, if the arm is fixed, the muscle will pull the trunk nearer to the arm and is thus an important muscle for climbers. It is also (surprisingly) a muscle that is attached to the lower limb girdle and that receives its nerve supply from the neck, so that in paraplegic patients it can be used to elevate one side of the pelvis (*hip-hitching*). It also prevents bedsores: if both hands are placed on the bed the bottom can be lifted from the bed and a more comfortable position sought.

## MUSCLES THAT ATTACH THE SCAPULA TO THE TRUNK

The only joint between the upper limb and the trunk is the sternoclavicular joint. The scapula is held in place by pieces of wire in mounted skeletons but entirely by muscles in real life. These muscles can move the scapula in various directions and thus augment movements at the shoulder joint, or they can act as fixators so that other muscles can act from a fixed base to move the humerus. One of the most important of these muscles has been mentioned already, namely trapezius. The others are *serratus anterior*, the *rhomboids* and *levator scapulae*, the first named being the one that you *must* remember.

### Serratus anterior

This arises from the upper eight ribs (the external oblique (p. 197) arises from the lower eight so they interdigitate on the middle four). It then passes as a flat sheet deep to the scapula and subscapularis to reach the medial border of the scapula (see Fig. 3.12). The fibres from the lower part of the origin converge to the inferior angle and form the strongest part of the muscle. These fibres pull the inferior angle forwards (Fig. 3.10) and thus rotate the scapula, which carries the humerus with

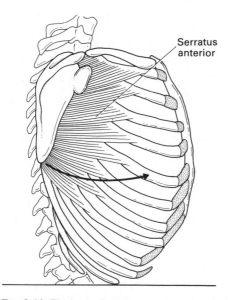

Serratus
anterior

**Fig. 3.10** The lower digitations of serratus anterior rotate the scapula by pulling the inferior angle forwards. NB The muscle passes *deep* to the scapula to be attached to its *medial* border.

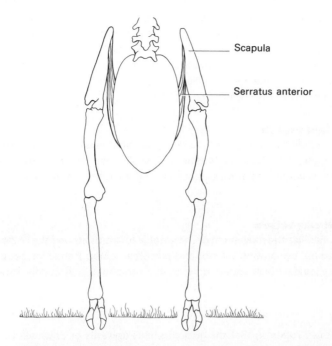

**Fig. 3.11** In quadrupeds, serratus anterior suspends the trunk between the forelimbs.

it, held in place by the rotator cuff muscles. Acting as a whole, the muscle will pull the scapula bodily forwards (*protraction*) and is used by boxers delivering a straight left. As is shown in Fig. 3.11, it also suspends the trunk between the scapulae, or, acting more powerfully, helps to increase the range of 'press-ups'. The nerve supply is the long thoracic nerve (of Bell), hence the old saying, '(C) 5 6 7, Bells of Heaven'. If the nerve is damaged, the trunk will fall forwards on that side when the weight is taken on the arms. The scapula thus appears to be projecting backwards ('winged scapula') but, in fact, the scapula is in its correct position and the trunk has fallen forwards.

### Rhomboids

These are two flat muscles that form a continuous sheet passing from the spines of some of the thoracic vertebra to the medial border of the scapula. They brace the scapula back.

### Levator scapulae

This arises from the transverse processes of the upper cervical vertebrae and is attached to the medial border of the scapula above the spine. It elevates the scapula as in shrugging the shoulders.

## MOVEMENTS OF THE SCAPULA AND CLAVICLE

The scapula is an extremely mobile bone, being held in place only by muscles. Its movements are closely associated with those of the upper limb and of the clavicle.

The chief function of the clavicle is to act as a prop, holding out the scapula, and therefore the upper limb, from the trunk. The coracoclavicular ligament is the main bond between the bones, so that when the clavicle is fractured medial to this ligament, the prop action is lost, the patient holds up his arm with the other hand and the broken ends of the bone usually overlap.

The movements of the scapula are as follows.

### Elevation and depression

The whole scapula can be elevated, carrying the lateral end of the clavicle with it, as in shrugging the shoulders. Muscles responsible are: trapezius and levator scapulae for elevation; pectoralis major and minor (plus gravity) for depression.

### Protraction and retraction

Protraction involves the movement of the scapula forwards around the chest wall and is carried out by serratus anterior and pectoralis minor. Retraction results in bracing the shoulders back and is done by the rhomboids and middle fibres of trapezius.

### Rotation

The scapula can rotate so that the glenoid points upwards or downwards. The former movement is carried out by the lower fibres of serratus anterior and the upper and lower fibres of trapezius acting as a couple (Fig. 3.10). The opposite movement is usually carried out by gravity but pectoralis minor, levator scapulae and the rhomboids can assist.

### Elevation of the upper limb

To summarize the muscles used in producing a full abduction of the upper limb, the initial stage (the first 15–20° or so) is done by supraspinatus, since deltoid is at a mechanical disadvantage. Deltoid then takes over (by means of its central multipennate fibres) and, shortly after, movement at the glenohumeral joint is supplemented by rotation of the scapula produced by the lower fibres of serratus anterior and the upper and lower fibres of trapezius.

## THE AXILLA

This is an important region because through it pass all the nerves, vessels and lymphatics of the upper limb and it also contains a number of groups of lymph nodes that drain the whole upper limb, a good deal of the trunk and, in particular, the breast. You can easily grasp the anterior and posterior walls of the axilla between your fingers and thumb and can also confirm by palpation that the medial wall consists of ribs with their associated muscles. A perpendicular line dropped from a point midway between the anterior and posterior axillary folds is called the *midaxillary line* and is an important landmark. The axilla itself is a truncated three-sided pyramid in shape (Fig. 3.12) with its apex consisting of the narrow triangular gap between the 1st rib, the clavicle and the scapula (see Fig. 2.2), through which the nerves and vessels have to pass.

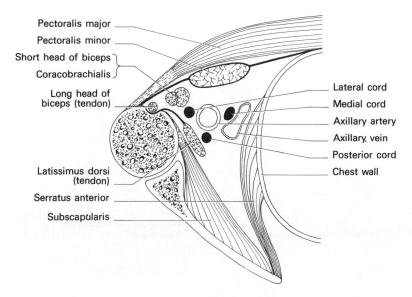

**Pectoralis major**
**Pectoralis minor**
**Short head of biceps**
**Coracobrachialis**
**Long head of biceps (tendon)**
**Latissimus dorsi (tendon)**
**Serratus anterior**
**Subscapularis**

**Lateral cord**
**Medial cord**
**Axillary artery**
**Axillary vein**
**Posterior cord**
**Chest wall**

**Fig. 3.12** A horizontal section through the axilla along the line shown in Fig. 3.13.

## The anterior wall of the axilla

This consists of two muscles, *pectoralis major* and *minor*, and a sheet of fascia called the *clavipectoral fascia* (Fig. 3.12).

### Pectoralis major

This is a large and powerful muscle upon which, in the female, most of the breast sits. It is attached by means of two heads to the front of the chest. The clavicular head arises from the medial half of the clavicle (opposite the clavicular head of sternocleidomastoid) while the sternocostal head is attached to the anterior surface of the sternum almost up to the midline and to the upper six costal cartilages. The rectus abdominis muscle is attached to the 5th, 6th and 7th costal cartilages (p. 199) and so some of pectoralis major arises from the rectus sheath. The muscle fibres converge to be inserted into the prominent lateral lip of the intertubercular sulcus. The upper (clavicular) fibres fold under the sternocostal fibres to form a thick bilaminar tendon so that the anterior wall of the axilla feels thick right up to its insertion. The nerve supply is from the medial and lateral pectoral nerves. Pectoralis major is an adductor and medial rotator of the arm and the clavicular fibres can flex and the sternocostal fibres extend the arm at the shoulder joint. Acting in the opposite direction it can pull the trunk upwards if the arms are fixed above the head as in grasping wall bars. The simplest way to make the muscle contract, which is necessary during palpation of a lump in the breast, is to ask the patient to put her hand on the 'hip' and press medially.

Since deltoid is attached to the lateral third of the clavicle and pectoralis major to the medial half, there is a small part of the clavicle (one-sixth of its length) to which neither muscle is attached and which forms the base of a triangular space

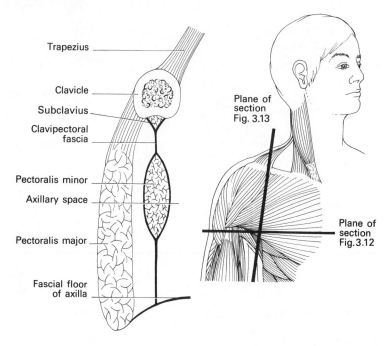

Trapezius

Clavicle

Subclavius

Clavipectoral fascia

Pectoralis minor

Axillary space

Pectoralis major

Fascial floor of axilla

Plane of section Fig. 3.13

Plane of section Fig.3.12

**Fig. 3.13** A coronal section through the anterior wall of the axilla to show its components.

(*deltopectoral triangle*) between the muscles. This is visible on the surface in thin people and it contains some lymph nodes and the termination of the cephalic vein. Its floor is formed by the clavipectoral fascia (*see* below), which is pierced by the cephalic vein (going into the axilla) and the lateral pectoral nerve and thoraco-acromial artery (coming out).

### Pectoralis minor (*see* Fig. 3.20)

This is a small muscle that passes obliquely from the anterior ribs and costal cartilages (usually 3, 4 and 5) to the coracoid process. It can bring the scapula forward around the ribs (protraction) and depress the tip of the shoulder or, at least, prevent it rising. Hence it holds the scapula in place when the weight is taken on the hands, as in leaning on a table. The nerve supply is from the medial and lateral pectoral nerves. Pectoralis minor is enclosed in the clavipectoral fascia, which extends as a continuous sheet from its upper border to the clavicle (where it encloses the unimportant subclavius (*see* also p. 317) and from its lower border to the floor of the axilla (Figs 3.13 & 17.4).

## The posterior wall of the axilla (Fig. 3.14)

This consists of *subscapularis* sitting on the scapula, *teres major* and *latissimus dorsi*. Two of these have already been described. The third is teres major.

### Teres major

This passes from the lateral border of the scapula to the medial lip of the intertubercular sulcus. The tendon of latissimus dorsi twists round its lower border,

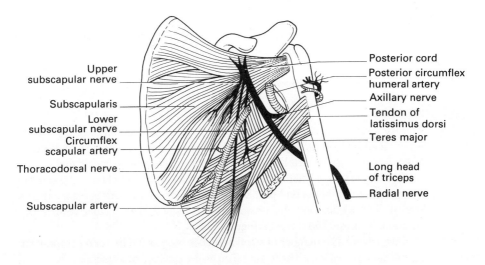

**Fig. 3.14** The muscles, nerves and vessels of the posterior wall of the axilla.

comes to lie in front of it and is inserted lateral to it. The nerve supply is the lower subscapular nerve and the muscle is an adductor and medial rotator of the humerus.

*Note*: (i) that the teres major tendon is the lowest of the three and its lower border officially marks the end of the axilla, i.e. the level at which, for example, the axillary artery becomes the brachial artery; (ii) that there is a triangular gap between subscapularis and teres major (see Fig. 3.7) through which the tendon of the long head of triceps can be seen; and (iii) that although teres major and teres minor both arise from the lateral border of the scapula, teres major passes in front of the long head of triceps and teres minor passes behind it. The former is a medial rotator and the latter a lateral rotator of the humerus. The large triangular gap mentioned in (ii) above is divided by triceps (long head) into a *quadrangular* and a *triangular space*. The boundaries of the former are subscapularis (infraspinatus posteriorly), triceps, teres major and the surgical neck of the humerus, while those of the latter are subscapularis, teres major and triceps. The quadrangular space transmits the axillary nerve and posterior circumflex humeral artery and the triangular space the circumflex scapular artery.

### The medial wall of the axilla

This consists of the upper part of serratus anterior, the upper ribs and the intercostal muscles.

### The lateral wall of the axilla

The lateral wall is almost non-existent, but as can be seen from Fig. 3.12 it is really formed by the floor of the intertubercular sulcus, to which is attached the tendon of latissimus dorsi. Running downwards are the long head of biceps (in the sulcus) and the short head of biceps and coracobrachialis, which are fused together in this region. They arise from the tip of the coracoid process.

## Contents of the axilla

The axilla contains a large amount of fat, the axillary artery and vein, the brachial plexus, and a number of important groups of lymph nodes. The axillary vessels and the brachial plexus are enclosed in a tube of fascia, *the axillary sheath*, which is a downward prolongation from the prevertebral fascia (p. 315).

### The axillary artery

This begins at the outer border of the 1st rib as a continuation of the subclavian artery. It ends by becoming the brachial artery at the lower border of teres major. It is divided into three parts by pectoralis minor, part two being behind that muscle. The axillary vein is medial to the artery and the cords of the brachial plexus are arranged around the second part of the artery, the lateral cord being lateral, the medial cord medial and the posterior cord behind. The medial cord is thus between the artery and the vein (Figs 3.15 & 3.18).

The artery has a number of small branches such as the *thoraco-acromial* and *lateral pectoral* arteries. The former pierces the clavipectoral fascia and breaks up into branches and the latter follows the lower border of pectoralis minor, giving

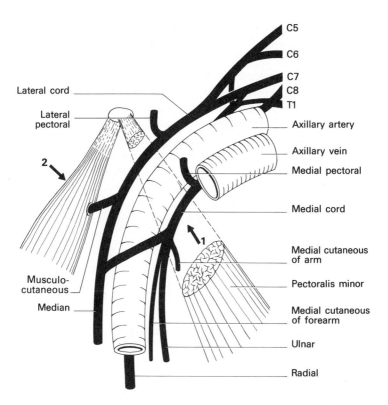

**Fig. 3.15** Diagram to show the relationships of the axillary artery and the brachial plexus. The posterior cord is hidden behind the axillary artery. The arrows 1 and 2 refer to the positions of the two sections shown in Fig. 3.18.

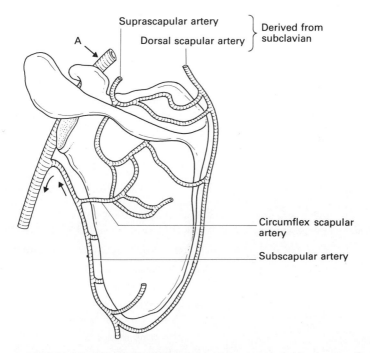

Suprascapular artery
Dorsal scapular artery } Derived from subclavian

A

Circumflex scapular artery

Subscapular artery

**Fig. 3.16** The scapular anastomosis. After an obstruction at A, blood can reach the upper limb by reversal of flow in the subscapular artery.

branches to the breast in the female. There are two larger branches, which may arise from a common stem. The *subscapular artery* follows the lateral border of the scapula, giving numerous muscular branches plus the *circumflex scapular*, which winds round to the back of the scapula (Fig. 3.16). These arteries anastomose with branches derived from the subclavian artery to form the scapular anastomosis (p. 318). The *posterior circumflex humeral* passes backwards through the quadrangular space with axillary nerve. It then winds round the surgical neck of the humerus, supplying the local muscles, principally deltoid.

### The axillary vein

This begins at the lower border of pectoralis minor as the continuation of the basilic vein (p. 64). It is joined by the venae comitantes of the brachial artery and by numerous tributaries corresponding to the branches of the axillary artery. Shortly before it becomes the subclavian vein at the outer border of the 1st rib, it receives the termination of the cephalic vein (p. 64).

## THE BRACHIAL PLEXUS

This extends from the neck into the axilla, so it will be convenient to describe the whole plexus in outline (Fig. 3.17). It arises by five *roots* from the ventral rami of C5, 6, 7 and 8 and T1. C5 and C6 unite to form the *upper trunk*, C8 and T1 unite to form the *lower trunk* and C7 continues unchanged to form the *middle trunk*.

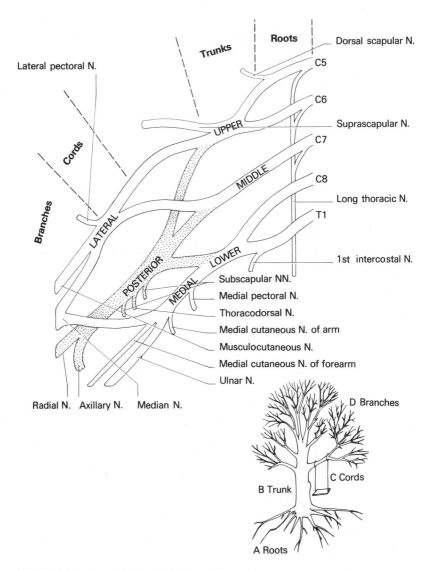

**Fig. 3.17** Diagram of the brachial plexus. The garden scene is to remind you that the order is roots, trunks, cords, branches.

The trunks lie in the posterior triangle of the neck and can be felt in the angle between sternocleidomastoid and the clavicle.

As they pass behind the clavicle, each trunk divides into anterior and posterior divisions. The three posterior divisions unite to form the *posterior cord*, the anterior divisions of the upper and middle trunks unite to form the *lateral cord* and the anterior division of the lower trunk continues as the *medial cord*. The cords are arranged around the second part of the axillary artery according to their names, i.e. the lateral cord is lateral and so on, and, below this level, each divides into a number of branches. There are also a number of supraclavicular branches which will be described in detail later. It helps in understanding the brachial plexus if you regard it as being composed of an anterior and a posterior cord, which supply skin and muscles on the front and back of the limb respectively. The anterior cord is subdivided into medial and lateral cords, which therefore, between them, supply skin and muscles on the front. The exceptions to this general rule are the *brachioradialis muscle* (p. 78) and the *skin of the backs of the finger tips* (p. 96). If you find it difficult to remember which are the roots, trunks, cords, etc., the diagram in Fig. 3.17 will help.

## Supraclavicular branches

The supraclavicular branches of the plexus arise directly from the roots or trunks. They are:

dorsal scapular (to the rhomboids)
long thoracic (to serratus anterior)
suprascapular (to supra- and infraspinatus)
branches to the scalene muscles.

These nerves will be described in Chapter 17.

The cords are lateral, medial and posterior to the second part of the axillary artery so their branches start off with a similar relationship (Fig. 3.18) but soon travel off in various directions.

## Lateral cord

### Lateral pectoral nerve

This passes forwards to enter pectoralis major and also sends a communicating loop to the medial pectoral nerve, through which it supplies pectoralis minor.

### Lateral root of the median nerve

The median nerve arises by medial and lateral roots, one from the lateral cord and one from the medial cord. The medial root crosses the axillary artery to join the lateral root so the median nerve itself is, at first, *lateral* to the artery.

### Musculocutaneous nerve

This passes from the lateral cord laterally to enter coracobrachialis, which it supplies, and then lies between biceps and brachialis, both of which it supplies (therefore known as the BBC nerve). It ends by becoming the lateral cutaneous nerve of the forearm.

## Medial cord

### Medial pectoral nerve

This passes forwards to pierce pectoralis minor, which it supplies, and then continues into pectoralis major. Thus *both* pectoral muscles are supplied by *both* pectoral nerves.

### Medial root of median nerve

See above.

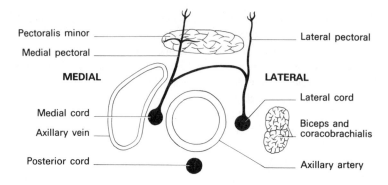

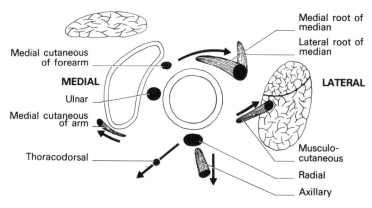

**Fig. 3.18** The axillary artery and brachial plexus of the *left* axilla, seen from below. The upper figure shows the second part of the axillary artery and the lower shows the third part.

### Ulnar nerve
This is a large branch that continues down the arm, lying at first between the axillary artery and vein, rather posteriorly.

### Medial cutaneous nerve of forearm
This is a surprisingly large nerve that also lies between artery and vein, in front of the ulnar nerve.

### Medial cutaneous nerve of arm
This is small and unimportant.

## Posterior cord

### Radial nerve
This passes down behind the axillary artery, travelling posteriorly and laterally. It gives branches to the long and medial heads of triceps *while still in the axilla.*

*Axillary nerve*

This and the radial nerve are the two largest branches of the posterior cord. It is inappropriately named because the first thing it does is to quit the axilla by passing backwards through the quadrangular space with the posterior circumflex humeral artery (see Fig. 3.14). It then winds round the surgical neck of the humerus, deep to deltoid, which it supplies (see Fig. 3.23). It also supplies teres minor and gives off the *upper lateral cutaneous nerve of the arm*, which supplies a small area of skin over the posterior border of deltoid. This small nerve is clinically important. The axillary nerve is vulnerable in fractures of the surgical neck and in dislocation of the shoulder joint (it is an inferior relation of the capsular ligament). In these cases it is obviously impossible to test deltoid by abducting the arm because of the severe pain, but the presence of a small area of lost or diminished sensation over the edge of deltoid will enable a diagnosis to be made.

*Upper and lower subscapular nerve*

These small nerves lie on, and supply, the posterior wall of the axilla, along with the thoracodorsal nerve, which arises between them (Fig. 3.14). The upper supplies subscapularis and the lower supplies subscapularis and then continues across its lower border to supply teres major.

*Thoracodorsal nerve*

This stays on the posterior wall of the axilla, runs down on latissimus dorsi and supplies it (p. 52).

## Two other nerves in the axilla

These are nerves that arise outside the axilla and then enter it.

*Intercostobrachial nerve*

This is a lateral branch of the second intercostal nerve that joins the medial cutaneous nerve of the arm and helps it to supply a small area of skin.

*Long thoracic nerve (of Bell)*

This has been mentioned already as a supraclavicular branch of the brachial plexus. It passes behind the neurovascular bundle in the axilla and then descends on the medial wall, which is formed by serratus anterior. It supplies this muscle only.

## Lymph nodes in the axilla

As well as the structures mentioned above, there are important groups of lymph nodes in the axilla. They are arranged in five main groups although they are not well demarcated one from another.

### Lateral group

These are arranged in the lateral part of the axilla along the course of the axillary vein. They drain almost the whole of the upper limb and their efferents pass to the *central* and *apical* groups.

**Anterior (or pectoral) group**

These lie deep to pectoralis major, along the lower border of pectoralis minor. They drain the front and side of the chest and abdomen, including part of the breast. Their efferents pass to the *central* and *apical* groups.

**Posterior (or subscapular) group**

These lie on the posterior wall of the axilla. Their area of drainage is, perhaps, unexpected but can best be visualized by thinking of the trapezius and latissimus dorsi muscles, i.e. they drain the back of the neck, and the whole of the back *down as far as the iliac crest*.

**Central group**

These are situated in the central part of the floor of the axilla. They receive afferents from the three groups mentioned already and their efferents pass to the *apical group*.

**Apical group**

These lie near the axillary vessels at the apex of the axilla. They receive afferents from all the other groups so that their efferents to the subclavian trunk (p. 23) represent the 'final common path' for the lymphatics of the upper limb.

Also communicating with the apical group are the efferents from a small number of nodes in the deltopectoral triangle (*infraclavicular nodes*).

## LYMPHATIC AND VENOUS DRAINAGE OF THE UPPER LIMB

Since the lymphatics and veins of the upper limb all end up in the axillary region, this will be an appropriate place to describe them.

### Venous drainage

The veins of the upper limb form *superficial* and *deep* systems. The deep veins consist of venae comitantes that accompany the arteries. They eventually drain into the brachial veins (or venae comitantes) and these, in turn, join the basilic vein to form the axillary vein. There are many communications between the deep and superficial veins. The superficial system comprises the *basilic* and the *cephalic* veins. The former arises from the venous network on the dorsum of the hand, runs up the medial side and then the front of the upper limb and finally pierces the deep fascia about halfway up the arm and joins the axillary vein. The cephalic vein begins from the dorsal venous network over the anatomical snuffbox (p. 90) and passes up the lateral side of the forearm, inclines to the front, travels up the groove between deltoid and pectoralis major and finally pierces the clavipectoral fascia to open into the axillary vein. The two superficial veins are joined by a *median cubital vein* over the cubital fossa and this is usually joined to the deep veins by a communicating branch.

### Lymphatic drainage

There are superficial and deep sets of lymph vessels in both upper and lower limbs.

In the superficial tissues of the upper limb, the fingers and the distal part of the palm drain into vessels on the dorsum of the hand. These pass up into the forearm, passing around the medial and lateral borders of the forearm to reach the front. On the front, the lymphatics tend to follow the superficial veins. Thus, from the proximal part of the palm, vessels ascend into the forearm where they follow the cephalic and basilic veins. The former group of vessels mostly enter the lateral axillary group of nodes but a few enter the infraclavicular group and thence to the apical group. The vessels following the basilic vein may be interrupted by some small *supratrochlear nodes* above the medial epicondyle but they mostly continue up to join the lateral axillary nodes.

The deep system, which communicates with the superficial system in places, follows the main vessels of the limb, radial, ulnar, brachial, etc., and ends up in the lateral group of nodes.

## THE BREAST

The male breast is rudimentary, although primitive ducts are present and carcinoma can occur.

The female breast is made up of fat, fibrous tissue and glandular tissue, fat predominating in the non-lactating breast. It rests mainly on pectoralis major, but overlaps the edge of this muscle to lie on serratus anterior and the external oblique aponeurosis. There is an extension superolaterally towards the axilla — the *axillary tail* (of Spence). The breast is separated from the muscles by a layer of loose areolar tissue — the *retromammary space* — which can be used for the insertion of a prosthesis via an incision in the crease below the breast. Advanced carcinoma of the breast may infiltrate this space and invade the muscle so that the tumour becomes fixed. On the surface, the nipple is surrounded by the *areola*, which contains sebaceous *areolar glands*. These become enlarged in pregnancy to form *Montgomery's tubercles*.

Fibrous tissue septa separate the lobules of the glandular tissue and these are attached to the skin and the nipple. They are well developed in the upper part of the breast, where they form the *suspensory ligaments* (of Astley Cooper). Carcinoma involving the ligaments may cause them to shrink, producing characteristic dimpling of the skin. The glandular tissue itself consists mainly of ducts until puberty, when the ends branch out to become potential alveoli; these do not become fully formed until pregnancy occurs. The glands are arranged to form 15–20 lobes, each of which drains into a *lactiferous duct*. These converge towards the nipple and each becomes dilated to form a *lactiferous sinus* beneath the areola. The ducts then become narrower and turn up to open on to the surface of the nipple. Incisions into the breast are made radially to avoid cutting across the line of the ducts. A blocked duct may become dilated during lactation to form a *galactocoele*.

The blood supply of the breast is from the branches of the axillary artery (especially the lateral thoracic) and from the intercostal and internal thoracic arteries.

The lymphatic drainage is important in relation to the spread of carcinoma of the breast (Fig. 3.19). The plexus of lymphatics around the lobules of the gland

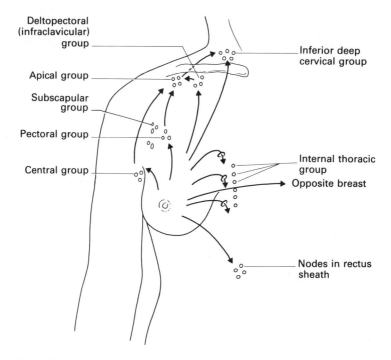

**Fig. 3.19** Lymphatic drainage of the breast.

communicates with the skin plexus, which is particularly dense around the nipple (the *subareolar plexus*). Drainage is mainly to the pectoral, subscapular and apical groups of axillary nodes and perhaps to nodes in the infraclavicular fossa. From the medial side of the breast lymphatics pass to parasternal lymph nodes along the internal thoracic vessels (although these are rarely affected in the absence of axillary node enlargement) to the opposite breast, and some may pass downwards to lymph nodes within the rectus sheath. By far the most important groups, however, are those in the axilla.

## THE ARM

The muscles of the arm (Fig. 3.20) are disposed around the humerus, brachialis in front and triceps behind being attached to very large areas of the shaft.

### Brachialis

The muscle arises from the front of the shaft of the humerus, crosses the anterior ligament of the elbow joint and is inserted into the tuberosity of the ulna by a powerful tendon. It flexes the elbow joint and its nerve supply is the musculo-cutaneous nerve, along with a small branch of the radial nerve.

### Coracobrachialis (Fig. 3.20)

This is a small muscle passing from the tip of the coracoid process to the medial

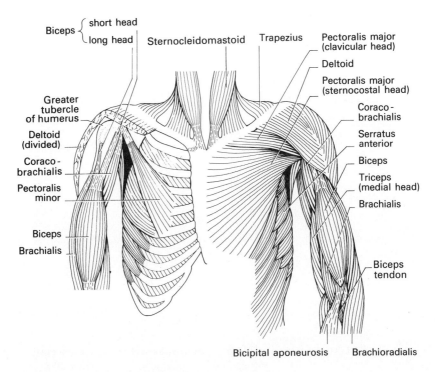

**Fig. 3.20** The muscles of the shoulder region and arm.

side of the humerus halfway down. It will thus move the arm forwards and medially as occurs during walking. The nerve supply is musculocutaneous.

### Biceps brachii

As its name implies, this muscle has two heads, the long head arising from above the glenoid fossa of the scapula and passing through the shoulder joint to emerge, surrounded by a tube-like sheath of synovial membrane, at the top of the intertubercular sulcus. It lies in the sulcus and then joins the short head, which arises in common with coracobrachialis from the coracoid process (Fig. 3.20). The muscle lies in front of brachialis with the musculocutaneous nerve sandwiched between the two. The thick tendon of insertion is attached to the radial tuberosity. An extension from the tendon, the *bicipital aponeurosis*, is a thin sheet of dense fascia that passes medially to blend with the deep fascia of the forearm (Fig. 3.20). It is very prominent when biceps contracts and must not be mistaken for the main tendon when eliciting a stretch reflex.

Biceps is supplied by the musculocutaneous nerve and is a powerful flexor and supinator of the forearm. Its long head helps in flexion at the shoulder.

### Triceps

This is the only muscle on the back of the arm and its three heads are unfortunately named *lateral, medial* and *long*. It would be much easier to understand if its heads

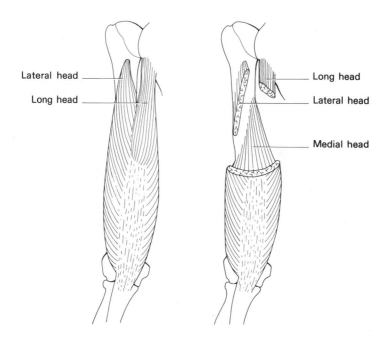

**Fig. 3.21** The three heads of triceps. The medial head is deep to the other two.

were named superficial (with two components) and deep since it is in two well-defined layers in the upper part of the arm, with the radial nerve lying between them (Figs 3.21 & 3.23).

The medial head (which one would prefer to be called deep) is attached to the whole of the lower part of the back of the humerus, overflowing on to the back of the medial and lateral intermuscular septa (see Fig. 1.6). The long and lateral heads arise, respectively, from the infraglenoid region and an oblique line on the back of the humerus. These two heads unite to form a 'superficial head' which blends with the medial head to be inserted into the olecranon at the upper end of the ulna. The medial head arises below the spiral groove and the lateral head above it.

Triceps is supplied by the radial nerve by means of branches that arises not only in the spiral groove (to the medial and lateral heads) but also in the axilla (to the long and medial heads). It is a powerful extensor of the elbow but remember that in the upright position extension is carried out by gravity so, when testing triceps, extension must be carried out against resistance.

## Nerves and vessels in the arm

### Musculocutaneous nerve

This has already been described. It supplies biceps, brachialis and coracobrachialis (BBC).

### Ulnar nerve

This is a non-stop traveller (i.e. no branches) above the elbow. In the axilla, it has been seen lying between the axillary artery and vein but in the arm it passes gradually backwards to pierce the medial intermuscular septum halfway down so that it comes to lie between the septum and the medial head of triceps (see Fig. 1.6). It then lies behind the medial epicondyle of the humerus, where it can be palpated (gently). If knocked accidentally, it gives rise to altered sensation in its cutaneous distribution (Fig. 3.22). Thus the 'funny bone' is not the 'humorous' but the ulnar nerve.

### Median nerve

This passes non-stop through the arm, crossing the brachial artery from lateral to medial halfway down the arm. At the elbow, therefore, it lies on the medial side of the artery (see Fig. 4.1).

### Radial nerve

After leaving the axilla, this nerve passes backwards, downwards and laterally, spiralling around the shaft of the humerus in the spiral (radial) groove. It enters the

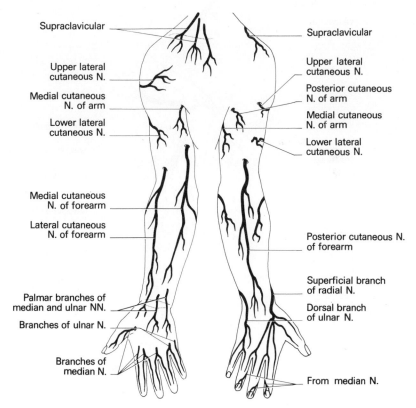

**Fig. 3.22** The cutaneous nerves on the front and back of the upper limb. The nerves, in life, are considerably smaller than shown here.

region between the upper part of the medial head and the long head (Fig. 3.23) and is then sandwiched between the medial and lateral heads. For a part of its course it is directly in contact with the bone and here it is vulnerable in fractures of the shaft. It pierces the lateral intermuscular septum so that it comes to lie in front of the lower end of the humerus and is therefore lateral to brachialis, and medial to two muscles of the forearm (brachioradialis and extensor carpi radialis longus) that are attached to the lateral supracondylar ridge. It is a little difficult to find in this region because it lies in a deep groove between the muscles. Whilst in the spiral groove it supplies the medial and lateral heads of triceps. If the nerve is damaged in the spiral groove, however, triceps will not be totally paralysed since branches to the long and medial heads arise in the axilla.

There are also three cutaneous branches, of which one, the *posterior cutaneous nerve* of the forearm, is large and supplies most of the skin of the back of the forearm (Fig. 3.22). However, there is so much overlap with adjacent nerve territories that sensory loss is slight in radial nerve lesions.

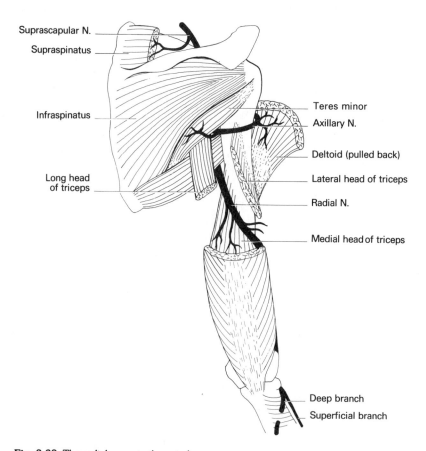

**Fig. 3.23** The radial nerve in the spiral groove.

### Brachial artery

This begins at the lower border of teres major and ends just below the elbow joint by dividing into radial and ulnar arteries. It is accompanied by venae comitantes (brachial veins), and supplies branches to the adjacent muscles, to the humerus and to the arterial anastomosis around the elbow joint. It has one large branch, the *profunda brachii*, which accompanies the radial nerve and takes part in the anastomosis around the elbow region.

# Chapter 4
# Forearm and Hand

## THE CUBITAL FOSSA

This is the region in front of the elbow joint and is triangular in shape. Its boundaries are formed by two of the muscles of the forearm that arise above the level of the elbow, viz. the *brachioradialis* and *pronator teres*, together with a line joining the two epicondyles of the humerus (Fig. 4.1). The floor of the fossa is formed by the *brachialis* muscle and the roof by deep and superficial fascia. In the latter lies the *median cubital vein*, which joins the cephalic and basilic veins (p. 64). The venous pattern, of course, is very variable but there is usually a vein in this region that is convenient for intravenous injections or for obtaining a blood sample. The medial and lateral cutaneous nerves of the forearm, derived from the medial cord and the musculocutaneous nerves respectively, lie near by.

There are three important structures in the fossa, whose position you need not memorize because in an examination you can palpate your own. The biceps tendon is thick and can be felt in the middle of the fossa between finger and thumb. Do not confuse it with the bicipital aponeurosis, which can be felt as a very sharp edge as it runs downwards and medially from the tendon of biceps. Just medial to the tendon, feel the pulsation of the brachial artery and, medial to this again, the median nerve may be rolled under the fingers (you will remember that it crosses the brachial artery, from lateral to medial, halfway down the arm).

At this stage, it will be helpful to revise the positions of the other two large nerves of the upper limb. The *radial nerve* lies deeply between brachialis and brachioradialis *in front* of the lateral epicondyle of the humerus and the *ulnar nerve* lies *behind* the medial epicondyle. This seems odd because the ulnar (and median) nerves supply chiefly the *front* of the forearm and hand while the radial nerve supplies the *back*.

## THE ELBOW JOINT

This is really a complex of joints between the humerus, radius and ulna. The elbow joint itself is between the trochlea of the humerus and the olecranon notch of the ulna; the coronoid process of the ulna engages with the coronoid fossa on the front of the humerus in full flexion, while the olecranon fits into the olecranon fossa of the humerus in full extension. During the movements of flexion and extension, the head of the radius glides on the capitulum of the humerus and, during full flexion, engages with the radial fossa on the front of the humerus. The elbow joint is therefore a hinge joint.

The *superior radio-ulnar joint* is a pivot joint between the head of the radius and a corresponding radial notch on the ulna. During pronation and supination the radial head rotates about an axis running through the centre of the head (see Fig. 2.6). This movement can easily be felt if you place your index finger on the

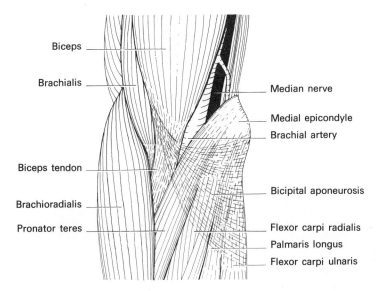

**Fig. 4.1** The right cubital fossa.

lateral epicondyle of the humerus and your middle finger alongside it. The middle finger now lies over the radial head, which can be felt rotating when the forearm is pronated and supinated.

The articulating surfaces are all covered by a layer of hyaline articular cartilage and all internal surfaces of the joint, except for the articulating surfaces, are lined by synovial membrane.

The joint is surrounded by a capsular ligament, which is strong except where it thins out anteriorly and posteriorly, as might be expected in a hinge joint. It is thickened by collateral ligaments at the sides. The ulnar collateral ligament is triangular and extends from the medial epicondyle of the humerus down to the olecranon and the coronoid process while the radial collateral ligament is from the lateral epicondyle of the humerus to the annular ligament (Fig. 4.2). The latter ligament surrounds the head of the radius, being attached medially to the edges of the radial notch of the ulna.

Many of the muscles that are attached to the bones in the region of the elbow joint are also attached to the ligaments.

## THE MUSCLES OF THE FRONT OF THE FOREARM

Most of these arise *above* the elbow joint so that they may have some flexor activity on this joint. They are best thought of as lying in three strata. The most superficial layer consists of four muscles, *pronator teres, flexor carpi radialis, palmaris longus* and *flexor carpi ulnaris*. They arise from a common flexor origin on the medial epicondyle and are partly fused together in the upper part of the forearm so that the muscle that forms the intermediate layer (*flexor digitorum superficialis*) cannot be seen. Towards the wrist, they narrow down to tendons so that the tendons of flexor superficialis can surface between them (Fig. 4.3).

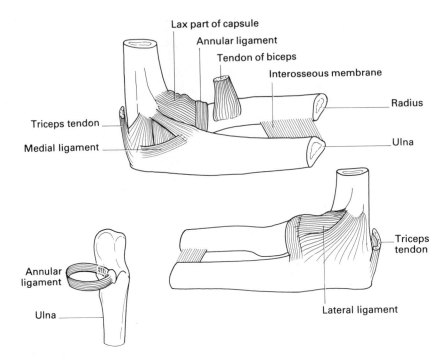

**Fig. 4.2** The ligaments of the elbow joint and superior radio-ulnar joint.

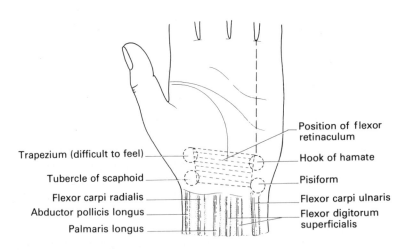

**Fig. 4.3** Palpable structures on the front of the wrist and hand. The four bony points are the attachments of the flexor retinaculum.

## Superficial layer (Fig. 4.4)

### Pronator teres

This is the only one of the four superficial muscles that only travels halfway down. It is inserted into a small area halfway down the radius at the summit of its convexity, thus gaining the maximum mechanical advantage in pronation. (It also has a deep head of origin from the coronoid process of the ulna).

### Flexor carpi radialis

This passes deep to the flexor retinaculum in its own compartment and is inserted into the bases of the 2nd and 3rd metacarpals. It is a flexor and abductor of the wrist.

### Palmaris longus

This is often absent and is unimportant except as a landmark. It is inserted into the palmar aponeurosis and flexor retinaculum.

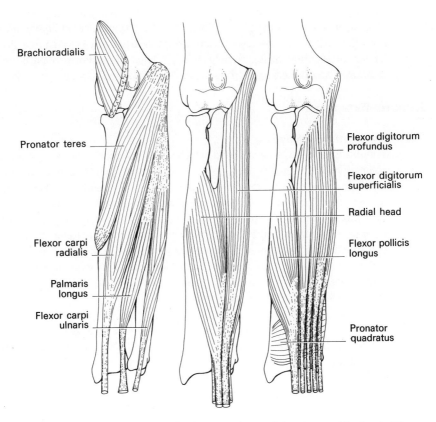

**Fig. 4.4** The muscles of the superficial, intermediate and deep layers of the front of the forearm.

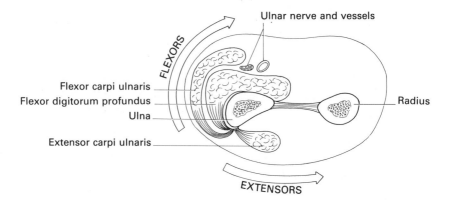

**Fig. 4.5** The palpable posterior border of the ulna (see Fig. 4.17) is the dividing line between flexors and extensors with the carpi ulnaris on each side.

### Flexor carpi ulnaris

This muscle has an extra origin by means of an aponeurosis from the subcutaneous posterior border of the ulna, which forms a 'pocket' in this region (Fig. 4.5). It is officially inserted into the pisiform bone but the pisometacarpal ligament passes from the pisiform to the base of the 5th metacarpal and is said to be the true tendon of insertion of flexor carpi ulnaris. Thus the pisiform bone is really a sesamoid bone in the tendon. This muscle is a flexor and an adductor of the wrist.

## Intermediate layer

### Flexor digitorum superficialis

This muscle also has an extra head, which is a thin sheet of muscle attached to the oblique line on the radius (Fig. 4.4). The two heads join to form an arch under which pass the median nerve and the ulnar artery. The tendons pass beneath the flexor retinaculum as two pairs, those to the ring and middle fingers in front and to the index and little fingers behind (Fig. 4.6). In the palm, they splay out and pass to the four fingers to enter the *fibrous flexor sheaths*. These are tough fibrous arches that pass across the palmar surface of the phalanges (although they are thinner and more flexible over the interphalangeal joints to allow movement). They thus form tunnels (see Fig. 4.15) which contain the long flexor tendons (*superficialis* and *profundus*) and their synovial sheaths.

As they enter the fibrous flexor sheaths, each tendon splits into two; each half twists through 180° and they then reunite, only to split up again to be attached to the sides of the middle phalanx (Fig. 4.7). The perforation in the tendon allows for the passage of flexor digitorum profundus, which is inserted into the distal phalanx. Superficialis is thus a flexor of all the joints that it crosses but it cannot flex the distal interphalangeal joint. This can be demonstrated by putting profundus out of action as follows. First, flex all your fingers into the palm. Now extend all but the ring finger. You will find that the distal phalanx of the ring finger is now paralysed.

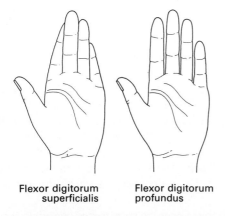

Flexor digitorum          Flexor digitorum
  superficialis             profundus

**Fig. 4.6** Diagram to help you remember the arrangement of the tendons of flexor digitorum superficialis and profundus. In the former, the tendons for the middle and ring fingers are in front and those for the index and little fingers are behind, but in the latter the four tendons are side by side.

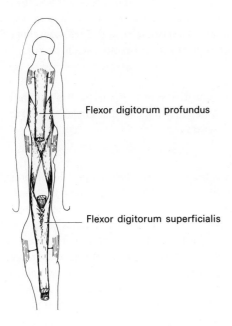

Flexor digitorum profundus

Flexor digitorum superficialis

**Fig. 4.7** The insertions of the long flexor tendons of the fingers.

## Deep layer

These three muscles lie in contact with the bones and the ligaments of the joints.

### Flexor digitorum profundus

This arises from the front of the ulna and the adjoining interosseous membrane and also by an aponeurosis from the posterior border of the ulna (Fig. 4.5). It splits into four tendons rather low down in the forearm, although the tendon for the index finger separates off from the other three quite proximally so that this finger enjoys a good deal of independence of movement. This is reflected in the distal palmar crease, which is the flexure line for the other three fingers. Note that it turns up towards the gap between the index and middle fingers so that the index finger is on its own. The tendons pass under the flexor retinaculum (lying side by side) and through the fibrous flexor sheaths and are inserted into the bases of the distal phalanges. They pass through the opening in the tendons of flexor digitorum superficialis. It is the low division of the muscle belly into its four tendons that stops flexion of the distal phalanx of the ring finger when the muscle is stretched by extending the other fingers (see above). *The four lumbrical muscles are attached to the four tendons of profundus. Do not confuse these with the interosseous muscles.*

This muscle can flex all the joints over which it passes, including the distal interphalangeal joint (but does not usually do so, see p. 91).

### Flexor pollicis longus

This is a unipennate muscle arising from the shaft of the radius and the adjacent interosseous membrane (see Fig. 4.4). It is inserted, like profundus, into a distal phalanx, namely that of the thumb, after passing beneath the flexor retinaculum. It can flex all the joints over which it passes (but does not usually do so).

### Pronator quadratus

This is the only muscle of the forearm that runs transversely; it is also the deepest. It passes from the lower few centimetres of the ulna to a similar area on the radius (see Fig. 4.4). It pronates the forearm.

### An oddity: brachioradialis

Although situated on the front of the forearm and although a flexor, brachioradialis really belongs to the extensor group in that it is supplied by the *radial* nerve and arises from the *lateral* side of the humerus, in fact, from the upper part of the supracondylar ridge. It is also odd in that, while most muscles cross the joint on which they act and immediately form an attachment to bone, brachioradialis crosses the elbow joint and then travels down to the lower end of the radius before inserting just above the styloid process. It is a flexor of the elbow joint when the forearm is semiprone (as in carrying a coat over the arm) and it can supinate the already pronated forearm and pronate it when it is already supinated.

### Nerve supply of the anterior forearm muscles

This has not been given for individual muscles because it is easier to remember that, excluding brachioradialis (see above), all the forearm muscles are supplied by the median nerve, directly or via its anterior interosseous branch, except for one and a half of them. These are flexor carpi ulnaris and the ulnar half of flexor digitorum profundus. The latter therefore has a dual nerve supply from the median and ulnar nerves and, in imagination, you should carry this 'division' in the muscle down to its tendons and their muscles so that the medial two lumbricals are supplied by the ulnar nerve and the lateral two by the median. *Do not confuse the lumbricals with the interossei. All the latter are supplied by the ulnar nerve.*

## THE FLEXOR RETINACULUM

All the tendons, as well as the main vessels and nerves, funnel down to the wrist region, where most, but not all of them, pass through the neck of the funnel which is the *carpal tunnel*. This is a rather narrow passage formed by the carpal bones and the flexor retinaculum. The carpal bones are shaped so as to form a curve (see Fig. 2.7) and only those at the sides of the curve come anywhere near the surface of the palm of the hand where they can be palpated. As you can see from Fig. 2.7, these are the *pisiform* and the *hook of the hamate* medially and the *scaphoid* and *trapezium* (the latter not easily felt) laterally. It is to these four bones that the flexor retinaculum is attached, thus completing the tunnel (Fig. 4.8). If you identify these bones in your own hand, you will not fall into the common error of thinking that the retinaculum lies under your wrist-watch strap. It is in the palm of the hand!

The structures that pass through the carpal tunnel (i.e. are deep to the retinaculum) are the median nerve and the tendons of flexor pollicis longus, flexors digitorum superficialis and profundus and flexor carpi radialis (in its own compartment). The tendons are enclosed in synovial sheaths. This makes a tight fit, so it is not surprising that no large arteries and veins pass through the tunnel as they might get compressed. This sometimes happens to the median nerve, which is always slightly flattened in this region. Compression may cause the *carpal tunnel syndrome*, in which there are motor and sensory changes in the structures supplied by this nerve.

The ulnar nerve and the ulnar artery pass superficial to the retinaculum so are exposed to possible injury but not to compression. If you cannot remember anything else about the flexor retinaculum, remember *ulnarover* and *mediunder*.

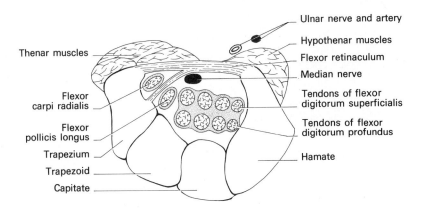

**Fig. 4.8** Cross-section through the carpal tunnel. Note particularly that the ulnar nerve is superficial and the median nerve deep to the flexor retinaculum.

## SYNOVIAL SHEATHS OF THE FLEXOR TENDONS

These surround parts of the flexor tendons in the manner described in Chapter 1. Flexor pollicis longus has its own sheath and flexors digitorum superficialis and profundus share one as they pass through the carpal tunnel (Figs 4.8 & 4.9). This sheath ends, however, in the palm of the hand except for the sheath for the little finger, but in the fibrous flexor sheaths (p. 76) the tendons for the index, middle and ring fingers have their own synovial sheaths. The relation of these sheaths to the skin creases in the palm are shown in Fig. 4.9. There is often a communication between the sheath of flexor pollicis longus and that of the long flexors behind the flexor retinaculum. The sheaths are important because, if an infection in the fingers (for example, a whitlow) is inadequately treated, the sheaths may become infected and this is an extremely serious condition, sometimes leading to restriction of finger movement.

## JOINTS OF THE WRIST AND HAND

### Wrist (radiocarpal joint)

This is a condyloid joint so that movements of flexion and extension, abduction and adduction can occur. No active rotation is possible but it is not needed since pronation and supination of the forearm replace it.

The bones taking part are the lower end of the radius and the scaphoid, the lunate and (indirectly) the triquetral. The head of the ulna is in contact with a triangular disc of fibrocartilage, which helps the radius to form a complete upper

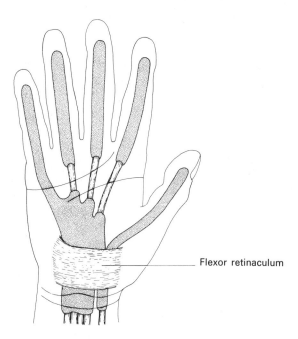

Flexor retinaculum

**Fig. 4.9** The synovial sheaths of the long flexor tendons in the palm.

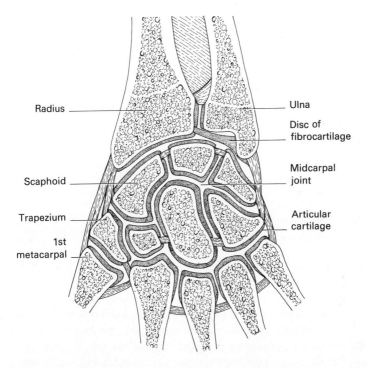

**Fig. 4.10** A section through the carpal joints. Note particularly the line of the wrist joint, the midcarpal joint between the proximal and distal rows of bones, and the first carpometacarpal joint.

articular surface (Fig. 4.10). The disc is attached to the edge of the ulnar notch of the radius and to the base of the styloid process of the ulna. It also extends over the styloid to blend with the ulnar collateral ligament. The joint has a capsular ligament, reinforced at the sides by radial and ulnar collateral ligaments.

### Intercarpal joints

These are complex but you should, at least, know about the *midcarpal joint*, which is between the proximal and distal rows of bones (Fig. 4.10). Movements at the midcarpal joint are similar to those at the wrist joint but, in some directions, even more extensive.

### Carpometacarpal joints

The most important of these is the first, that between the base of the 1st metacarpal and the trapezium. It is a sellar or saddle-shaped joint, does not communicate with any other joints and has movements similar to those of a ball and socket joint, i.e. flexion, extension, abduction and adduction, and medial and lateral rotation. The thumb, in a position of rest, lies more or less at right angles to the fingers and so, therefore, do the planes of its movements. Thus, in abduction of the fingers, the edge of each finger leads the way and, in abduction of the thumb,

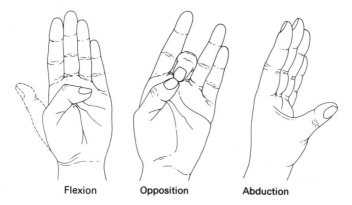

Flexion          Opposition          Abduction

**Fig. 4.11** Movements of the thumb (see text for description).

the same is true so the thumb is moved away from the palm of the hand (Fig. 4.11). Similarly, in flexion, the thumb is moved across the palm and, in extension, the nail leads the way (Fig. 4.11). Rotation is difficult to display as it is usually combined with flexion, but can be appreciated if you grip the head of the metacarpal and twist it medially and laterally. The combination of flexion and medial rotation is *opposition*, so called because the pad of the thumb can be opposed to the pads of the fingers, especially the index and middle fingers. This will be discussed further later in this section.

The other carpometacarpal joints are less important. Those of the index and middle fingers show only slight gliding movement but, if you grip the head of the ring or of the little finger, you will find quite a range of flexion and extension (see p. 91).

### Other joints of the fingers and thumb

The metacarpophalangeal joints are condyloid joints and have flexion and extension, abduction and adduction but no rotation. Abduction is used in a special sense for the fingers, meaning movement away from the middle finger, while adduction is towards the middle finger. The deep transverse palmar ligaments run between the thickened palmar ligaments of the metacarpophalangeal joints of the four fingers. The interphalangeal joints are hinge joints.

### MUSCLES OF THE HAND (Fig. 4.12)

These are arranged in two main groups around the 1st and the 5th metacarpals, forming the *thenar* and *hypothenar eminences*, along with a number of deeper muscles in the central 'hollow' of the palm. The skin in the centre of the palm is particularly thick and cannot easily be pinched up. This is because it is attached deeply to the *palmar aponeurosis*, a thick triangular mass of fascia that is attached proximally to the flexor retinaculum, blending in with the tendon of palmaris longus (if present). Distally, it splits up into slips for the four fingers and blends with the fibrous flexor sheaths.

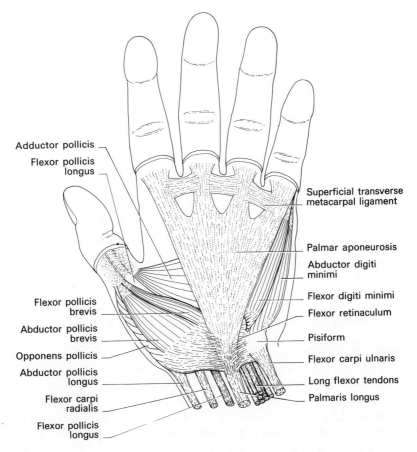

Adductor pollicis
Flexor pollicis longus
Flexor pollicis brevis
Abductor pollicis brevis
Opponens pollicis
Abductor pollicis longus
Flexor carpi radialis
Flexor pollicis longus

Superficial transverse metacarpal ligament
Palmar aponeurosis
Abductor digiti minimi
Flexor digiti minimi
Flexor retinaculum
Pisiform
Flexor carpi ulnaris
Long flexor tendons
Palmaris longus

**Fig. 4.12** The superficial muscles of the palm. Not all the tendons shown at the wrist are palpable.

## Thenar muscles

### Abductor pollicis brevis
This extends from the flexor retinaculum and some carpal bones to the lateral side of the proximal phalanx of the thumb.

### Flexor pollicis brevis
Its attachments are similar to those of the abductor.

### Opponens pollicis
The origin is similar but it is inserted into the shaft of the 1st metacarpal.

### Adductor pollicis
This muscle lies deeper than the other three. It has an *oblique head* from some of the carpal bones and a *transverse head* from the shaft of the 3rd metacarpal

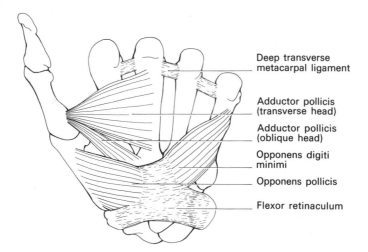

**Fig. 4.13** The opponens muscles and adductor pollicis.

(Fig. 4.13). It is inserted into the medial side of the base of the proximal phalanx of the thumb, along with the 1st palmar interosseous (see below).

### Insertion of the thenar muscles

To sum up the insertions of the four muscles attached to the sides of the proximal phalanx, abductor and flexor pass to the lateral side and adductor and the 1st palmar interosseous to the medial side (Fig. 4.14). A sesamoid bone is found in the combined tendon on each side and the flexor pollicis longus tendon passes between them on its way to the distal phalanx.

## Hypothenar muscles

*Abductor digiti minimi*
This arises from the pisiform and is inserted into the ulnar side of the base of the proximal phalanx of the little finger.

*Flexor digiti minimi*
This arises from the flexor retinaculum and is inserted into the ulnar side of the proximal phalanx.

*Opponens digiti minimi*
This arises from the flexor retinaculum and is inserted into the shaft of the 5th metacarpal.

### Actions and nerve supply of thenar and hypothenar muscles

The names of the muscles indicate their actions but note that since the muscles often work together (for example opposition of the thumb involves both flexion and medial rotation) the muscle bellies are often fused. Opposition of the little

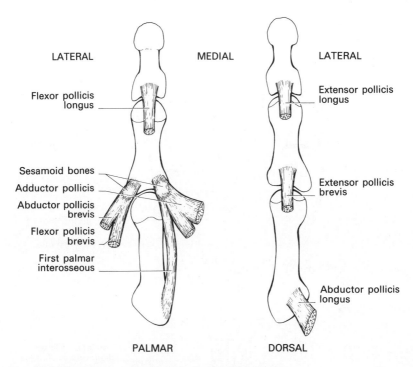

LATERAL    MEDIAL    LATERAL

Flexor pollicis longus

Extensor pollicis longus

Sesamoid bones
Adductor pollicis
Abductor pollicis brevis
Flexor pollicis brevis
First palmar interosseous

Extensor pollicis brevis

Abductor pollicis longus

PALMAR    DORSAL

**Fig. 4.14** The tendons inserted into the bones of the thumb.

finger is slight but important in gripping (p. 91). The nerve supply of the hypothenar muscles is the ulnar nerve and that of the thenar muscles is often said to be the median nerve. However, flexor pollicis brevis *usually* has an extra branch from the ulnar nerve and opponens *often* has a similar branch. If you want to test the median nerve, therefore, you should test *abduction* of the thumb.

### Lumbrical muscles

These are four slender muscles which arise from the tendons of flexor digitorum profundus (Fig. 4.15). Their tendons are inserted into the radial side of each of the proximal phalanges of the fingers and into the dorsal extensor expansion (see below). They have similar actions to those of the interossei in flexing the metacarpophalangeal joints without flexing the interphalangeal joints. The nerve supply is similar to that of flexor digitorum profundus.

### The interosseous muscles

These are eight muscles that arise from the shafts or bases of the metacarpals. If you remember that the dorsal arise from two metacarpals and the palmar from only one and also that the dorsal abduct the fingers and the palmar adduct (hence the classical mnemonic D.AB and P.AD), you can easily work out their insertions, which are shown in Fig. 4.16. You will see that whichever way the middle finger moves to the side it will be called abduction, so it has two dorsal interossei attached; similarly the middle finger cannot be adducted (it is there already) and so

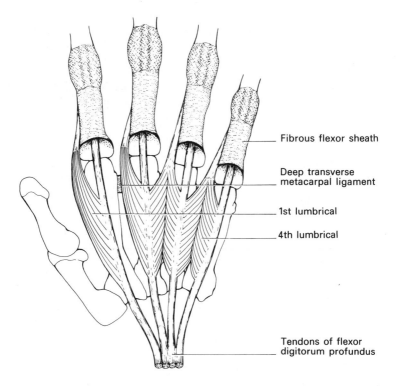

**Fig. 4.15** The lumbrical muscles.

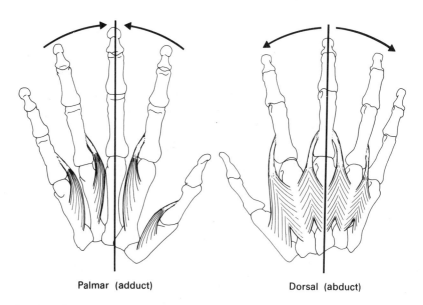

Palmar (adduct)          Dorsal (abduct)

**Fig. 4.16** The palmar and dorsal interossei.

has no palmar interossei. In addition to their insertions into the sides of the proximal phalanges, the interossei are also inserted into the dorsal extensor expansion (see below).

*All* the interossei are supplied by the deep branch of the ulnar nerve. Their functions will be described below.

## THE MUSCLES OF THE BACK OF THE FOREARM

The subcutaneous posterior border of the ulnar is a watershed between the extensors and the flexors, particularly between flexor and extensor carpi ulnaris (see Fig. 4.5). If you start from this and travel medially you will find flexor carpi ulnaris, palmaris longus, flexor carpi radialis and pronator teres (see Fig. 4.4).

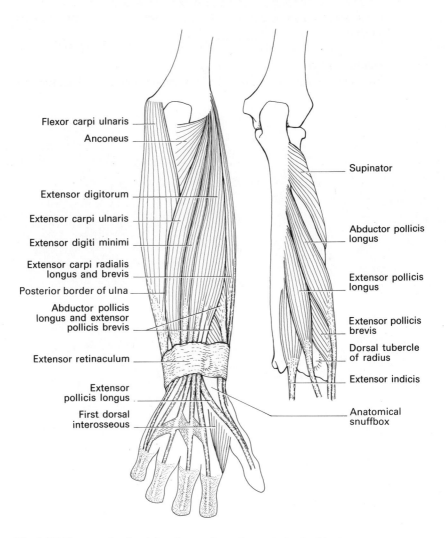

**Fig. 4.17** The superficial and deep layers of muscles on the back of the forearm.

Travelling laterally you will meet extensor carpi ulnaris, extensor digiti minimi, extensor digitorum, etc. (Fig. 4.17, p.87). The extensor muscles can be divided into superficial and deep groups.

## Superficial group of extensor muscles

These arise from a common extensor origin on the lateral epicondyle of the humerus.

### Extensor carpi ulnaris

Like flexor carpi ulnaris, this has an additional aponeurotic origin from the posterior border of the ulna and it is inserted into the base of the 5th metacarpal. It is an extensor and adductor of the wrist joint. *Note*: both flexor and extensor carpi ulnaris are attached to the 5th metacarpal, the former via the pisometacarpal ligament.

### Extensor digiti minimi

This is small and unimportant; it joins the tendon of extensor digitorum to the little finger.

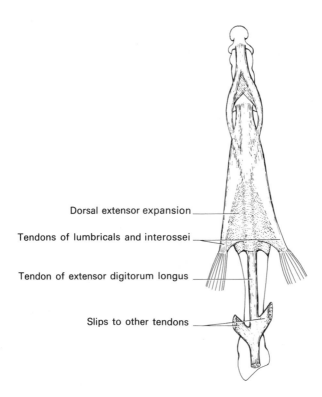

Dorsal extensor expansion

Tendons of lumbricals and interossei

Tendon of extensor digitorum longus

Slips to other tendons

**Fig. 4.18** The insertions of the lumbricals, interossei and extensor digitorum longus.

### Extensor digitorum

The muscle belly divides into four tendons, one for each of the fingers. On the base of each finger, the tendon spreads out to form a triangular *dorsal digital expansion* (Fig. 4.18). The base of the triangle receives part of the insertion of the appropriate interossei and lumbricals so that these muscles can extend the interphalangeal joints by pulling on the extensor tendon. The apex of the triangle splits into three slips, of which the middle one is attached to the base of the middle phalanx while the other two unite to be attached to the base of the distal phalanx. The dorsal extensor expansion is bound down at the sides to the deep transverse palmar ligament. The tendons of extensor digitorum are bound together on the back of the hand by tendinous slips (Fig. 4.17), which makes it difficult to extend the ring finger alone from a clenched fist.

Extensor digitorum extends the metacarpophalangeal and interphalangeal joints simultaneously as in opening a clenched fist.

### Extensor carpi radialis longus and brevis

These muscles arise from the lateral supracondylar ridge rather than the common extensor origin. You will remember that flexor carpi radialis was inserted into the bases of the 2nd and 3rd metacarpals. The corresponding extensor is split into two parts, *longus* and *brevis*, which are inserted respectively into the bases of the 2nd and 3rd metacarpals. They extend and abduct the wrist.

## Deep group of extensor muscles

These cannot be seen properly until the superficial muscles are retracted or removed, although their tendons come to the surface near the wrist.

### Supinator (Fig. 4.19)

Just as, on the flexor surface, pronator teres does not have a tendon that reaches the wrist, supinator, on the extensor surface, is also restricted to the upper forearm. It arises from the supinator crest of the ulna (see Fig. 2.6) as well as the common extensor origin and the annular ligament, and then wraps around the back of the radius to be attached to the upper third of the shaft. It supinates (Fig. 4.19) and, along with biceps, is used in powerful movements such as screwing in screws and turning door knobs.

### Abductor pollicis longus, extensor pollicis brevis and extensor pollicis longus

These muscles are best dealt with together. They arise, along with extensor indicis, from the backs of the radius, or ulna, or both (plus the interosseous membrane) and are each inserted into the base of a different bone of the thumb: the metacarpal, proximal phalanx and distal phalanx in the order shown above (see Fig. 4.14). The tendons of abductor pollicis longus and extensor pollicis brevis come to the surface by emerging from under cover of the lateral border of extensor digitorum and winding round the lateral border of the forearm (the so-called 'outcropping' muscles), while extensor pollicis longus is easily recognized because it uses the dorsal tubercle of the radius as a pulley, thus changing its direction slightly (see Fig. 4.17). Identify these tendons on your own wrist. They form the

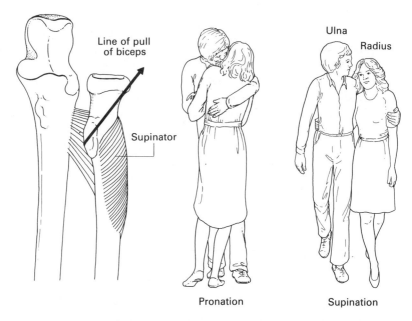

**Fig. 4.19** The supinator arises from the supinator crest of the ulna and (not shown) from the lateral epicondyle of the humerus. It wraps around the radius and the two students on the right are enjoying demonstrating how it supinates.

boundaries of the *anatomical snuffbox*. In its floor can be felt the styloid process of the radius and the scaphoid and also the pulsation of the radial artery (p. 93). Tenderness in the snuffbox is an important sign of fracture of the scaphoid.

### Extensor indicis

This is small and unimportant; its tendon joins the tendon of extensor digitorum to the index finger.

All the muscles mentioned above, except for supinator, have synovial tendon sheaths in the region of the wrist, where they pass under the extensor retinaculum, but there are no synovial sheaths on the dorsal aspect of the fingers.

### Nerve supply of the extensor muscles

All are supplied by the deep branch of the radial nerve (posterior interosseous nerve), so that a lesion of this nerve will paralyse the extensors and give rise to 'wrist drop'. More important than this, however, is the effect on grasp.

### Grasp

The versatility of the human hand is evident if one compares a person wielding a sledge-hammer with a student carrying out a careful dissection of the nerves of the orbit. In the former case, the fingers and thumb are wrapped around the handle and the muscles, both flexor and extensor, of the forearm contract powerfully. In

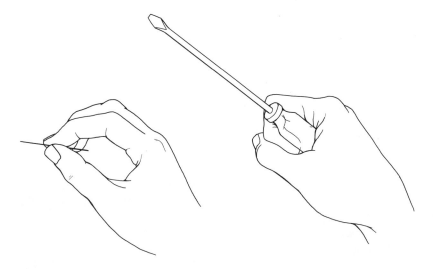

**Fig. 4.20** The 'precision grip' and the 'power grip'.

the latter, the thumb is opposed to the pads of the fingers with the forceps between and most of the work is done by the small muscles of the hand. These two types of grip are known respectively as the *power grip* and the *precision grip* (Fig. 4.20).

### Power grip

This involves a flexion of all joints of the fingers and thumb and the muscles used as prime movers are principally flexor pollicis longus and flexors digitorum superficialis and profundus. All these muscles, however, cross the wrist joint and could flex it, so this is prevented by synergistic (p. 15) contraction of the wrist extensors. If the extensors are paralysed by a radial nerve injury, a power grip is impossible because the wrist flexes as well as the fingers, so the fingers are pulled open by the relative shortness of the extensors (try gripping with your wrist fully flexed).

In gripping a rounded object, flexion of the carpometacarpal joints of the little and ring fingers (p. 82) helps by closing down the medial side of the palm on to the object.

### Precision grip

Now try doing up a button without using your thumb. *The thumb is by far the most important digit on the hand* by virtue of its ability to be brought round so that the pad of the thumb comes to be opposite the pads of the fingers (*opposition*). This is done by opponens pollicis along with flexor and adductor. Now try doing up a button with the interphalangeal joints flexed (by flexor profundus). The difficulty here is that the thumb can only be opposed to the finger nails; what is needed is a movement of flexion at the metacarpophalangeal joints combined with extension at the interphalangeal joints. This can be done by the interossei and lumbricals because of their insertion into the dorsal digital expansion, *and this is their most important action* (*not* the abduction and adduction mentioned on p. 85). A lesion of the ulnar nerve that paralyses the interossei and the medial two lumbricals will produce the opposite effect, namely extension at the metacarpophalangeal joints and flexion at the interphalangeal joints. This position is the so-called '*ulnar claw hand*'. The index and middle fingers are less affected than the other two because they still have their lumbricals, supplied by the median nerve.

## THE ARTERIES OF THE FOREARM AND HAND

The brachial artery divides into radial and ulnar arteries at the level of the head of the radius. All the major arteries in this region give one or more branches to the arterial anastomosis around the elbow joint but they will not be described in detail. Note that the radial and ulnar arteries originate near the midline while the radial and ulnar nerves enter the forearm respectively in front of the lateral epicondyle and behind the medial epicondyle. The arteries are therefore 'inside' the nerves (Fig. 4.21).

### Radial artery

This is a superficial artery in the forearm, being overlapped only by the edge of brachioradialis. It therefore lies, in turn, on all the muscles or tendons that are attached to the radius (Fig. 4.22), — 'To Swim Properly, Flex Forearm, Pronate

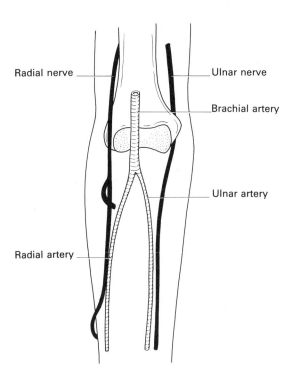

**Fig. 4.21** This shows why the radial nerve is lateral to the radial artery but the ulnar nerve is medial to the ulnar artery.

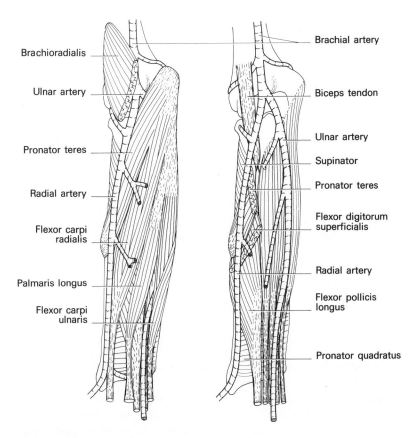

**Fig. 4.22** The main arteries of the forearm.

Radius': tendon of biceps, supinator, pronator teres, flexor superficialis (radial head), flexor pollicis longus, pronator quadratus, radius. While it lies on the lower end of the radius its pulsation may easily be felt ('the pulse'). It gives branches to the surrounding muscles and finally winds backwards deep to the two lateral tendons of the snuffbox (abductor pollicis longus and extensor pollicis brevis) to lie in the snuffbox itself before passing between the two heads of the first dorsal interosseous muscle to enter the palm. Here it gives a branch to the thumb (*princeps pollicis*) and to the lateral side of the index finger (*radialis indicis*) before joining up with the deep branch of the ulnar artery to form the *deep palmar arch* (Fig. 4.23). The arch gives off three palmar metacarpal arteries which join the common palmar digital arteries (see below).

#### Ulnar artery

This is a deep artery in the forearm, although rarely it may pass superficial to the forearm muscles and if it is mistaken for a vein an 'intravenous' injection of, say, an anaesthetic may have spectacular consequences. Normally, though, the artery

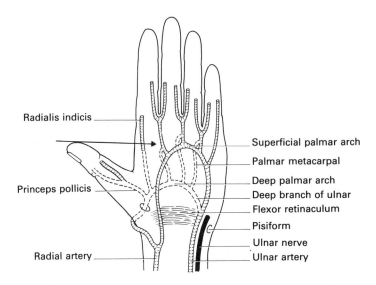

Radialis indicis

Superficial palmar arch

Palmar metacarpal

Deep palmar arch

Princeps pollicis

Deep branch of ulnar

Flexor retinaculum

Pisiform

Ulnar nerve

Radial artery

Ulnar artery

**Fig. 4.23** The principal arteries of the palm. The arrow indicates that the superficial palmar arch is at the level of the extended thumb.

passes deep to pronator teres and deep to the arch of flexor digitorum superficialis. It therefore comes to lie on flexor profundus and one-third of the way down the arm it joins the ulnar nerve (Fig. 4.21) and the two run down together. They are overlapped by flexor carpi ulnaris but emerge to lie immediately lateral to that tendon and its insertion into the pisiform. It then divides into *superficial* and *deep branches*. The former forms the *superficial palmar arch*, which crosses the palm, deep to the palmar aponeurosis, at the level of the extended thumb. The arch is completed by a small branch from the radial artery or one of its branches. The arch gives off three *common palmar digital arteries* (Fig. 4.23), which are directed towards the clefts between the 2nd, 3rd, 4th and 5th fingers (see below). They join the palmar metacarpal arteries and the resulting vessel divides to give palmar digital arteries along the sides of the fingers. These lie slightly anterior to the phalanges and their pulsations can be felt.

### Common interosseous artery

This is the largest branch of the ulnar artery, the others being relatively unimportant. It divides into anterior and posterior interosseous arteries, which run down the forearm anterior and posterior to the interosseous membrane and accompanying respectively the *anterior interosseous* branch of the median nerve and the deep (*posterior interosseous*) branch of the radial nerve. Note, however, that the posterior interosseous artery enters the back of the forearm by passing between the radius and ulna while the nerve winds round the radius to reach this position. See also p. 145. Branches are given to surrounding structures. The venous drainage of the upper limb has already been described on p. 64.

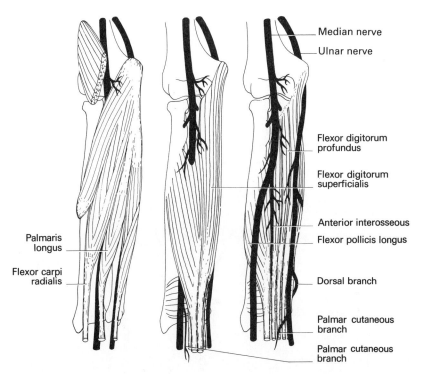

**Fig. 4.24** The median and ulnar nerves on the front of the forearm. Compare with Fig. 4.4.

## NERVES OF THE FOREARM AND HAND

### Median nerve

This has been seen at the elbow lying on branchialis and just medial to the brachial artery. It passes between the two heads of pronator teres (so is separated from the ulnar artery by the deep head) and under the arch of flexor digitorum superficialis (Fig. 4.24). It is adherent to the deep surface of superficialis until it emerges on the lateral side of that muscle, between it and flexor carpi radialis and just lateral to palmaris longus, if present (Fig. 4.25). It then runs deep to the flexor retinaculum to enter the palm of the hand, where it divides into branches. High in the forearm it gives off an anterior interosseous branch that runs down on the interosseous membrane to supply two and a half of the three muscles of the deep layer, i.e. flexor pollicis longus, pronator quadratus and half of flexor digitorum profundus. The main trunk supplies all the other muscles on the front of the forearm except for one (see ulnar nerve below).

In the hand, a large and important branch is given off just as the nerve emerges from the carpal tunnel. This nerve turns back and enters the thenar eminence (Fig. 4.25). For this reason incisions should not be made along the whole length of the palmist's 'line of life' (the flexure line of the thumb) or this nerve is bound to be

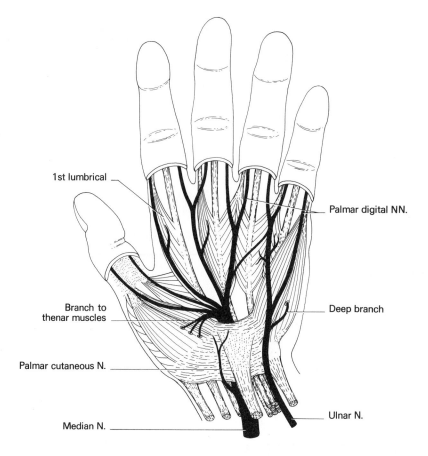

1st lumbrical

Palmar digital NN.

Branch to
thenar muscles

Deep branch

Palmar cutaneous N.

Ulnar N.

Median N.

**Fig. 4.25** The median and ulnar nerves in the palm of the hand. Note that the important branch of the median nerve to the thenar muscles is given off at the distal margin of the flexor retinaculum.

cut. It supplies the three muscles of the thenar eminence although two of them are liable to receive additional branches from the ulnar nerve (p. 85).

The median nerve then gives branches to the two lateral lumbricals and divides into palmar digital nerves that supply adjacent sides of the medial three and a half digits. At the tips of the fingers these nerves bend dorsally and supply the nail beds and parts of the dorsal surfaces of these fingers.

Lesions of the median nerve produce an extensive area of sensory loss and inability to flex the terminal phalanges of the index and middle fingers. The thumb is held in adduction and extension, giving an 'ape-like' hand.

## Ulnar nerve

After passing behind the medial epicondyle of the humerus, the ulnar nerve inclines forwards and passes between the two heads of flexor carpi ulnaris to enter the forearm. It seems odd that the nerve should enter the posterior compartment above the elbow (p. 69) and then return to the front of the forearm below the elbow, but this is useful in cases of suture of a damaged ulnar nerve as the nerve may be transplanted to the front of the elbow, thus gaining some extra length of nerve and preventing stretching.

In the forearm it lies on flexor digitorum profundus and is overlapped by flexor carpi ulnaris until it emerges at the wrist to lie just lateral to the pisiform alongside the ulnar artery. It ends in the region of the hypothenar eminence by dividing into superficial and deep branches.

In the forearm it supplies branches to one and a half muscles (flexor carpi ulnaris and half of flexor profundus) and also gives a dorsal branch that supplies the dorsal surface of the medial one and a half digits. The terminal superficial branch supplies the palmar surfaces of these digits as well as the region of the nail bed. The deep branch is more important and supplies the hypothenar muscles as it passes through them and then arches across the palm with the deep palmar arch, lying on the metacarpals and interosseous muscles. It supplies all these interossei and the two medial lumbricals and ends up by supplying adductor pollicis.

Lesions of the ulnar nerve produce noticeable wasting of the interosseous muscles, especially the 1st dorsal, which normally produces the bulge on the dorsum of the hand between the 1st and 2nd metacarpals. It also gives rise to a claw hand (p. 92) and some weakness in adduction of the wrist.

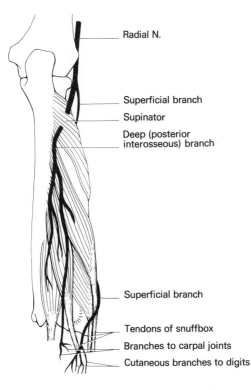

Radial N.

Superficial branch

Supinator

Deep (posterior interosseous) branch

Superficial branch

Tendons of snuffbox

Branches to carpal joints

Cutaneous branches to digits

**Fig. 4.26** The radial nerve in the back of the forearm.

**Radial nerve**

From its position in front of the elbow, deep between the brachialis and brachioradialis muscles, the radial nerve enters the forearm and promptly divides into superficial and deep branches. The superficial branch is purely cutaneous and follows the radial artery for part of its course (see Fig. 4.21) before passing backwards deep to brachioradialis, and splitting up into dorsal digital nerves for the supply of the lateral three and a half digits on their dorsal aspects. The deep branch (posterior interosseous) is purely motor. It winds round the upper part of the radius, passing through the supinator muscle, which it supplies. It then descends between the extensor muscles and supplies them (Fig. 4.26).

Lesions of the radial nerve produce 'wrist drop' and inability to use a power grip (p. 91). The interphalangeal joints, however, can be extended by using the interossei and lumbricals. The sensory loss is not very important, owing to extensive overlap of areas supplied by cutaneous nerves and also because sensation on the dorsum is much less important than that on the palmar surface of the hand.

# Chapter 5
# The Bones of the Lower Limb

## COMPARISON OF THE LIMB BONES

The bones of the upper and lower limbs are constructed on the same basic plan but there are important differences between them, mainly because the upper limb is constructed for mobility (and thereby sacrifices strength) while the lower limb is built mainly for its weight-bearing powers (and so sacrifices mobility). Whereas the upper limb girdle only articulates with the trunk at its sternoclavicular joint, the lower limb girdle is composed of the very solid hip bones, which have an extremely strong and almost immovable joint with the sacrum. Most students are surprised to find that this is a synovial joint since the surfaces involved (the auricular surfaces of the ilium and sacrum) are irregular (see Fig. 5.2). The femur is extremely strong and the ball at its upper end is deeply embedded in the acetabulum. In the lower part of the limbs there is an important difference in that the fibula plays no part in the formation of the knee joint (except for providing an attachment for the fibular collateral ligament) and is not a weight-bearing bone. The bones of the hand and feet are also comparable — obviously so in the case of the metatarsals and phalanges but less obviously in the case of the other bones.

## THE HIP BONE (Figs 5.1 & 5.2)

The hip bone is made up of the ilium, ischium and pubis, joined, in childhood, by hyaline cartilage, which becomes ossified after puberty. The junctions between the bones are indicated by dashed lines in Figs 5.1 & 5.2. The ilium is the bone that is mainly responsible for distributing the weight from the spine to the femoral heads and is therefore thickened along the line of weight transmission. The pubis and ischium act as ties or struts between the ilia (see Fig. 13.1) in much the same way as the clavicle acts as a prop holding the shoulder out, but they are not homologous with the clavicle in any way. In sitting, the weight is transmitted via the ischial tuberosity; between this and the seat is only the skin, a dense mass of fibrous tissue and perhaps a bursa. Although gluteus maximus covers the ischial tuberosity when the hip is extended, it fortunately slips upwards off the tuberosity when it is flexed — fortunately, because vascular muscular tissue would not stand up to being sat on whereas fibro-fatty tissue, with its low requirement of blood, accepts this insult with equanimity. For the same reason, the long muscles of the sole of the foot stop short of the tubercles of the calcaneum, which are separated from the ground by a dense mass of fibro-fatty tissue.

## THE FEMUR (Fig. 5.3)

Massive compared with the humerus, the femur has many resemblances to it except that the portion corresponding to the anatomical neck is elongated so that it acts as a lever for rotation at the joint and, as it is a common site for fracture, it

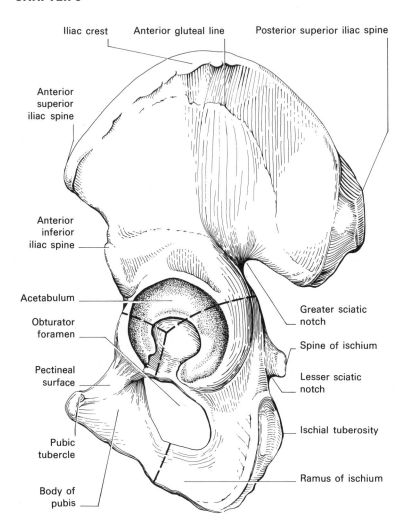

**Fig. 5.1** The left hip bone, lateral aspect.

therefore corresponds also to the surgical neck. At the lower end, the femur articulates with only one long bone (plus the patella). The articular surface at the lower end is more complex than it may seem at first sight and since it is important in the locking of the knee joint in full extension (p. 140) it will now be described in more detail. Figure 5.3 shows the lower end of the femur and you should note two facts. Firstly, the lateral condyle projects further forward than the medial condyle (compare with an end-on view of the fist, whose shape is affected by the flexion that can take place at the carpometacarpal joint of the little finger). This prevents lateral displacement of the patella (p. 121). Secondly, the areas for articulation with the medial and lateral condyles of the tibia differ considerably. The anterior area is for articulation with the patella; the area for the lateral condyle is more or less circular and that for the medial condyle is oval. Therefore the shapes of the

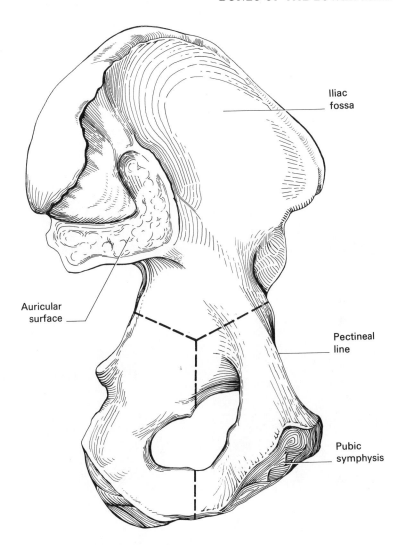

Iliac
fossa

Auricular
surface

Pectineal
line

Pubic
symphysis

**Fig. 5.2**  The left hip bone, medial aspect.

lateral and medial menisci are circular and oval respectively, as are the facets on the tibial condyles. When the knee joint is extended, this means that the lateral condyle will come to the end of its run before the medial; extension will stop on the lateral side but will continue on the medial side so that the tibia rotates laterally. This has the effect of tightening all the ligaments and producing a very stable joint. More details of this important mechanism are given in Chapter 6.

## The blood supply of the femur

The femur provides a good example of the blood supply of bones, which has been described in Chapter 1. There is a nutrient foramen in the middle of the shaft which is directed away from the growing end (the lower end). There are also

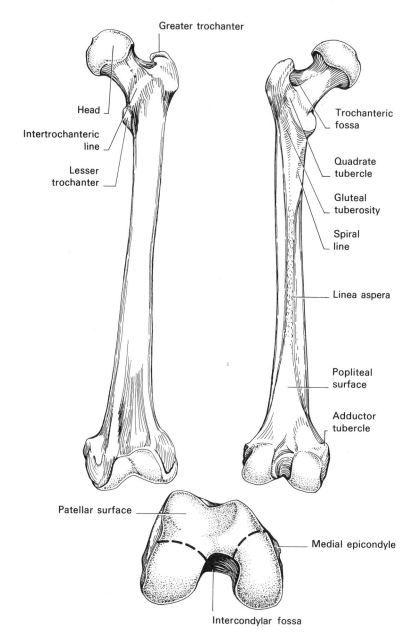

**Fig. 5.3** The left femur, anterior and posterior aspects and the lower end. The dashed lines indicate the limits of the areas for articulation with the lateral (circular) and medial (oval) condyles of the tibia.

nutrient foramina at the ends. At the upper end these fall into two groups: a series of large foramina around the base of the head which belongs to vessels that run up to the head in the retinacula (p. 114), and some smaller foramina in the pit or *fovea* in the head through which pass vessels that reach the head through the ligamentum teres. In the adult there is little or no anastomosis between these two sets of vessels so that if the retinacula are interrupted by a fractured neck of femur the head may die (*avascular necrosis*).

## THE PATELLA

The anterior surface is roughened for the attachment of part of the quadriceps tendon. The posterior surface is smooth since it is covered with hyaline cartilage in life. It is divided into lateral and medial facets for articulation with the corresponding areas of the femur. The lateral facet is the larger, which explains the old trick of 'siding' a patella — if placed on the table with the apex pointing away from you, it will fall to the side to which it belongs.

## THE TIBIA (Fig. 5.4)

Note the peculiar shape of the upper epiphysis of the tibia, shown in Fig. 5.5. Weight transmission is through the expanded upper end of the tibia (the *tibial*

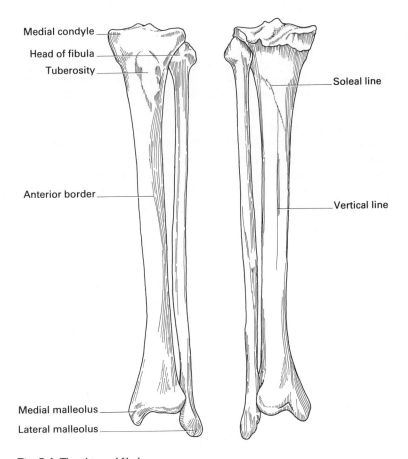

**Fig. 5.4** The tibia and fibula.

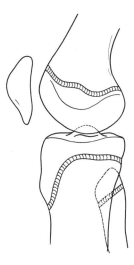

**Fig. 5.5** The positions of the epiphysial lines around the knee. Note the odd shape of the upper tibial epiphysis (traced from an X-ray).

*plateau*) and then through the lower end to the talus. The medial malleolus forms the medial part of the mortise that holds the talus in place. The anterior border and medial surface of the tibia are subcutaneous and related only to overlying structures at the upper and lower ends. At the upper end are the insertions of the flattened tendons of three muscles, often with one or more bursae between them. These are *sartorius*, *gracilis* and *semitendinosus*. At the lower end lies the *great saphenous vein*, which can normally be seen, and always located, since it crosses the tibia at a point 2 cm above and in front of the medial malleolus. On the posterior surface, the *soleal line* marks off a triangular upper area to which the popliteus muscle is attached, while, below the line, a faint *vertical line* divides the rest of the posterior surface into medial and lateral areas. Numerous other muscles and tendons are attached to the tibia but there are no muscular attachments to the lower third.

## THE FIBULA (Fig. 5.4)

To tell to which side a fibula belongs, first identify the lower end by the triangular facet (the upper facet is more or less circular). Behind the facet is a pit, or fossa. Remember the Pit is Posterior (Fig. 5.6). Since the facet faces medially, the side to which the bone belongs is obvious. A common error is to believe that the facet is for articulation with the tibia; in fact, it is for the talus. The inferior tibiofibular joint is marked by a rough area above the facet, since this joint is a syndesmosis.

The relation between tibia and fibula is best understood by remembering that, in bones that are roughly triangular in cross-section, the surfaces are named in accordance with the border that lies opposite, for example the anterior surface is the one opposite the posterior border. The tibia has *anterior*, *medial* and *lateral* (or interosseous) borders and its surfaces are therefore called posterior, lateral and

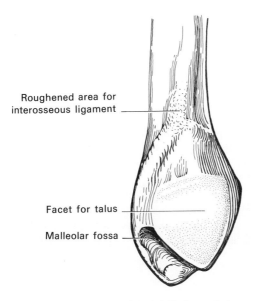

Roughened area for interosseous ligament

Facet for talus

Malleolar fossa

**Fig. 5.6** The lower end of the left fibula, medial aspect.

medial. The fibula has *anterior*, *posterior* and *medial* (or interosseous) borders so its surfaces are posterior, anterior and lateral (Fig. 5.7). In the case of the fibula, however, the surfaces have very different areas. Identify the anterior border by following up the two sides of the triangular subcutaneous area at the lower end of the fibula. Travelling medially, the next border you come across is the medial (interosseous) border. It is not impressive, and is very near to the anterior border, so that the anterior surface is extremely narrow. Now travel laterally from the anterior border and you will find the rather rounded posterior border. Between the

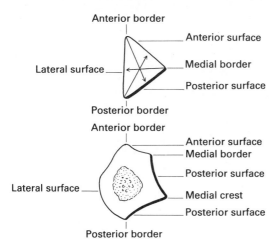

Anterior border

Anterior surface

Lateral surface

Medial border

Posterior surface

Posterior border

Anterior border

Anterior surface

Medial border

Posterior surface

Lateral surface

Medial crest

Posterior surface

Posterior border

**Fig. 5.7** The borders of the fibula, diagrammatically and actually. The posterior surface is the tricky one so it is drawn as a thick line in both pictures.

two is the lateral surface, which, if followed downwards, spirals round to become posterior at the lower end of the fibula because the tendons of the peroneal muscles use this slight groove as a pulley, round which they run to enter the sole of the foot (see Fig. 7.2). Everything between the posterior and medial borders is the posterior surface and this is confusing because it has upon it the *medial crest*, which is sharp and prominent, so that part of the true posterior surface faces anteriorly.

It is now possible to make sense of the attachments of muscles and fascial layers (see Fig. 7.7). The extensor group of muscles are attached to the anterior surface, and the peronei to the lateral surface. The interosseous membrane stretches between the two interosseous borders and separates the extensors from the deep layer of flexors. These arise from tibia, or fibula, or both, and they are covered by a rather dense layer of fascia. Behind this are the two powerful plantar flexors, *gastrocnemius* and *soleus*, and the unimportant one, *plantaris*. Soleus arises from both tibia (soleal line) and fibula.

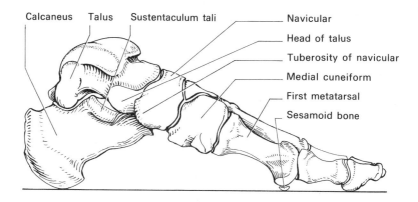

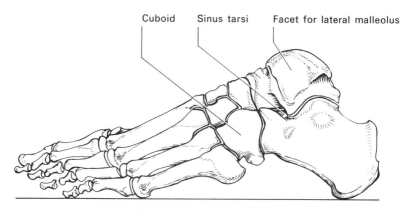

**Fig. 5.8** The skeleton of the left foot, medial and lateral aspects.

The fibula is obviously not a weight-bearing bone and its main functions are to provide muscular attachments and to help to form the ankle joint by means of its lower end, the *lateral malleolus*.

## THE TARSAL BONES

These are shown in Figs 5.8 & 5.9. The whole weight of the body is transmitted from the tibia to the talus, which is held firmly in position between the medial and lateral malleoli. From the talus, the weight is transmitted backwards to the calcaneus and forwards to the metatarsal heads, in the normal standing position.

## TALUS

The upper surface has a large facet for articulation with the tibia and facets on the medial and lateral sides for the medial and lateral malleoli respectively. It is important in examining fractures in this region to remember that the lateral

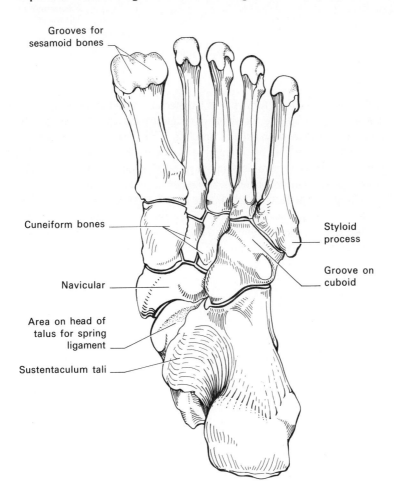

**Fig. 5.9** The plantar surface of the skeleton of the left foot.

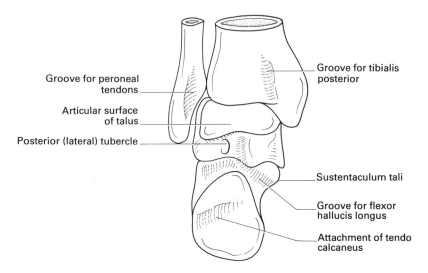

Groove for peroneal tendons

Articular surface of talus

Posterior (lateral) tubercle

Groove for tibialis posterior

Sustentaculum tali

Groove for flexor hallucis longus

Attachment of tendo calcaneus

**Fig. 5.10** The ankle bones from behind. Notice the grooves for tendons on each of the four bones.

malleolus comes down further than the medial; hence the medial facet is comma-shaped and restricted to the upper part of the talus, while the lateral is triangular and occupies most of the lateral side (Fig. 5.8). Note that the superior articular surface is wider in front than behind so that, when the foot is dorsiflexed, the talus is wedged between the malleoli but in plantar flexion there is a little play so that a minor degree of inversion and eversion can occur at the joint.

The groove at the back of the talus (Fig. 5.10) is for the *flexor hallucis longus* tendon. It lies just alongside the posterior (also called lateral) tubercle, which is ossified from a separate centre and may not fuse with the rest of the talus. In this case it is called an *os trigonum*.

It may avoid some errors if you remember that there are no muscular or tendinous attachments to the talus although, of course, a number of ligaments are attached to it.

## CALCANEUS

The medial side is easily recognized because of the sustentaculum tali, which is cantilevered out to support the head of the talus. There are two facets for talocalcaneal joints, a saddle-shaped posterior facet and a curved anterior one which includes the sustentaculum tali (Fig. 5.11). Often the latter is divided into two separate facets. The posterior surface can be divided into three areas (Fig. 5.10). The middle roughened area marks the insertion of the tendo calcaneus. The upper area is smooth and is the site of a bursa between the tendon and the bone, while the lower area is covered by part of the fibro-fatty pad that forms the heel (see Fig. 7.6). The under-surface has medial and lateral tubercles, of which the former is the larger and is the weight-bearing part of the heel. The superficial muscles of the sole of the foot are covered by the plantar aponeurosis and the site

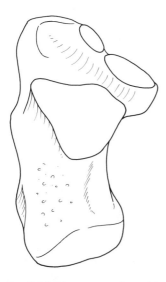

**Fig. 5.11** The upper surface of the calcaneus, showing the large and the two small facets for the talus.

of attachment of this to the medial and lateral tubercles is obvious. The muscles, therefore, are only found in front of the weight-bearing part of the calcaneus.

## THE CUBOID

This articulates with the anterior facet on the calcaneus and, like the hamate in the hand, supports the two postaxial metatarsals (p. 111). Its under-surface has a groove for the tendon of *peroneus longus* (see Fig. 5.9).

## NAVICULAR

This is supposed to be boat-shaped but is more like a coracle. It articulates with the head of the talus behind and with the three cuneiforms in front. On the under-surface of the tarsus there is a triangular gap between the navicular and the sustentaculum tali (see Fig. 5.9) and the head of the talus appears to be unsupported here. In life the *spring ligament* (p. 161) spans between the two bones and provides a resilient support.

## CUNEIFORMS

Cuneus means a wedge (Fig. 5.12). The shape of the cuneiform bones therefore helps to maintain the transverse arch of the foot.

## METATARSALS

The 1st metatarsal is very obviously thicker than the others because it takes most of the body weight during stepping off. The 2nd metatarsal is the longest and thinnest and, if the mechanism of the foot is upset by muscle injury or fatigue, may find itself taking over the job of the 1st metatarsal, in which case it is liable to break (*march fracture*). As in the hand (metacarpals), the 1st metatarsal articulates with

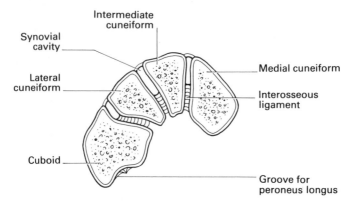

**Fig. 5.12** The wedge shape of the cuneiform bones is largely responsible for the transverse arch of the foot (actually it is only half an arch).

only one bone (the medial cuneiform) while the 2nd metatarsal takes on three (the three cuneiforms). The 1st metatarsal head is large and marked by two grooves on its under aspect. Two sesamoid bones articulate with these (p. 153).

# Chapter 6
# The Thigh

There are many similarities between the lower and upper limbs but these cannot easily be understood without a brief reference to their development. They appear as paired *limb buds* at an early stage of development (see Fig. 1.12), and are at first simple flipper-like appendages so that the upper and lower limbs are similar in appearance. Each has a *dorsal* and a *ventral surface* and a *pre-axial* and *postaxial* border, the former being the border on the headward edge of the limb. Later in development, the ends become expanded and flattened to form *hand* and *foot plates*, in which the digits develop. The digits nearest to the pre-axial borders are the thumb and the big toe. The limbs then rotate in the direction of the arrows in Fig. 1.12 so that the pre-axial border (and thumb) of the upper limb are lateral and the pre-axial border (and big toe) of the lower limb are medial. Thus the dorsal surfaces of the thigh and leg face anteriorly and the ventral surfaces posteriorly. The course of the great saphenous vein is a rough guide to the pre-axial border. The cubital fossa corresponds to the popliteal fossa and the extensor group of muscles are on the back of the forearm and the front of the leg. *Flexion at the ankle* is therefore the position adopted by a ballet dancer on her pointes. This is also known as *plantar flexion*, the opposite movement of extension usually being known as *dorsiflexion*.

Before beginning a detailed description of the thigh, it is necessary to understand the basic form of the region. It is firstly very important to realize that the femur does not run straight down the middle of the thigh. The heads of the two femora are separated by the width of the pelvis while their lower ends may be almost in contact. The shaft of the femur is therefore very oblique, particularly in the female with her wider pelvis (Fig. 6.1). The four parts of the quadriceps muscle are wrapped around the femur so they, too, are oblique, while the adductors fill in the space between quadriceps and the medial side of the thigh which is marked out by gracilis. The adductors are inserted into the linea aspera and so are on a more posterior plane than quadriceps; hence there is a depression between the two groups of muscles which forms the floor of the femoral triangle in the upper part of the thigh but which is bridged over by sartorius to form the *adductor (subsartorial) canal* in the lower part (Fig. 6.1). Behind the adductor group, adductor magnus in particular, are the three hamstring muscles.

The deep fascia of the lower limb is dense and strong and has many tendons and muscles attached to it (p. 34). It is particularly strong down the lateral side of the thigh, where it is thickened by longitudinal fibres to form the *iliotibial tract*. This is attached above to the iliac crest, superficial and deep to the tensor fasciae latae muscle, which is inserted into it, as is three-quarters of gluteus maximus. Below, it is attached to the lateral condyle of the tibia. It serves as a useful source of long fascial strips that are used as a darning material in the surgical repair of hernias.

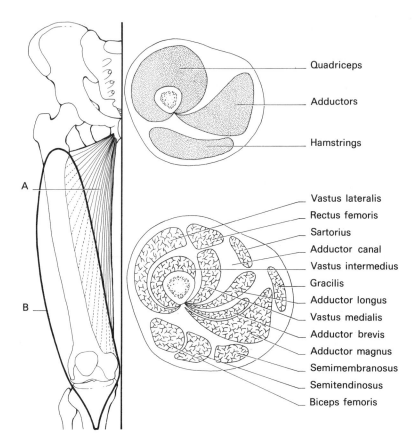

A – Adductors

B – Outline of main quadriceps mass

**Fig. 6.1** Diagrams to show the main groups of muscles in the thigh. Note the obliquity of the femur and hence of the quadriceps femoris.

## THE HIP JOINT

This is a ball and socket joint and is very much more stable than the shoulder joint because of the very deep socket, its very strong ligaments and the powerful muscles that surround it. The weight-bearing part of the acetabulum is covered with articular cartilage and is horseshoe-shaped (see Fig. 5.1); the lower part of the acetabulum is occupied by a fat pad. The whole is deepened by a *labrum acetabulare*, which is carried across the acetabular notch by a *transverse ligament* (Fig. 6.2). The articular surface of the head of the femur is similarly covered with cartilage except for the area of attachment of the ligamentum teres (see below). The *capsular ligament* encloses the joint, being attached along the trochanteric line anteriorly but to the posterior aspect of the femoral neck behind. It has three thickenings which pass between the femur and each of the three components of

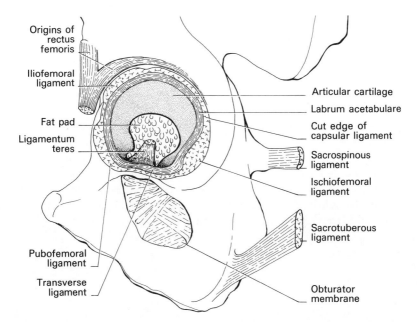

**Fig. 6.2** Lateral view of the acetabulum after division of the capsular ligament and removal of the femur.

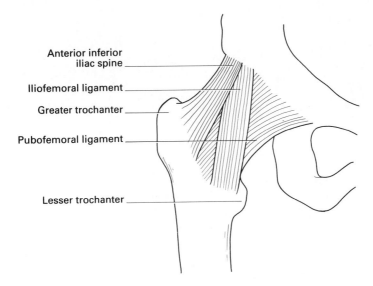

**Fig. 6.3** The hip joint, shown from the front because this is where the most powerful ligaments are.

the hip bone so that they are named *iliofemoral, ischiofemoral* and *pubofemoral* ligaments (Fig. 6.2). The iliofemoral ligament is the strongest; it passes from the anterior inferior iliac spine to the intertrochanteric line and is Y-shaped (Fig. 6.3).

Some of the fibre bundles of the capsular ligament pass up along the neck of the femur as fibrous bands called *retinacula*. They convey blood vessels to and from the head of the femur, entering via the small holes around the base of the head (see also Chapter 5). The synovial membrane has the usual disposition. Movements that can take place at the joint are flexion and extension, abduction and adduction and medial and lateral rotation. Circumduction is a combination of these movements.

In examining the range of movement at the hip joint it is important to exclude simultaneous movement of the pelvis. Thus, in testing abduction, the pelvis must be held firmly to prevent it tilting; when examining a standing patient the position of the pelvis must always be checked (Fig. 6.4). Similarly, in testing for flexion and extension the movements of the pelvis may cause problems. Figure 6.5 shows how a hip that is held in flexion may be made to appear normal by tilting the pelvis and increasing the lumbar curve of the spine.

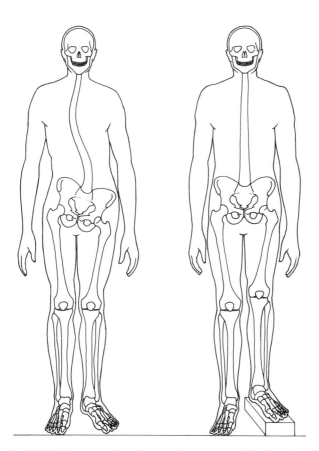

**Fig. 6.4** A shortening of one leg may be compensated for by a tilting of the pelvis. It may thus not be apparent unless the examiner investigates the relative levels of the anterior superior iliac spines.

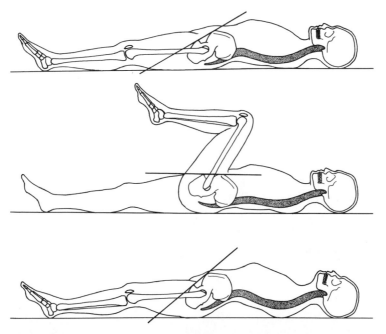

**Fig. 6.5** Flexion of the hip joint can tilt the pelvis backwards and flatten the lumbar spine. The lowest diagram shows how a flexion deformity of the hip joint can be compensated for by a forward tilt of the pelvis and an increase in the lumbar curvature.

## THE FEMORAL TRIANGLE

The boundaries are the inguinal ligament, the lateral border of sartorius and the *medial* border of adductor longus. The latter muscle, therefore, lies in the floor of the triangle (Fig. 6.6). The other muscles of the floor are (from medial to lateral) pectineus, psoas tendon and iliacus. Covering the muscles is the thick deep fascia of the thigh (*fascia lata*), in which lies the *saphenous opening*. The latter is covered by the *cribriform fascia*, which is pierced by the great saphenous vein, some lymphatics and some small arteries that are branches of the femoral artery; the accompanying veins drain into the great saphenous. The site of entry of the great saphenous vein into the femoral vein is 4 cm downwards and laterally from the pubic tubercle, an important landmark in the surgery of varicose veins (Fig. 6.7).

The principal contents of the femoral triangle are the femoral nerve, artery and vein (from lateral to medial), and their branches and tributaries, and the superficial and deep inguinal lymph nodes. The femoral nerve is separated from the hip joint by the iliacus muscle, the femoral artery by psoas and the femoral vein by pectineus. Medial to the femoral vein is the *femoral canal*, a potential opening between the abdomen and the thigh through which a hernia may occur (Fig. 6.8 and see p. 205). In the axilla, a tube of fascia from the prevertebral fascia in the neck is pushed down into the axilla to form the *axillary sheath* (p. 315). In the same way, the lining fascia of the abdominal cavity (transversalis fascia and psoas

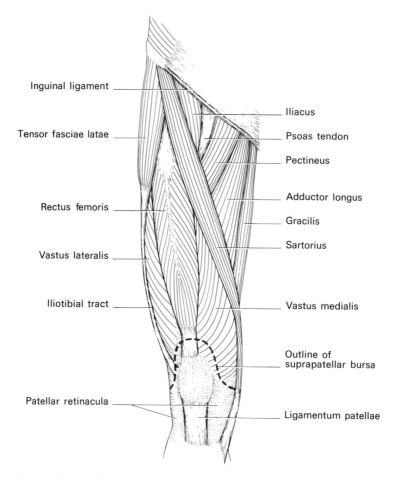

**Fig. 6.6** The muscles of the front of the thigh. The dashed line shows the outline of the suprapatellar bursa.

fascia) is evaginated under the inguinal ligament to form the *femoral sheath*. But the femoral nerve is outside (behind) the psoas fascia (see Fig. 9.7) and is thus outside the femoral sheath, which therefore encloses the artery, the vein and the canal. Incidentally, when a nerve, artery and vein run side by side they are usually in this order: NAV or VAN — for example, in the femoral triangle, in the intercostal spaces and in the pelvis (obturator nerve, artery and vein). An obvious and important exception to this general rule is in the popliteal fossa where the order is: AVN from deep to superficial.

### The inguinal lymph nodes

These are arranged in two groups, *superficial* and *deep* (to the deep fascia). The superficial nodes form a T-shape since one row of nodes runs parallel to and below the inguinal ligament while another row is arranged vertically along the great

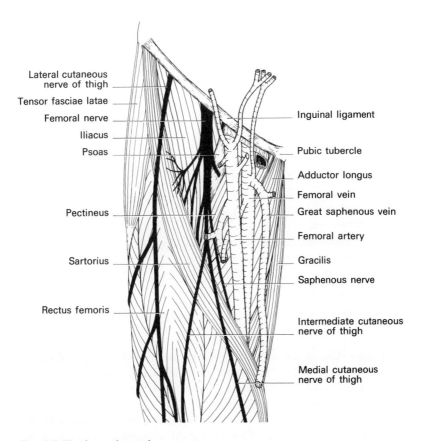

Lateral cutaneous nerve of thigh

Tensor fasciae latae

Femoral nerve

Iliacus

Psoas

Pectineus

Sartorius

Rectus femoris

Inguinal ligament

Pubic tubercle

Adductor longus

Femoral vein

Great saphenous vein

Femoral artery

Gracilis

Saphenous nerve

Intermediate cutaneous nerve of thigh

Medial cutaneous nerve of thigh

**Fig. 6.7** The femoral triangle.

saphenous vein. The deep nodes lie along the femoral vein. The most important thing to remember about the superficial nodes is that they drain the superficial tissues not only of the whole of the lower limb (including the buttocks) but also of the lower abdominal wall, the external genitalia, the perineum, the lower part of the anal canal and, to some extent, the uterus by means of lymphatics that follow the round ligament.

## The muscles of the femoral triangle

### Iliacus

This arises from most of the iliac fossa of the hip and is inserted, along with the psoas tendon, into the lesser trochanter and the region just below it. It is supplied, inside the abdomen, by branches of the femoral nerve.

### Psoas major

This arises from the transverse processes and sides of the bodies and intervertebral discs of the five lumbar vertebrae and from a series of fibrous arches that span the

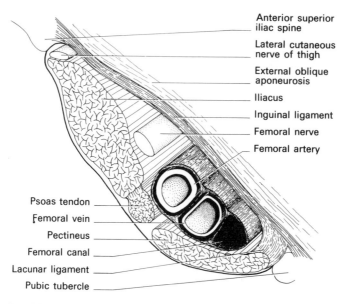

Anterior superior
iliac spine

Lateral cutaneous
nerve of thigh

External oblique
aponeurosis

Iliacus

Inguinal ligament

Femoral nerve

Femoral artery

Psoas tendon

Femoral vein

Pectineus

Femoral canal

Lacunar ligament

Pubic tubercle

**Fig. 6.8** The structures that pass below the inguinal ligament.

concave sides of the vertebral bodies. It passes with iliacus under the inguinal ligament (Fig. 6.8) and is inserted into the lesser trochanter, partly fusing with iliacus so that the two muscles are often referred to as *iliopsoas*. The nerve supply is from L1, 2 and 3 inside the abdomen.

Iliopsoas is a powerful flexor of the hip joint, both in flexing the lower limb and in raising the trunk from the lying to the sitting position. Since the greater part of the muscle is intra-abdominal (Fig. 6.9), in conditions such as acute appendicitis the hip is often held slightly flexed to relax the muscle.

### Pectineus

This arises from the superior ramus of the pubis and is inserted into the femur just below the lesser trochanter. It is supplied by the femoral and obturator nerves and is a flexor and adductor of the thigh.

### Adductor longus

This arises *by a tendon* from the pubis just below the pubic tubercle and is inserted into the linea aspera. It is supplied by the obturator nerve and is an adductor of the thigh. The insertions of iliopsoas, pectineus and adductor longus form a continuous line (Fig. 6.10) so that the muscles are more or less edge to edge. Figure 6.10 thus shows the floor of the femoral triangle from behind.

### Sartorius

This arises from the anterior superior iliac spine and is inserted into the medial surface of the tibia just below the knee. The tendon is flat and just behind it lie the equally flat tendons of gracilis and semitendinosus, with (usually) a bursa between

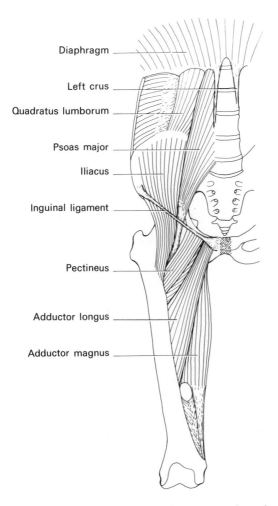

**Fig. 6.9** Psoas, iliacus and the adductor group of muscles.

Labels on figure:
Diaphragm
Left crus
Quadratus lumborum
Psoas major
Iliacus
Inguinal ligament
Pectineus
Adductor longus
Adductor magnus

the two last-named muscles. Hence, 'Say Grace Before Tea'. Sartor = tailor, and the muscle puts the lower limb into the cross-legged seated position; it produces abduction and lateral rotation at the hip and flexion at the knee. The nerve supply is femoral.

## Nerve supply of the muscles around the hip

If you find the nerve supplies of the muscles above (and below) confusing, it helps to classify them according to their action on the hip. *Flexors* are supplied by the femoral nerve, *extensors* by the inferior gluteal and sciatic, *abductors* by the superior gluteal, and *adductors* by the obturator (Fig. 6.11). Pectineus is a flexor and adductor, so is supplied by two nerves, as is adductor magnus, which is an extensor and an adductor.

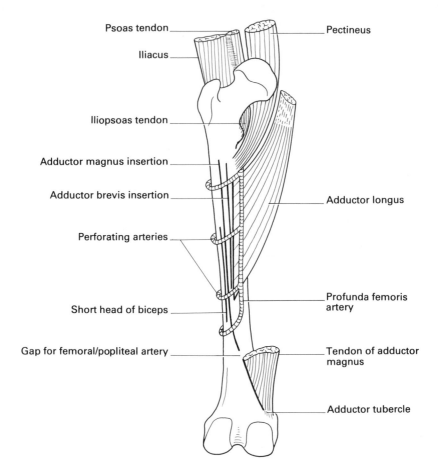

**Fig. 6.10** The muscles of the floor of the femoral triangle *seen from behind*. The attachments of other muscles to the linea aspera are shown so that you can see which muscles are perforated by the perforating arteries.

## THE FRONT AND MEDIAL SIDES OF THE THIGH

The muscles of the front of the thigh lie on a plane in front of the muscles that are inserted on to the back of the femur (some of these are mentioned above) and are shown in Fig. 6.6.

*Tensor fasciae latae* (often mistakenly written and pronounced as 'tensor fascia lata') This small muscle arises from the iliac crest and is inserted into the iliotibial tract. It is an abductor of the hip and therefore supplied by the superior gluteal nerve. Via the iliotibial tract it can also extend the knee.

### Quadriceps femoris

As its name implies, this muscle has four components; it is often confused with *quadratus* femoris. Rectus femoris is the 'odd man out' since it is the only one of

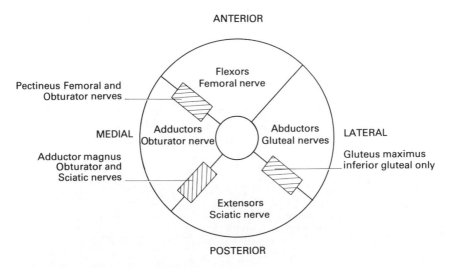

**Fig. 6.11** Nerve supply of the muscles that act on the hip joint.

the components that crosses the hip joint and is a weak flexor of this joint. It arises from the anterior inferior iliac spine and also has an oblique head from just above the acetabulum. Vastus medialis and vastus lateralis arise from the corresponding lips of the linea aspera and wrap around the femur, thus burying vastus intermedius, which arises from the shaft (see Figs 6.1 & 6.6). Rectus femoris lies in front of these three muscles (see Fig. 6.7). All four muscles converge and fuse to be (officially) inserted into the patella, which is, itself, connected to the tibial tubercle by the *ligamentum patellae*. In fact, the ligamentum is the true tendon of the insertion so that the patella is a sesamoid bone in the tendon. There are fibrous expansions, the *patellar retinacula*, from the sides of the tendon which partly enclose the knee joint, blending with its capsular ligament and helping to strengthen it (see Fig. 6.6). Deep to quadriceps and extending for a hand's-breadth above the patella is the *suprapatellar bursa* (see Fig. 6.6). This communicates freely with the knee joint and when there is a joint effusion the patella floats on a cushion of fluid. A *deep infrapatellar bursa* separates the ligamentum patellae from the upper part of the tibial tubercle (see Fig. 6.29) and there are also superficial *prepatellar* and *infrapatellar* bursae. Before the days of handy mops, these became filled with fluid in the condition of *housemaid's knee*. Quadriceps is supplied by branches from the femoral nerve and, in accordance with Hilton's law, the nerve to rectus gives a branch to the hip joint and the nerves to the vasti give branches to the knee joint. Thus disease of the hip joint may give rise to referred pain in the knee.

Quadriceps femoris is the principal extensor of the knee joint, not only in such actions as kicking but also in raising the body from a sitting to a standing position. Since quadriceps is moulded around the femur, its line of pull is oblique and tends to dislocate the patella laterally. It can, in fact, do so during a sudden violent contraction (it can even pull the patella apart!) but this is counteracted by the lower fibres of vastus medialis, which extend further distally than those of vastus

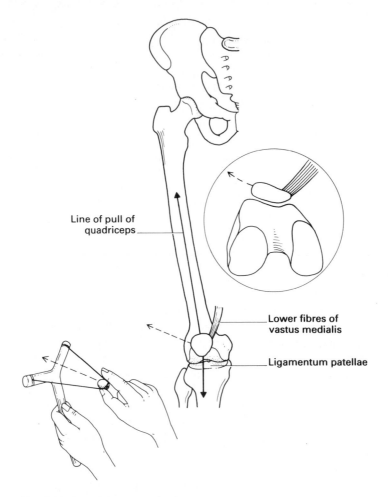

**Line of pull of quadriceps**

**Lower fibres of vastus medialis**

**Ligamentum patellae**

**Fig. 6.12** A sudden contraction of quadriceps can catapult the patella laterally. The two restraining mechanisms are shown.

lateralis (Fig. 6.12), and also by the forward projection of the lateral condyle. In full extension of the knee, quadriceps is responsible for the locking mechanism (p. 141). Its tone also contributes greatly to the stability of the joint by means of the patellar retinacula.

## The adductor muscles

These comprise *adductors longus, brevis* and *magnus*, along with gracilis and pectineus. They lie on a plane posterior to quadriceps (see Fig. 6.1) since they are attached to the linea aspera. The three adductor muscles lie one behind the other, with longus in front forming part of the floor of the femoral triangle. It has already been described, as has pectineus. All the adductors receive a nerve supply from the obturator nerve.

### Adductor brevis

This extends from the inferior ramus of the pubis to the linear aspera, overlapping the insertions of pectineus and adductor longus. It can therefore be seen from the front by separating these two muscles.

### Adductor magnus

This muscle is difficult to visualize and to draw, since it has a twist. Its origin is from the whole lengh of the inferior ramus of the pubis and the ramus of the ischium (so that its origin is a line running posterolaterally), whereas its insertion is into the whole length of the linea aspera, which is a vertical line (see Fig. 6.10). The most anterior fibres from the pubis run to the uppermost part of the linea aspera (i.e. the gluteal tuberosity) so that the muscle has a free horizontal upper order, which can just be seen in Fig. 6.13. The most posterior fibres stretch from the ischial tuberosity to the adductor tubercle and are therefore vertical. The latter part of the muscle is often called the *hamstring portion* because: (i) it arises from the ischial tuberosity; (ii) it has vertical fibres; (iii) it is supplied by the sciatic nerve (as opposed to the rest of the muscle, which is supplied by the obturator nerve, Fig. 6.11); and (iv) it is an extensor of the hip joint. The attachment of the muscle to

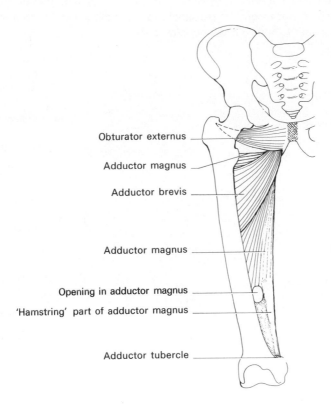

Obturator externus

Adductor magnus

Adductor brevis

Adductor magnus

Opening in adductor magnus

'Hamstring' part of adductor magnus

Adductor tubercle

**Fig. 6.13** Adductor brevis and magnus (longus has been removed).

the linea aspera is interrupted to form a gap (Figs 6.9 & 6.13) through which the femoral vessels pass.

### Gracilis

This passes down the whole length of the medial side of the thigh from the pubis and ischium to the medial side of the tibia (Say Grace Before Tea).

### Actions of the adductor muscles

All are, of course, adductors and are well developed in riders. Their main function, however, is as postural muscles. In addition, adductor longus, because of its anterior position, is also a weak flexor while adductor magnus (hamstring portion), because of its posterior position, is also an extensor. They have been known collectively as 'custodes virginitatis'.

### The adductor canal (subsartorial canal or Hunter's canal)

This is a passage, triangular in cross-section, on the medial side of the thigh (Fig. 6.14). Its floor is formed by adductor longus, but this is replaced by adductor magnus in the lower part of the thigh. Its lateral wall is vastus medialis and it is roofed over by a dense layer of fascia reinforced by sartorius. It contains the lower part of the femoral vessels and some nerves (see below). The part of the floor formed by adductor magnus is perforated to allow the femoral vessels to pass through the floor to appear in the popliteal fossa.

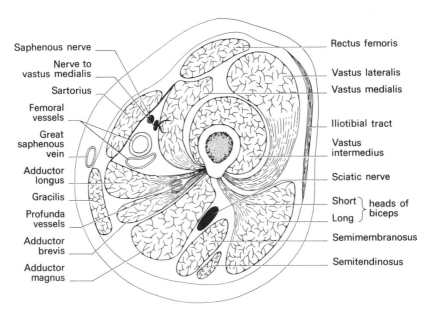

**Fig. 6.14** A cross-section through the thigh to show the subsartorial (adductor) canal.

## The vessels and nerves of the front of the thigh

### Femoral artery

This enters the thigh under the inguinal ligament halfway between the anterior superior iliac spine and the midline (the *femoral point*). It passes down to the apex of the femoral triangle, enters the adductor canal and at the lower border of adductor longus passes through the opening in adductor magnus, whereupon it changes its name to the *popliteal artery*. The femoral vein is at first medial to it but then crosses behind it (see Fig. 6.7) and at the top of the popliteal fossa the vein is lateral to the artery (see Fig. 6.25).

The artery gives off small superficial branches in the inguinal region to the lower abdomen, lateral thigh and external genitalia, but the branch you must remember is the *profunda femoris* (profunda means deep). Whereas the femoral artery passes superficial to the adductor longus muscle, the profunda artery passes deep (see Fig. 6.10). The profunda vein is superficial to the artery so that at the apex of the femoral triangle the vascular structures from front to back are: femoral artery, femoral vein, profunda vein, profunda artery. All could be damaged by a bullet or stab wound in this region (Fig. 6.15). Vascular surgeons often refer to the femoral and profunda arteries as the *common femoral, superficial femoral* and *deep femoral*.

The profunda artery gives off *medial* and *lateral circumflex femoral arteries which pass round to the back of the thigh, as do the four perforating arteries*. These arise from the profunda and wind round the shaft of the femur, perforating any muscles that they meet, i.e. the muscles attached to the linea aspera (Figs 6.10 & 6.16). The one muscle that *all* the perforating arteries perforate is adductor magnus, since this extends along the whole length of the linea aspera (see Fig. 6.10). As can be seen in Fig. 6.16 these vessels, along with the superior and inferior gluteal vessels, form a longitudinal anastomosis which can maintain the circulation if the femoral artery is occluded, at least in a young person.

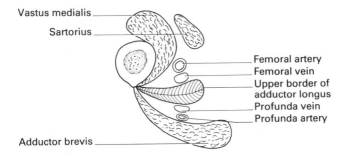

**Fig 6.15** Cross-section of some of the thigh muscles near the apex of the femoral triangle. The four major vessels of the lower limb lie one behind the other.

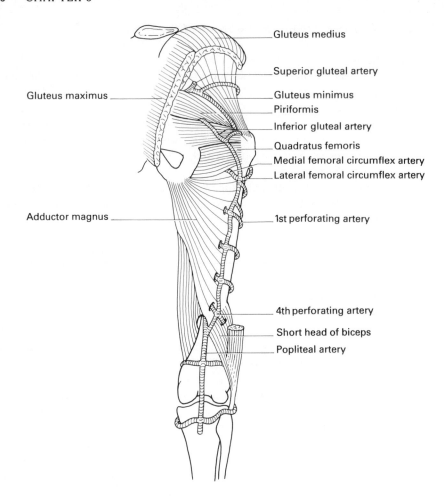

**Fig. 6.16** The longitudinal arterial anastomosis on the back of the thigh as it would appear if the femoral artery were obstructed.

### Femoral vein

Its tributaries correspond to the branches of the femoral artery, but it also receives the great saphenous vein, into which the small superficial veins of the inguinal region drain.

### Femoral nerve

The femoral nerve and the obturator nerve both come from the same lumbar nerves: L2, 3 and 4. Having read the introduction to this chapter, you will understand why the branches of these nerves that form the femoral nerve lie on a plane more dorsal than those of the obturator nerve. The femoral nerve is formed in the abdomen and emerges under the inguinal ligament *outside* the femoral sheath and lying on iliacus. Within the abdomen, the femoral nerve or branches from its roots supply psoas and iliacus. In the thigh it gives a branch to pectineus and then divides into *anterior* and *posterior divisions*. The former has one muscular and two cutaneous branches and the latter two muscular and one cutaneous. The anterior division gives the *nerve* (or nerves) to *sartorius* and the *medial* and *intermediate cutaneous nerves of the thigh*. The posterior division gives the *nerve to rectus femoris*, the nerves to the *three vasti* and the *saphenous nerve* (Fig. 6.17). The saphenous nerve and the nerve to vastus medialis are confusable since they are of similar size and both enter the adductor canal. Only the saphenous nerve comes out again, to accompany the great saphenous vein and supply the adjacent skin right down to the medial side of the foot.

### Lateral cutaneous nerve of the thigh

*This is a separate nerve* that enters the region beneath the lateral part of the

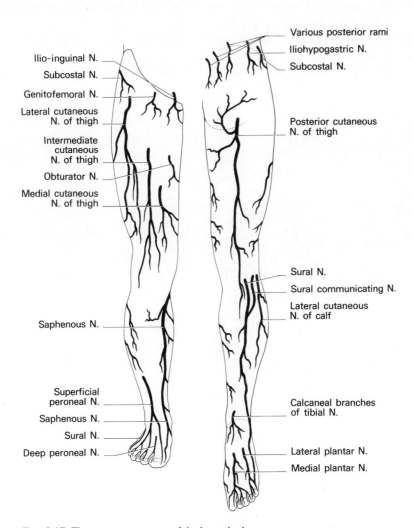

**Fig. 6.17** The cutaneous nerves of the lower limb.

inguinal ligament (Fig. 6.17). It has to pass through the very thick deep fascia in this region and sometimes becomes compressed as it does so, giving rise to pain and a tingling sensation in its area of distribution (*neuralgia paraesthetica*).

## Obturator vessels and nerve

These enter the region through the obturator foramen. The arterial branches encircle the foramen and anastomose freely with the local arteries. The nerve divides into anterior and posterior branches (Fig. 6.18), which lie respectively in front of and behind adductor brevis. The anterior division supplies adductor brevis, longus and gracilis. The posterior division supplies obturator externus, adductor brevis and magnus (also supplied by the sciatic nerve) and usually pectineus (also supplied by the femoral nerve, see Fig. 6.11). The obturator nerve also supplies both hip and knee joints so that pain in the hip joint may be referred to the knee.

### *Obturator externus* (see Fig. 6.13)

This muscle arises from the obturator membrane and the pubis and ischium

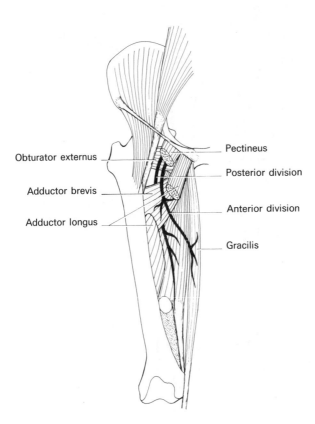

**Fig. 6.18** The obturator nerve in the thigh, shown by removing adductor longus and pectineus.

around it, dips under the hip joint and is inserted into the trochanteric fossa. It is a lateral rotator of the thigh.

## THE GLUTEAL REGION AND BACK OF THIGH

This gluteal region is limited above by the iliac crest, which, in the male, forms a watershed between the gluteal muscles and the muscles of the back. In the female, however, the fat of the region is continuous with the fat of the lumbar region so that the iliac crest is not a prominent feature. The gluteal fat is supported below by attachment of the skin to the deep fascia, thus causing a marked gluteal fold in the skin between the region and the back of the thigh. This does *not* represent the lower border of gluteus maximus.

### The gluteal muscles

Only the three main gluteal muscles will be dealt with in detail.

### Gluteus maximus

This is the largest muscle in the body although its nerve supply, the inferior gluteal nerve, is not particularly large. The muscle must have huge motor units (p. 13) and is not, therefore, renowned for its delicate and precise movements. It arises from the posterior part of the ilium and the adjacent part of the sacrum and is partly inserted into the gluteal tuberosity of the femur. This bony attachment, however, only receives about one-quarter of the fibres, the remaining three-quarters being inserted into the iliotibial tract along with tensor fasciae latae (Fig. 6.19). The two muscles have sometimes been known collectively as the deltoid of the hip joint and, acting together, they can abduct the thigh. The lower part of the muscle covers the ischial tuberosity (Fig. 6.20) but fortunately slips up to expose it when the thigh is flexed, so that it does not get sat upon (p. 99). The muscle is a powerful extensor, being used especially in such activities as climbing stairs and rising from a sitting position. It is also an abductor and lateral rotator and via its insertion into the iliotibial tract it helps to balance the trunk on the tibia and femur.

### Gluteus medius and minimus

Both extend from the outer surface of the ilium to the greater trochanter, the former muscle covering the latter (Fig. 6.21) and both being partly covered by gluteus maximus. They are supplied by the superior gluteal nerve and are abductors and medial rotators of the hip. The action you *must* remember is their tilting action on the pelvis (Fig. 6.22). When standing on the *right* leg, the *right* glutei contract and prevent the pelvis sagging to the left. Thus the glutei contract alternately during walking. Interference with the muscles or with the length of angulation of the lever they work on (the neck of the femur) will cause the pelvis to drop to the opposite side when the weight is taken on the affected leg (*Trendelenberg's sign*).

### The smaller muscles

These are shown in Fig. 6.21. They are lateral rotators. Piriformis arises from the

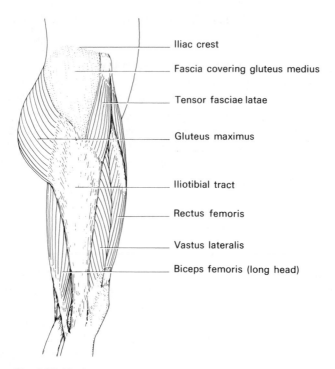

**Fig. 6.19** The lateral side of the thigh. Note the two muscles inserted into the iliotibial tract.

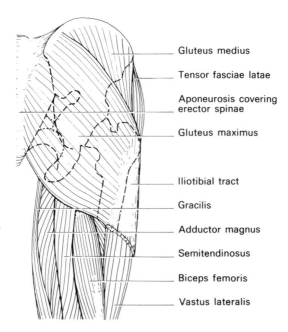

**Fig. 6.20** The superficial muscles of the gluteal region.

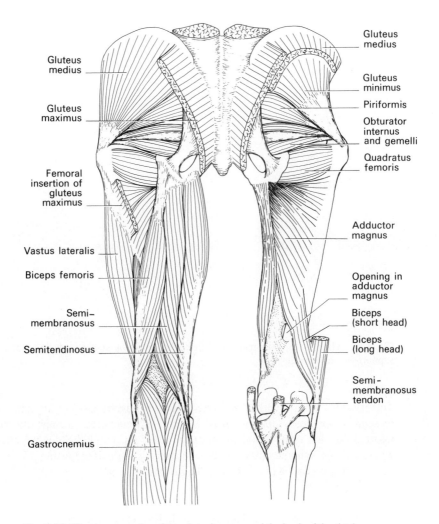

**Fig. 6.21** The deep muscles of the gluteal region and the back of the thigh.

front of the sacrum and so forms part of the pelvic wall. it is also useful as a landmark (see below).

### The hamstring muscles

These include the three true hamstrings and the 'hamstring' portion of adductor magnus (see above). Adductor magnus, in fact, is attached to almost the whole length of the linea aspera and, together with the short head of biceps and the popliteal surface of the femur, forms the floor of the back of the thigh (Fig. 6.21). The opening in adductor magnus forms a communication between the adductor canal and the back of the thigh and this is the route taken by the main vessels. You should imagine the hamstrings as being superimposed upon this floor. The hamstrings, *including* the 'hamstring' part of adductor magnus, are all attached to the ischial tuberosity and all, *except* for adductor magnus, cross two joints.

**Fig. 6.22** When the weight is taken on the *right* leg, the *right* gluteus medius and minimus support the pelvis on that side.

### Semitendinosus

This arises in common with biceps but remains on the medial side, lying superficial to semimembranosus. Its tendon is thin and prominent but flattens out to be inserted into the tibia behind sartorius (p. 118). Its muscle belly is uppermost whereas that of semimembranosus is towards the lower part of the thigh. Thus the two muscles are packed head to tail, like sardines in a tin.

### Biceps femoris

The tendon of biceps is folded around the fibular collateral ligament and, like that ligament, is attached to the upper end of the fibula. It thus crosses the thigh (and the sciatic nerve) from medial to lateral.

The hamstring muscles extend the hip and flex the knee but in many people they are not long enough to allow full flexion of the hip with extension of the knee. When testing the range of flexion at the hip joint it is therefore necessary first to flex the knee. The hamstrings are used powerfully to raise the body from bending forward. The semimembranosus and semitendinosus can medially rotate the knee, while biceps can laterally rotate it.

## Nerves and vessels of the gluteal region

These are best remembered in relation to piriformis (Fig. 6.23). Above this muscle, the *superior gluteal artery* and *nerve* (along with veins) emerge from the pelvis through the greater sciatic foramen. Below, the largest feature is the *sciatic*

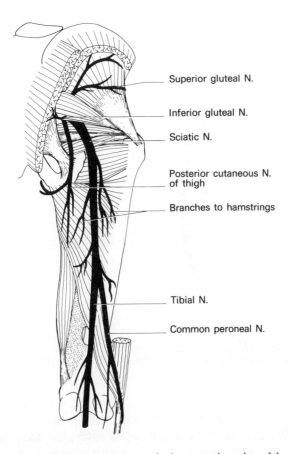

**Fig. 6.23** The sciatic nerve and other major branches of the sacral plexus.

nerve (sometimes this is replaced by two nerves, see below), the *inferior gluteal nerve* and vessels and the *posterior cutaneous nerve of the thigh*. There is also the very small nerve to *quadratus femoris* and an artery lying between two nerves: the *internal pudendal artery*, the *nerve to obturator internus* and the *pudendal nerve*. These three are simply taking the easiest route from the pelvis to the perineum by crossing the spine of the ischium (p. 275) and will be described in Chapter 13.

### Superior gluteal nerve

Its two branches lie between gluteus medius and minimus, supply them both and end in the tensor fasciae latae. It is the *nerve of the abductors of the hip*.

### Inferior gluteal nerve

This supplies gluteus maximus only. Note the relative sizes of the nerve and muscle.

### Sciatic nerve

The gluteal region is often used for intramuscular injections and the sciatic nerve

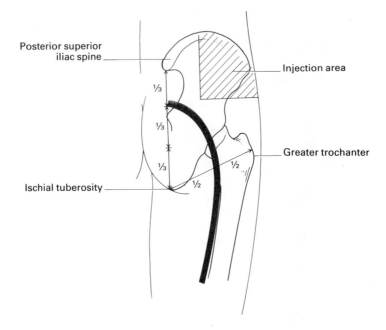

**Fig. 6.24** Surface markings of the sciatic nerve. Intramuscular injections are given into the upper and outer quadrant of the buttock (or into the lateral side of the thigh).

must be avoided at all costs. Its surface markings are shown in Fig. 6.24 and the shaded area indicates that injections should be made into the upper lateral quadrant of the buttock (but the lateral side of the thigh is better).

The sciatic nerve really consists of two nerves, the *tibial* and the *common peroneal*, bound together to form the largest nerve in the body. Not uncommonly, however, the two nerves leave the pelvis separately and, when this happens, the common peroneal usually pierces piriformis. The normal point of division, however, is near the apex of the popliteal fossa. The nerve passes over the short muscles of the gluteal region (Figs 6.21 & 6.23) before emerging from under cover of gluteus maximus. The hamstrings arise from the ischial tuberosity, which is medial to the sciatic nerve. The only one of the hamstrings that crosses to the lateral side, the long head of biceps, therefore has to cross the sciatic nerve. In its course, the sciatic nerve gives branches to the hamstrings (including the 'hamstring' part of adductor magnus). These all come from the tibial component except for the branch to the short head of biceps, which is from the common peroneal component.

### The posterior cutaneous nerve of the thigh

This nerve is surprisingly large and is distributed to a wide area of skin over the back of the thigh and the popliteal fossa. It also supplies perineal branches and branches than turn up to supply the skin of the buttocks.

### The arteries of the back of the thigh

These are normally rather inconspicuous, the femoral artery providing the major blood supply to the lower limb. The superior and inferior gluteal arteries supply local muscles and anastomose with each other, with the medial and lateral circumflex femoral arteries and with the lst perforating artery. This is the so-called *cruciate anastomosis* but the arteries do not form a cross as is popularly supposed. When the femoral artery is occluded, these anastomoses, along with the chain of anastomoses that joins the perforating arteries, can carry blood to the popliteal artery.

## THE POPLITEAL FOSSA

This diamond-shaped area is bounded by the hamstrings above and by the two heads of gastrocnemius below (Fig. 6.25). It is roofed over by dense deep fascia, which is pierced by the *short saphenous vein* on its way to join the popliteal vein (not shown in Fig. 6.25). Its floor is formed largely by the bare popliteal surface of the femur, below which is the posterior part of the capsule of the knee joint and, below that again, the dense fascia covering the popliteus muscle. It contains a great deal of fat and a small group of popliteal lymph nodes. The sciatic nerve usually divides near the upper end of the fossa into tibial and common peroneal nerves. The former passes straight down the middle while the latter follows the tendon of biceps to the fibula and winds obliquely round the neck of that bone. The tibial nerve forms a triad with the popliteal vessels, i.e. the femoral artery and vein, which changed their names as they passed through the gap in adductor magnus.

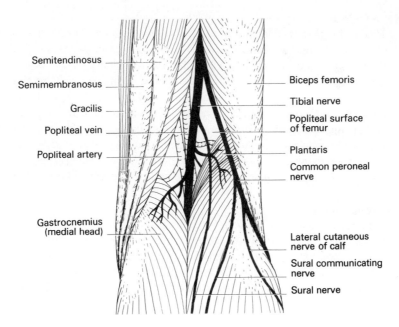

**Fig. 6.25** The right popliteal fossa.

The nerve is the most superficial and the artery is the deepest, lying directly on the bone, where it can be damaged by supracondylar fractures.

The popliteal artery gives a series of branches (*genicular arteries*) that form an anastomosis around the knee joint. It then divides into *anterior* and *posterior tibial arteries*. The former passes between the tibia and fibula to reach the front of the leg while the latter continues down the back. The vein has corresponding tributaries and also receives the short saphenous vein.

### Tibial nerve

In the fossa this gives articular (*genicular*), muscular and cutaneous branches. There are several genicular nerves to the knee joint, muscular branches supply gastrocnemius, soleus, plantaris and popliteus; the cutaneous branch is the *sural*, which is joined by the *sural communicating branch* from the common peroneal and then supplies skin along the lateral side of the leg and foot (see Fig. 6.17).

### The common peroneal nerve

This can be palpated just medial to the biceps tendon and again on the neck of the fibula. It gives genicular branches to the knee joint, the *sural communicating nerve* and also the *lateral cutaneous nerve of the calf*. It ends by dividing into *superficial* and *deep peroneal nerves*, the former supplying the structures on the lateral side of the leg and the latter passing to the front of the leg, where it joins the anterior tibial artery. Note especially that the nerve reaches the front of the leg by passing round the lateral bone, while the artery reaches it by passing between the bones (see Fig. 7.4). The same is true of the back of the arm (p. 94).

## KNEE JOINT

This is a hinge joint which can also undergo some rotation. The articular surfaces of the femur and the tibia are covered by articular cartilage, as is the posterior surface of the patella. The patella articulates with the femur; it does *not* cover the gap between the bones. Within the knee joint lie the *medial* and *lateral menisci*. The shapes of the articular surfaces are shown in Fig. 6.26. Those of the femur are not immediately obvious but close inspection will reveal faint grooves that demarcate the upper limit of the surfaces, when it can be seen that the articular area on the lateral condyle is roughly circular while that of the medial condyle is oval. The menisci and the articular facets of the tibia have a corresponding shape. *The fibula plays no part in the formation of the joint and is not a weight-bearing bone.*

### The ligaments

The capsular ligament encloses the joint in the usual way and has two major openings in it. In front, the synovial membrane of the joint is continuous, through a large opening, with the *suprapatellar bursa*. This is a bursa that lies deep to quadriceps and extends upwards for three finger-breadths above the top of the patella (Figs 6.6 & 6.27). Posterolaterally, there is an opening for the passage of the tendon of popliteus (Fig. 6.27).

On the medial side, the *tibial collateral ligament* is a wide thickening in the capsular ligament, extending a surprisingly long way down the tibia (Fig. 6.28). On its deep surface, it is firmly attached to the medial meniscus. The *lateral*

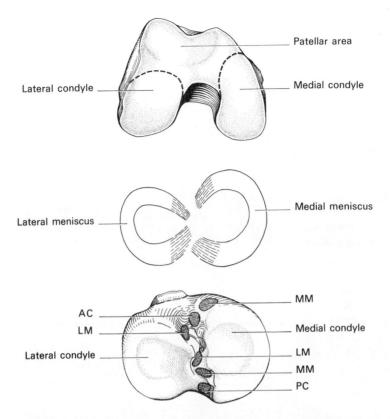

**Fig. 6.26** The lateral articular surface on the femoral condyle, the lateral meniscus and the lateral articular surface of the tibia are all circular. The corresponding surfaces on the medial side are all approximately oval. LM and MM: lateral and medial menisci; AC and PC: anterior and posterior cruciate ligaments. The ligaments themselves are shown (in the other knee) in Fig. 6.29.

*ligament*, on the other hand, is a firm cord-like structure extending from the lateral epicondyle of the femur to the head of the fibula; it is largely covered over by the biceps tendon but because it stands away from the joint capsule it is not related to the lateral meniscus (Fig. 6.27). Posteriorly, the capsular ligament is thin (as might be expected in a hinge joint) but is reinforced by some minor thickenings, one of which, the *oblique popliteal ligament,* is an extension from the semimembranosus tendon (see Fig. 6.32).

The *cruciate ligaments* are within the joint capsule. The anterior cruciate passes from the front of the intercondylar area of the tibia to the medial side of the lateral femoral condyle while the posterior runs from the back of the intercondylar area to the lateral side of the medial condyle (Figs 6.29 & 6.30). In case you find these attachments difficult to memorize, here is a lamp to guide your footsteps:

**L**ateral condyle receives the
**A**nterior cruciate.
**M**edial condyle receives the
**P**osterior cruciate.

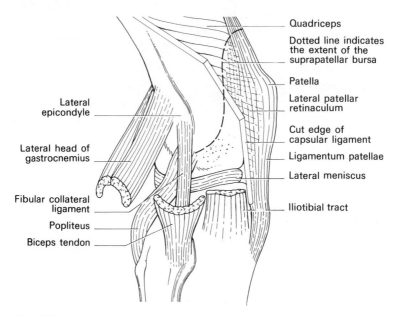

Quadriceps

Dotted line indicates
the extent of the
suprapatellar bursa

Patella

Lateral patellar
retinaculum

Cut edge of
capsular ligament

Ligamentum patellae

Lateral meniscus

Iliotibial tract

Lateral
epicondyle

Lateral head of
gastrocnemius

Fibular collateral
ligament

Popliteus

Biceps tendon

**Fig. 6.27** Knee joint, lateral aspect after removing part of the capsular ligament.

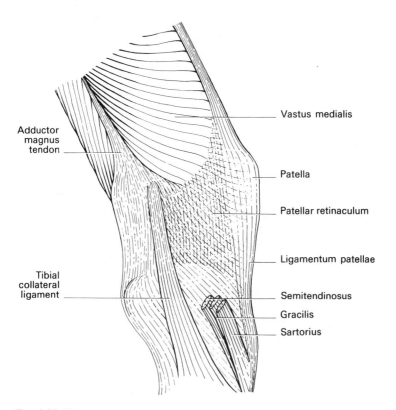

Vastus medialis

Adductor
magnus
tendon

Patella

Patellar retinaculum

Ligamentum patellae

Tibial
collateral
ligament

Semitendinosus

Gracilis

Sartorius

**Fig. 6.28** Knee joint, medial aspect.

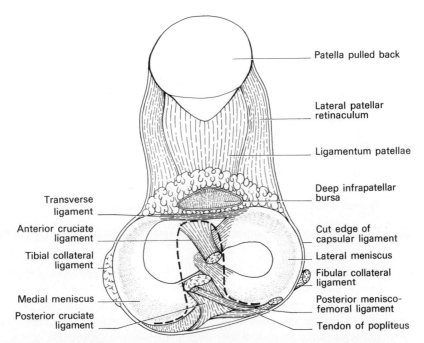

**Fig. 6.29** The articular surface of the right tibia after disarticulation at the knee joint and retraction of the patella. The dotted line indicates the synovial membrane in the vicinity of the cruciate ligaments.

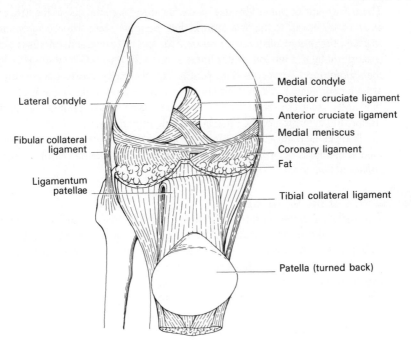

**Fig. 6.30** Anterior view of the flexed knee joint after division of quadriceps and retraction of the patella.

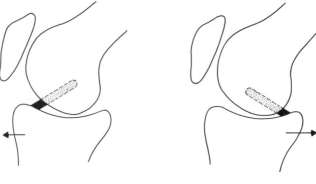

Anterior cruciate ligament          Posterior cruciate ligament

**Fig. 6.31** The cruciate ligaments prevent forward and backward displacement of the tibia.

Although inside the capsular ligament, the cruciates are outside the synovial cavity, since they appear to invaginate the synovial cavity from behind (Fig. 6.29). Their main function is to prevent forward and backward movement of the tibia (Fig. 6.31) but they also help to keep the bones together (they are taut in all positions of the joint) and they help to prevent medial rotation of the tibia, a movement which tends to twist them around each other.

### The menisci

These are made of rather fibrous fibrocartilage and are attached to the tibia at each end by their *horns*, to the tibia around their periphery by *coronary ligaments* and to each other anteriorly by a *transverse ligament*. The lateral meniscus is grooved posteriorly by the tendon of popliteus and is also attached to the femur by two *meniscofemoral ligaments* (Fig. 6.32). The menisci serve to deepen the tibial articular surfaces, to fill in dead space and thus help the distribution of synovial fluid and to adapt the tibial surfaces to the changing curvature of the femur during flexion and extension of the knee by opening out or closing.

Tears in the menisci are a common form of knee injury — most commonly the medial meniscus because it is the less mobile, being attached firmly to the tibial collateral ligament.

### Movements at the knee joint

During flexion and extension the femoral condyles carry out a rolling movement in relation to the upper surface of the tibia but this is necessarily combined with gliding so that the bones remain aligned (try the movement with a tibia and femur to understand this). Full flexion is limited by contact of the soft parts and full extension by tension in the muscles and ligaments behind the joint. Rotation of the tibia can best be carried out when the knee is flexed to a right angle but lateral rotation of the tibia (or medial rotation of the femur) also occurs involuntarily during full extension. Figure 6.26 shows that the anteroposterior dimension of the femoral articular surface is greater on the medial than on the lateral condyle. Thus,

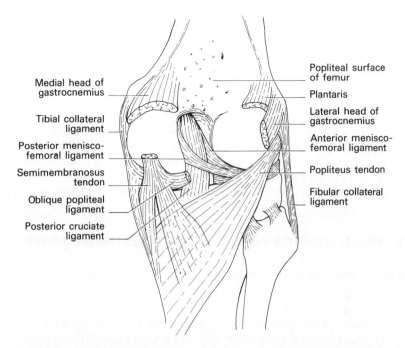

Medial head of gastrocnemius

Tibial collateral ligament

Posterior menisco-femoral ligament

Semimembranosus tendon

Oblique popliteal ligament

Posterior cruciate ligament

Popliteal surface of femur

Plantaris

Lateral head of gastrocnemius

Anterior menisco-femoral ligament

Popliteus tendon

Fibular collateral ligament

**Fig. 6.32** Posterior aspect of the knee joint after removal of the capsular ligament.

during extension, the lateral condyle comes to the end of its run before the medial, so extension stops on the lateral side. Continuing contraction of quadriceps has the effect of continuing the movement on the medial side so that a lateral rotation of the tibia occurs to produce a 'screwing home' effect. In this position (the close-packed position), all the ligaments are taut and the joint is locked and very stable. Weakness of quadriceps affects this movement and the knee cannot then be rigidly extended.

# Chapter 7
# The Leg and Foot

Before reading this chapter you should read again the introduction to Chapter 6 so that you appreciate the meaning of flexion and extension at the ankle joint and thus the position of the flexor and extensor muscles. Once you have realized that the front of the leg and dorsum of the foot correspond to the back of the forearm and hand, you will find very many points of comparison between the upper and lower limbs.

## EXTENSOR ASPECT OF LEG AND DORSUM OF FOOT

### Extensor group of muscles (Fig. 7.1)

These three muscles (four if you include the unimportant peroneus tertius) all dorsiflex the foot and all are supplied by the deep peroneal nerve. The muscles are important in the 'carry through' stage of walking, i.e. they keep the toes clear of the ground when the non-weight-bearing leg is being carried forward ready for the next step. A lesion of the nerve will cause *foot drop* (cf. *wrist drop* resulting from a lesion of the posterior interosseous nerve) so that the toes drag along the ground and the tips of the shoes are scuffed. This nerve lesion can be diagnosed by an inspection of the shoes.

### *Tibialis anterior*

This arises from the shaft of the tibia, and its thick tendon is easily recognized on the medial side of the front of the ankle region as it passes down to the base of the 1st metatarsal and the medial cuneiform. As well as being a dorsiflexor, it is therefore also an invertor of the foot.

### *Extensor digitorum longus and extensor hallucis longus*

Both arise from the very narrow anterior surface of the fibula (see Figs 5.7 & 7.7), digitorum from the upper three-quarters and hallucis from the middle half, the hallucis muscle belly being between digitorum and tibialis anterior. The tendons are inserted into the distal phalanges of the toes, extensor hallucis longus into the big toe and extensor digitorum longus into the other four toes. The latter tendons also have dorsal extensor expansions that receive lumbricals and interossei and they have additional attachments to the middle phalanges in an exactly similar way to the corresponding tendons in the hand (p. 88). The muscles extend the toes and dorsiflex the foot. The lowermost fibres of extensor digitorum longus may give rise to a tendon that is attached to the base of the 5th metatarsal (*peroneus tertius*).

Note that no muscle attachments on the front (or on the back) encroach on the lower third of the tibia so that this part of the bone has a poor blood supply.

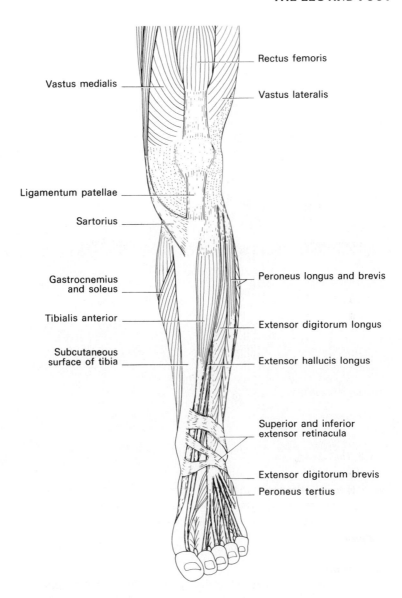

**Fig. 7.1** The extensor (dorsiflexor) group of muscles.

Labels on figure:
Rectus femoris
Vastus medialis
Vastus lateralis
Ligamentum patellae
Sartorius
Gastrocnemius and soleus
Peroneus longus and brevis
Tibialis anterior
Extensor digitorum longus
Subcutaneous surface of tibia
Extensor hallucis longus
Superior and inferior extensor retinacula
Extensor digitorum brevis
Peroneus tertius

## The peroneal muscles

These two muscles, *peroneus longus* and *brevis*, arise from the lateral surface of the fibula, the former from the upper two-thirds and the latter from the lower two-thirds. Their tendons use the lower end of the fibula as a pulley and pass forwards on the lateral side of the foot, peroneus brevis being the uppermost (Fig. 7.2). Brevis is attached to the base of the 5th metatarsal but longus turns medially, through the groove on the cuboid (converted into a tunnel by the long plantar

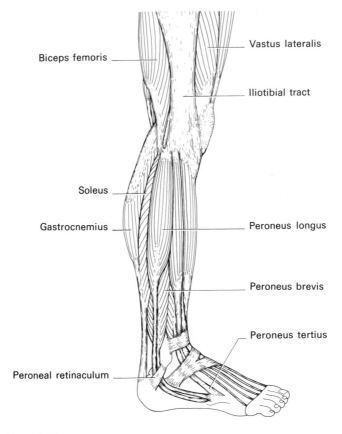

Biceps femoris

Vastus lateralis

Iliotibial tract

Soleus

Gastrocnemius

Peroneus longus

Peroneus brevis

Peroneus tertius

Peroneal retinaculum

**Fig. 7.2** The peronei.

ligament) and is attached to the lateral sides of the base of the lst metatarsal and the medial cuneiform, opposite to tibialis anterior. They are evertors of the foot and peroneus longus may act as a dorsiflexor. The nerve supply is superficial peroneal.

### Extensor digitorum brevis

This is a small muscle attached to the upper surface of the calcaneus. Its four tendons join those of the main extensors to the *medial four* toes. That to the big toe is often called *extensor hallucis brevis*. The nerve supply is deep peroneal.

## Extensor and peroneal retinacula

These are thickenings in the already thick deep fascia of the leg, the collagen fibres running transversely. Their function is to hold the tendons down and prevent them bowstringing. The *superior extensor retinaculum* (Fig. 7.1) is a simple band passing from tibia to fibula. The *inferior retinaculum* is Y-shaped with a stem that is attached to the calcaneus and two limbs that are attached medially to the medial malleolus and the deep fascia on the medial side of the foot, including the plantar aponeurosis. The peroneal retinaculum holds down the tendons of the peroneal muscles and runs from the fibula to the calcaneus. The tendons are enclosed in synovial sheaths beneath and in the vicinity of the retinacula.

## Nerves and vessels on the front of the leg

As in the back of the forearm, the main artery passes between the two bones while the nerve winds round the lateral bone (see Fig. 7.4).

### Anterior tibial vessels

The artery is accompanied throughout by a pair of venae comitantes. The artery is a terminal branch of the tibial artery and passes above the upper border of the interosseous membrane and then descends, lying deeply between the muscles which it supplies. In front of the ankle joint it lies midway between the two malleoli and just lateral to the tendon of extensor hallucis longus, an important point to examine for pulsation when investigating occlusive arterial disease of the lower limb. It then changes its name to *dorsalis pedis* and travels forwards to the level of the base of the metatarsals. It passes down to the sole of the foot between the two heads of the first dorsal interosseous muscle, but as it does so it gives the *1st dorsal metatarsal artery* to the cleft between the big and second toes. Among its other branches is the *arcuate artery*, which arches across to the lateral side of the foot and gives the remaining three dorsal metatarsal arteries, which divide to supply the sides of the toes.

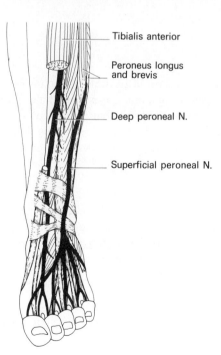

Tibialis anterior

Peroneus longus and brevis

Deep peroneal N.

Superficial peroneal N.

**Fig. 7.3** The nerves on the front of the leg. They are, in fact, not as large as shown in the diagram.

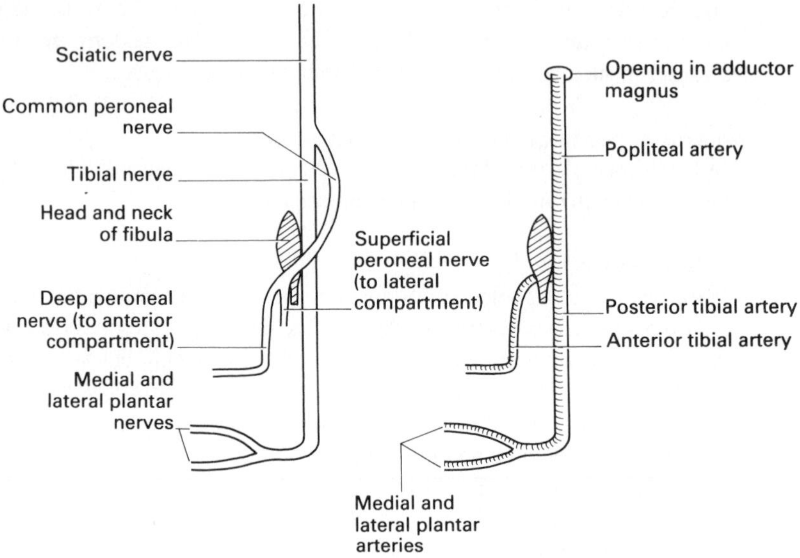

**Fig. 7.4** The principal nerves and arteries of the leg, viewed from the lateral side. Remember that the nerve to the extensor aspect (front) of the leg winds round the lateral bone but the artery passes between the bones; the same arrangement is seen in the upper limb.

*Deep peroneal nerve*

This is a branch of the common peroneal nerve that winds round the neck of the fibula and pierces the origin of extensor digitorum longus to reach the front of the leg (Fig. 7.3, p. 145). It runs in company with the anterior tibial artery, supplies the extensor muscles and, after reaching the dorsum of the foot, supplies extensor digitorum brevis and gives a cutaneous branch to the cleft between the big and second toes (see Fig. 6.17).

*Superficial peroneal nerve*

This is the other branch of the common peroneal. It runs down between the peronei, supplies them and then becomes superficial on the front of the leg, where it breaks up into branches to supply most of the dorsum of the foot and the toes (Fig. 7.3). Since the branches of the sciatic nerve are so commonly confused, they are shown in outline form in Fig. 7.4.

## FLEXOR ASPECT OF LEG AND SOLE OF FOOT

There are two groups of muscles in the calf. The most superficial group comprise the three muscles that, in some animals, are of equal size and are known as triceps surae. In man, one of them, plantaris, is rudimentary, leaving only gastrocnemius and soleus.

## Gastrocnemius

This muscle arises by two heads from the medial and lateral condyles of the femur (Fig. 7.5). They join to form the main belly of the muscle, which is inserted in common with soleus.

## Soleus

This muscle has two origins, one from the soleal line on the tibia (see Fig. 5.4) and one from the upper third of the posterior surface of the fibula. The muscle fibres on the deep surface of the muscle are arranged in a multipennate manner but the superficial fibres run longitudinally. Soleus and gastrocnemius are inserted along with plantaris (a small muscle arising with the lateral head of gastrocnemius) into the tendo calcaneus (or *Achilles tendon*, so called because Achilles' mother held him by this region of the foot when she dipped him into the river Styx to make him

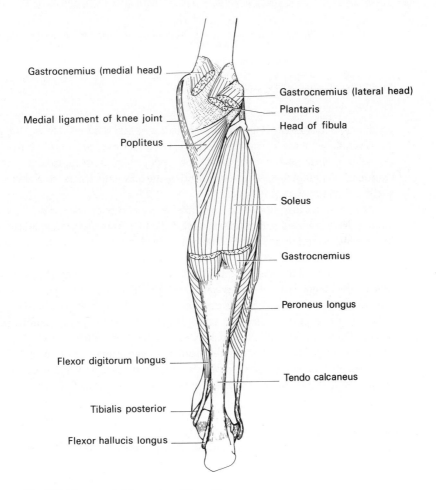

**Fig. 7.5** The superficial muscles of the calf.

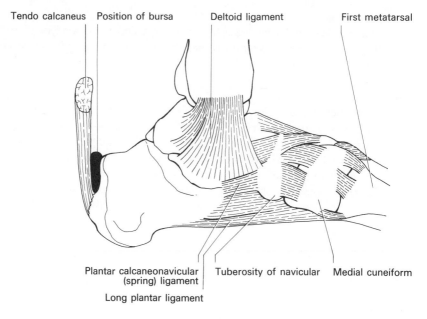

Tendo calcaneus    Position of bursa        Deltoid ligament                First metatarsal

Plantar calcaneonavicular | Tuberosity of navicular   Medial cuneiform
(spring) ligament
Long plantar ligament

**Fig. 7.6** Ankle joint, medial aspect.

invulnerable to wounds. It later proved to be a pity that she forgot to wet the part she was holding him by). The tendo calcaneus is attached to the middle third of the posterior surface of the calcaneus. The upper third is occupied by a bursa (*retrocalcaneal bursa*) and the lower third by the fibro-fatty tissue which takes the weight when standing (Fig. 7.6).

Both muscles are supplied by the tibial nerve and both are powerful plantar flexors. They are used particularly in raising the heel from the ground in walking, in standing on tiptoe and, when the foot is off the ground, in such activities as pedalling a bicycle. The calcaneus acts as a lever in these actions, since it projects a long way behind the ankle joint. The deep, multipennate fibres of soleus have only a small range but are very powerful (p. 14). This part of soleus is therefore used as a postural muscle. Soleus does not cross the knee joint, so when the knee is bent (so that gastrocnemius is relaxed) plantar flexion is only performed by soleus. These two muscles, particularly soleus, have an extensive venous plexus within the body of the muscle and they form the main *muscle pump* (p. 21). These veins are also a common site of deep venous thrombosis, which may occur after surgery, and, in this condition, passive dorsiflexion gives rise to calf pain (Homan's sign).

### The deep muscles of the back of the leg

There are four muscles in this group, all of which are supplied by the tibial nerve, the uppermost being the small *popliteus*. This muscle arises from the tibia above the soleal line (Fig. 7.5) and passes deep to the fibular collateral ligament, grooving the back of the lateral meniscus as it does so. It is inserted into a fossa on

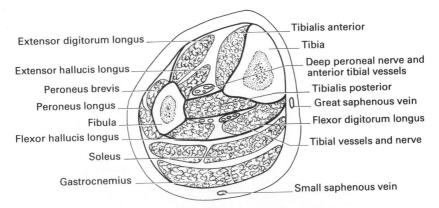

Extensor digitorum longus

Extensor hallucis longus

Peroneus brevis

Peroneus longus

Fibula

Flexor hallucis longus

Soleus

Gastrocnemius

Tibialis anterior

Tibia

Deep peroneal nerve and anterior tibial vessels

Tibialis posterior

Great saphenous vein

Flexor digitorum longus

Tibial vessels and nerve

Small saphenous vein

**Fig. 7.7** Cross-section of the leg, to be studied in conjunction with Fig. 5.7.

the lateral side of the lateral condyle of the femur. It is supplied by the tibial nerve and its action is to reverse the locking of the knee joint that is performed by quadriceps (p. 141), i.e. it is a flexor and a medial rotator of the tibia.

### Tibialis posterior

Arising from the posterior surfaces of the tibia and fibula and the interosseous membrane, its tendon runs in a groove behind the medial malleolus, turns forward across the deltoid ligament (Fig. 7.5) and is inserted into the tubercle of the navicular and most of the other tarsals and metatarsal bases but not the talus (*the talus has no muscle attachments*). It is separated from the tibialis anterior tendon by a roughly triangular hollow forming a sort of 'anatomical snuffbox' which has the head of the talus in its floor. It is an invertor and plantar flexor of the foot and helps to hold up the medial arch.

### Flexor digitorum longus

This arises from the posterior surface of the tibia lateral to tibialis posterior (Fig. 7.7) and its tendon crosses that of tibialis posterior behind the medial malleolus (Fig. 7.8), turns forwards across the sustentaculum tali (see Fig. 7.18), receives the attachment of flexor digitorum accessorius (see Fig. 7.11) and breaks up into four tendons for each of the four lesser toes. Its tendons resemble closely those of flexor digitorum profundus in the hand, i.e. they enter a fibrous flexor sheath, pass through a complex opening in a more superficial tendon (flexor digitorum brevis) and are inserted into the terminal phalanx.

### Flexor hallucis longus

Arising from the posterior surface of the fibula behind tibialis posterior, its tendon passes in a groove behind the talus and another below the sustentaculum tali (see Fig. 5.10). In the sole of the foot it is crossed superficially by the tendon of flexor digitorum longus (Fig. 7.8) and it is inserted into the distal phalanx of the big toe.

Flexors digitorum and hallucis longus flex the toes and help in plantar flexion of the foot. It helps to remember that the tendon of flexor digitorum longus *is*

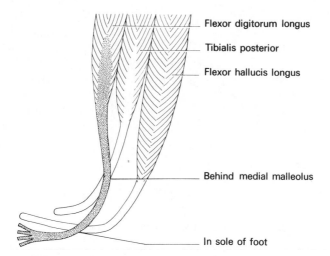

Flexor digitorum longus

Tibialis posterior

Flexor hallucis longus

Behind medial malleolus

In sole of foot

**Fig. 7.8** If you are confused over the relations of the tendons of the flexors, remember that the flexor digitorum longus tendon is superficial twice, once behind the medial malleolus and once in the sole of the foot.

*superficial twice*; it crosses superficial to tibialis posterior behind the medial malleolus and crosses superficial to flexor hallucis longus in the sole of the foot (Fig. 7.8).

## Nerves and vessels of the back of the leg

### Posterior tibial artery

This begins at the lower border of popliteus where the popliteal artery divides, and it travels down the back of the leg deep to gastrocnemius and soleus, supplying the muscles of the region. It ends by passing medially, between the medial malleolus and the heel (where its pulsation may be felt), to enter the sole of the foot deep to abductor hallucis. Here it divides into *medial and lateral plantar arteries*. Its main branch is the *peroneal artery*, which lies lateral to, and parallel with, the posterior tibial artery (of the radial and ulnar arteries in the forearm). This has a *perforating branch* that pierces the interosseous membrane and *may replace the dorsalis pedis artery*.

### Tibial nerve

This runs along with the posterior tibial artery, giving branches to the deep surface of soleus and gastrocnemius and to the three deep flexor muscles. It finally enters the sole of the foot deep to abductor hallucis and divides into medial and lateral plantar nerves.

## THE SOLE OF THE FOOT

You will find it very helpful to combine the study of the palm of the hand and the sole of the foot, remembering that the big toe and the thumb are both the pre-axial digits. The muscles of the sole are usually described as being in four layers. This is

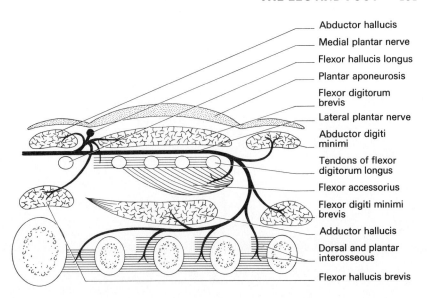

**Fig. 7.9** A highly diagrammatic cross-section of the sole to show the various planes.

not strictly true, but it is helpful in description. It is also not strictly true, but helpful as an *aide-mémoire*, that the four layers consist respectively of: three muscles; two muscles and two tendons; three muscles; and two muscles and two tendons (Fig.7.9). It is fairly obvious that the superficial muscles must have a longer span than the deeper muscles so that the first layer spans the whole length of the sole of the foot, from calcaneus to phalanges, while the third and fourth layers are restricted to the forefoot (that part in front of the talonavicular and calcaneocuboid joints). Their nerve supplies will not be mentioned in the following account, since it is easier to remember these in conjunction with those of the hand muscles (see Fig. 7.17).

### The plantar aponeurosis

Below (i.e. superficial to) the muscles of the sole of the foot is the plantar aponeurosis, which is very similar to the palmar aponeurosis. It has a thick and strong central part which covers the central muscle of the first layer, flexor digitorum brevis, and is immediately deep to the superficial fascia of the sole. It is attached posteriorly to the calcaneus just behind the muscle attachments, while anteriorly it blends with the skin at the creases at the bases of the toes and also sends five slips, one to each toe. Each of these splits into two which pass deeply, one on each side of the flexor tendons to that toe, and finally fuse with the *deep transverse metatarsal ligaments*. These are tough ligaments that link the plantar ligaments of the metatarsophalangeal joints of *all five* toes. Thus the big toe does not have the freedom of movement of the thumb.

### *The first layer (three muscles)*

This consists of a flexor between two abductors (Fig. 7.10). *Abductor hallucis* extends from the plantar surface of the calcaneus, just in front of the weight-

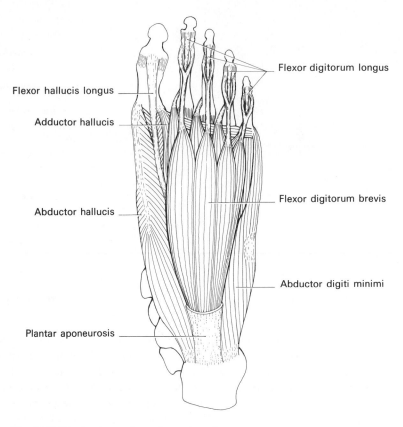

Flexor digitorum longus

Flexor hallucis longus

Adductor hallucis

Flexor digitorum brevis

Abductor hallucis

Abductor digiti minimi

Plantar aponeurosis

**Fig. 7.10** The first layer of muscles in the sole.

bearing part, to the medial side of the first phalanx of the big toe. It can abduct the first metatarsophalangeal joint (at least in people who are not used to wearing shoes). Many people, however, cannot do this effectively and it is probably more important in helping to maintain the medial longitudinal arch. *Abductor digiti minimi* has similar attachments (to the little toe) on the lateral side. *Flexor digitorum brevis* is attached to the calcaneus close to the other two muscles and divides into four tendons, one for each of the lesser toes. It is very similar in its insertion to flexor digitorum superficialis in the upper limb, i.e. each tendon, after entering a fibrous flexor sheath, is perforated in a similar way to that of superficialis and the tendon of flexor digitorum longus passes through it. It is then attached to the sides of the *middle* phalanx.

### The second layer (two muscles, two tendons)

The two tendons are those of *flexors digitorum* and *hallucis longus*, whose insertions into the distal phalanges have already been mentioned. Associated with them are *flexor digitorum accessorius* and the *lumbricals* (Fig. 7.11). The former arises by two heads that arise from the under-surface of the calcaneus, straddling the attachment of the long plantar ligament. It is inserted into the side of the tendon of flexor digitorum longus and can straighten out the oblique pull of that

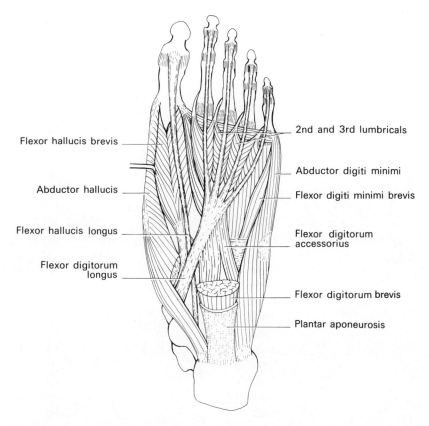

Flexor hallucis brevis

2nd and 3rd lumbricals

Abductor hallucis

Abductor digiti minimi

Flexor digiti minimi brevis

Flexor hallucis longus

Flexor digitorum accessorius

Flexor digitorum longus

Flexor digitorum brevis

Plantar aponeurosis

**Fig. 7.11** The second layer of muscles in the sole after removal of flexor digitorum brevis.

tendon; it can also hold the toes flexed when flexor digitorum longus relaxes, as occurs when simultaneous dorsiflexion of the foot occurs. The lumbricals are similar to those in the hand except for the second, which arises from *two* tendons (Fig. 7.11). Its nerve supply is different too (see Fig. 7.17).

### The third layer (three muscles)

This consists of an adductor between two flexors. The adductor is *adductor hallucis* (Fig. 7.12), which, like adductor pollicis has an oblique head, from some of the tarsal bones. Its transverse head is small, from metacarpal heads. It is inserted into the lateral side of the base of the phalanx of the big toe. The two flexors, of the hallux and of the little toe, arise from tarsal bones and the base of the 5th metatarsal respectively. *Flexor digiti minimi brevis* (no one seems to know why 'brevis' is used since there is no longus) is inserted into the lateral side of the proximal phalanx, while *flexor hallucis brevis* splits into two and is inserted into both sides of the proximal phalanx in company with the adductor (medially) and the abductor (laterally). The conjoined tendons each contain a sesamoid bone and the tendon of flexor hallucis longus runs between them. (Compare with the thumb, see Fig. 4.14, which is slightly different.)

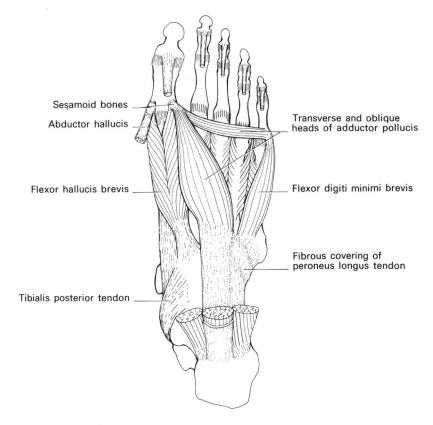

Sesamoid bones

Abductor hallucis

Flexor hallucis brevis

Tibialis posterior tendon

Transverse and oblique heads of adductor pollucis

Flexor digiti minimi brevis

Fibrous covering of peroneus longus tendon

**Fig. 7.12** The third layer of muscles in the sole.

### The fourth layer (two muscles, two tendons)

*The two tendons are those of peroneus longus*, which runs across the tarsus to reach the 1st metatarsal and medial cuneiform, and *tibialis posterior*, which has slips to most of the tarsal bones. The two muscles are the *dorsal* and *plantar interossei*. These are similar to the corresponding muscles in the hand, with two exceptions. Adduction and abduction take place in reference to the *second toe* (the *middle finger* in the hand). There is no plantar interosseous in the first interspace. You should now be able to work out the attachments of the interossei (Fig. 7.13). The second toe, since it can only be abducted (see also p. 85), has two dorsal interossei, attached to its proximal phalanx, but no plantar interossei. As well as abducting and adducting (not important), the interossei, assisted by the lumbricals, flex the metatarsophalangeal joints while extending the interphalangeal joints. The long flexors sometimes have the effect of buckling the toes (Fig. 7.14), in the same way that the links of a short length of bicycle chain will buckle if the ends are pushed together. This is counteracted by the interossei. They also help to maintain the transverse arch by bunching the metatarsals together.

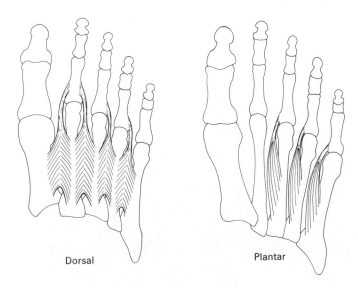

Dorsal                                  Plantar

**Fig. 7.13** The fourth layer of muscles in the sole.

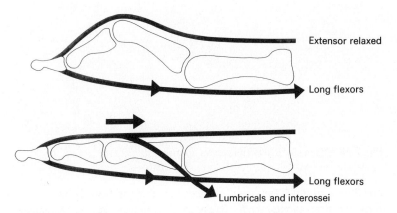

Extensor relaxed

Long flexors

Long flexors

Lumbricals and interossei

**Fig. 7.14** An important action of the lumbricals and interossei in the foot is to prevent buckling of the toes.

## Nerves and vessels of the sole

These are derived from the tibial nerve and the posterior tibial artery. They enter the foot deep to abductor hallucis and divide into medial and lateral plantar branches.

### Arteries (accompanied by venae comitantes)

The *medial plantar artery* runs forwards towards the big toe, giving branches to muscles and the medial toes. The latter join branches of the lateral artery (Fig. 7.15). The *lateral plantar artery* crosses to the lateral side between flexor accessorius and flexor digitorum brevis and, near the base of the 5th metatarsal, divides into *superficial* and *deep branches*. The superficial branch supplies the lateral toes

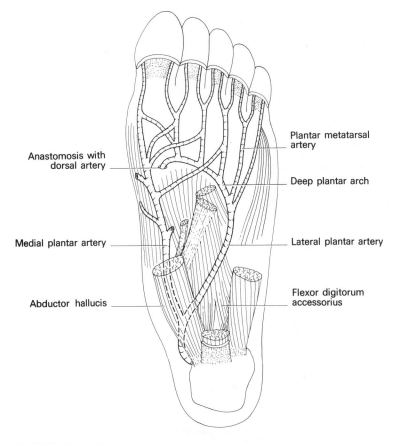

Anastomosis with
dorsal artery

Plantar metatarsal
artery

Deep plantar arch

Medial plantar artery

Lateral plantar artery

Abductor hallucis

Flexor digitorum
accessorius

**Fig. 7.15** The medial and lateral plantar arteries.

and the deep branch runs deeply between the third and fourth layers of muscles to form the *deep plantar arch*. The arch is completed by the termination of the dorsalis pedis, which has come through from the dorsum of the foot (compare with the deep *palmar* arch, which is completed by the radial artery). The deep plantar arch gives off the *plantar metatarsal arteries*, some of which are joined by branches of the medial plantar artery, to supply the toes.

**Nerves** (Fig. 7.16)

The *medial plantar nerve*, accompanied by its artery, runs forwards between abductor hallucis and flexor digitorum brevis and supplies cutaneous branches to three and a half toes and muscular branches to four muscles (Fig. 7.17). The *lateral plantar nerve* follows the lateral plantar artery to the base of the 5th metatarsal, where it divides into *superficial* and *deep branches*. These supply cutaneous branches to the lateral one and a half toes and muscular branches to the remaining muscles (see Fig. 7.9). Both medial and lateral plantar nerves also give branches to supply the skin of most of the sole of the foot.

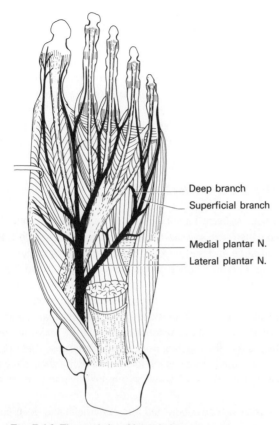

Deep branch
Superficial branch

Medial plantar N.
Lateral plantar N.

**Fig. 7.16** The medial and lateral plantar nerves.

| PRE-AXIAL NERVES | | POSTAXIAL NERVES | |
|---|---|---|---|
| MEDIAN | MEDIAL PLANTAR | ULNAR | LATERAL PLANTAR |
| Abductor pollicis brevis | Abductor hallucis | Abductor dig. minimi | Abductor dig. minimi |
| Flexor pollicis brevis | Flexor hallucis brevis | Flexor dig. minimi | Flexor dig. minimi |
| Opponens pollicis | — (unnecessary) | Opponens dig. minimi | — (even more (unnecessary)) |
| Lumbricals 1 and 2 | Lumbrical 1 | Lumbricals 3 and 4 | Lumbricals (2), 3 and 4 |
| Flexor digitorum superfic. | Flexor digitorum brevis | Adductor pollicis | Adductor hallucis |
| 3½ digits | 3½ digits | ALL interossei | ALL interossei |
| | | 1½ digits | 1½ digits |
| | | + | + |
| | | Palmaris brevis. (Unimportant and only included for symmetry) | Flexor dig. accessorius |

**Fig. 7.17** Nerve supply of the muscles of the palm and the sole. They are identical except for the 2nd lumbrical and one additional muscle supplied by the postaxial nerve in each case.

### Comparison of the nerves of the hand and foot

You will recall that the *pre-axial* nerves of the palm of the hand and the sole of the foot are the median and the medial plantar nerves, while the *postaxial* nerves are the ulnar and lateral plantar. The nerve supplies are best learnt together. As you will see from Fig. 7.17, they are identical except that there are no opponens muscles in the foot (for obvious reasons), that the nerve supply of the 2nd lumbrical is different and that there is no flexor accessorius in the hand. You will realize that flexor digitorum superficialis in the upper limb (with its perforated tendons) is equivalent to flexor digitorum brevis in the foot (with its perforated tendons).

## LYMPHATIC AND VENOUS DRAINAGE OF THE LOWER LIMB

The lymphatic and venous systems have been mentioned here and there in the preceding chapters and they will now be discussed as a whole (since this is a popular examination question and is also extremely important clinically).

The two major groups of lymph nodes have already been described. They are the *inguinal* and *popliteal* nodes. The deep lymphatics follow the major vessels, being interrupted in the back of the leg by the popliteal lymph nodes. They all finally end in the inguinal nodes. The superficial lymphatics tend to run with the superficial veins, the great and small saphenous veins, with the majority draining to the medial side. They end in the inguinal nodes, including those that drain the gluteal region and the perineum (as well as the lower part of the abdominal wall). From the inguinal nodes, vessels pass under the inguinal ligament to enter the abdomen.

The venous system may also be subdivided into superficial and deep components. The deep veins are venae comitantes until they become large enough to be dignified by their own names, the *popliteal* and *femoral veins*. The superficial veins are usually two in number, with occasional extras. The *short saphenous vein* drains the lateral side of the foot towards the calf, ascends to pierce the deep fascia in the lower or middle third of the calf (*not* in the popliteal fossa) and enters the popliteal vein. The *great saphenous vein* drains most of the dorsum of the foot, passes about 2 cm above and in front of the medial malleolus, ascends the medial side of the leg about a hand's breadth behind the patella and then, on the front of the thigh, pierces the deep fascia via the saphenous opening about 4 cm below and lateral to the pubic tubercle and enters the femoral vein. The position of this opening, however, is very variable. As it does so, it receives a number of other superficial veins from the medial and lateral sides of the thigh, the external genitalia and the lower part of the abdominal wall.

The superficial veins contain valves, but since they are situated in the superficial fascia they are not directly affected by the muscle pump. How then does blood travel in them when standing? The answer is that most of it doesn't. There are a number of important communicating veins (*'perforators'*) that connect the superficial and deep venous systems, particularly above the medial malleolus and above and below the knee. The valves in these are directed towards the deep veins. When the muscles contract, the valves prevent the blood being pushed into the superficial veins and when they relax blood is drawn into the deep veins from the superficial. If the valves become incompetent, 'blow-outs' occur and the superficial veins become varicose.

## JOINTS OF THE ANKLE REGION AND FOOT

### The tibiofibular joints

The superior tibiofibular joint is a plane synovial joint. The inferior joint is more important and is a syndesmosis, the bones being held together by a powerful *interosseous ligament* situated just above the triangular facet at the lower end of the fibula. This ligament is of vital importance since it holds together the tibia and fibula and thus the medial and lateral malleoli, which form the socket for the talus.

### The ankle joint

This is a hinge joint between the tibia and fibula and the talus. Most of its more powerful ligaments, however, pass down to the calcaneus, so that the talus may be thought of as being held between the other bones (Fig. 7.18). As it is a hinge joint, the most powerful ligaments are at the side. There is another important ligament posteriorly, the *posterior tibiofibular ligament*, which extends from the pit at the lower end of the fibula to the posterior edge of the tibia. When the foot is plantar-flexed during the take-off stage in walking, there is naturally a tendency for the tibia and fibula to slide forward on the talus under the influence of gravity (Fig. 7.19). This is prevented by two factors. Firstly, the socket is deepened at the back by a posterior lip on the tibia (sometimes called the *third malleolus*) and by the posterior tibiofibular ligament. Secondly, the articular surface of the talus is wider

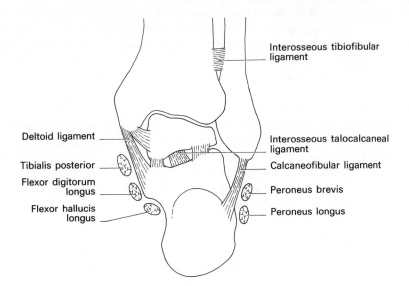

**Fig. 7.18** Ankle joint: diagram to show how the talus is held by ligaments between the tibia and fibula above and the calcaneus below.

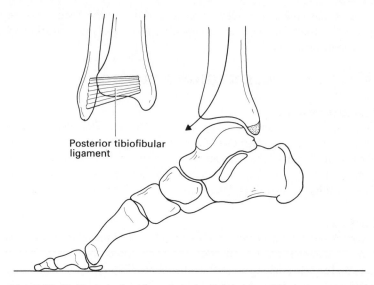

**Fig. 7.19** On lifting the heel from the ground the tibia and fibula are prevented from sliding forwards by the back of the tibia and the posterior tibiofibular ligament.

in front than behind. Even so, if one of the malleoli is fractured so that the socket is splayed, this *forward* movement of the leg on the talus may occur but is usually interpreted as a *backward* displacement of the foot (third-degree Pott's fracture).

As in all synovial joints, the articular surfaces are covered with articular cartilage and the synovial membrane lines everything in the joint except the articular surfaces. As in all hinge joints, the anterior and posterior parts of the capsular ligament are relatively thin, but remember that anteriorly the capsule extends forwards on to the neck of the talus. If you drop a knife in the dissecting room, its point could therefore enter the ankle joint, so take no chances.

### Medial collateral (deltoid) ligament

This is delta ($\triangle$)-shaped although its deep part consists only of a tough band of vertical fibres passing from the medial malleolus to the rough area below the comma-shaped facet on the talus (Figs 7.6 & 7.18). The superficial part is attached above to the medial malleolus and fans out to be attached to (from front to back) the tuberosity of the navicular, the spring ligament (see below), the sustentaculum tali and the back of the talus.

### Lateral collateral ligament

This has three components: *anterior and posterior talofibular ligaments* and a powerful, cord-like *calcaneofibular ligament* (Fig. 7.20). The latter runs downwards and backwards and so helps to prevent forward displacement of the socket on the talus.

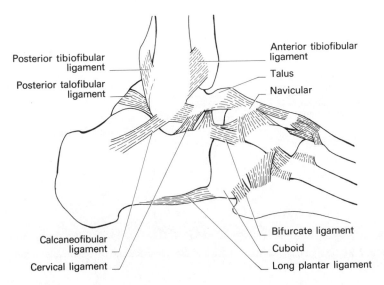

**Fig. 7.20** Ankle joint, lateral aspect after removal of the capsular ligament.

## Movements

As a hinge joint, flexion and extension (dorsiflexion and plantar flexion) are permitted, but in plantar flexion the narrower part of the talus is between the malleoli so that a little inversion and eversion can occur here.

### The subtalar and talocalcaneonavicular joints

These are sometimes known collectively as the *subtalar joint* since it is at this compound joint that inversion and eversion principally occur. The articular facets on the calcaneus are shown in Figs 5.11 & 7.21). The talus articulates with the posterior facet to form the *subtalar joint* and with the two anterior facets (which are frequently fused to form one joint surface) to contribute to the talocalcaneonavicular joint. The latter is a complicated joint as it also involves the navicular and the upper surface of the spring ligament. The ligaments of the subtalar joint comprise a capsular and medial and lateral ligaments. There is also an *interosseous talocalcaneal ligament*, which occupies the sinus tarsi, and a *cervical ligament* (Fig. 7.20). The talocalcaneonavicular joint has a weak *capsular ligament* and is reinforced by a *bifurcate ligament* (Fig. 7.21). The spring ligament is officially called the *plantar calcaneonavicular ligament* and it spans across between the sustentaculum tali and the tuberosity of the navicular, thus forming a support for the head of the talus.

The movements of inversion and eversion are complex but a useful way of thinking about movements of the foot is as follows. In dorsiflexion and plantar flexion, the talus is part of the foot and it moves on the tibia and fibula. In inversion and eversion, it forms part of the leg and the other bones of the foot rotate around it.

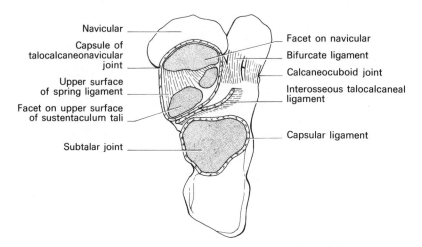

Navicular

Capsule of
talocalcaneonavicular
joint

Upper surface
of spring ligament

Facet on upper surface
of sustentaculum tali

Subtalar joint

Facet on navicular

Bifurcate ligament

Calcaneocuboid joint

Interosseous talocalcaneal
ligament

Capsular ligament

**Fig. 7.21** The joints between the talus and the calcaneus and navicular, after removal of the talus.

There is a plane synovial joint between the front of the calcaneus and the cuboid (Fig. 7.21), and this joint and the talonavicular joint are known collectively as the *transverse tarsal joint*. Movements here contribute to inversion and eversion.

The other joints of the tarsus and those of the metatarsal bases are of little interest except to specialists. There are altogether six synovial cavities in the tarsus and, except for those movements mentioned above, only slight gliding movements occur. The metatarsophalangeal and interphalangeal joints are very similar to those in the hand, but movements are more restricted.

### The arches of the foot

The foot is usually described as having two longitudinal arches: a *medial arch*, consisting of the calcaneus, talus, three cuneiforms and the medial three metatarsals, and a *lateral arch*, comprising the calcaneus, the cuboid and the lateral two metatarsals. There is also a *transverse arch*, formed by the cuneiforms, metatarsals and distal tarsal bones. In normal standing, the weight is thus taken mainly on the posterior part of the calcaneus and the metatarsal heads. The medial arch is the most prominent of the arches. Its summit is the talus, which transfers the body weight to the foot; the weight tends to force apart the calcaneus and the navicular and this is resisted by the tension of the spring ligament.

The arches are not rigid in the normal foot. They have a little 'give' and can be flattened by the body weight: slightly in most people but more or less completely in babies or in people who have never worn shoes (see Fig. 2.5). The arches are held up by *bone shape,* by *ligaments,* by *muscle action,* and by *muscles and tendons holding up the arches from above.* The shape of the bones is most important in the transverse arch, where the cuneiform bones are wedge-shaped. (This is the meaning of cuneiform. See, for example, the cuneus in the brain and the cuneiform script of certain ancient peoples, such as the Sumerians.) The ligaments are

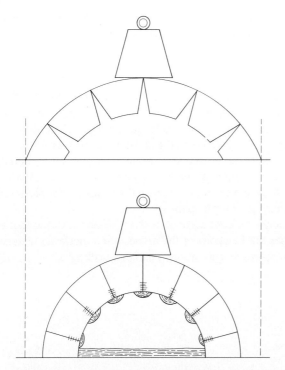

**Fig. 7.22** The main ligaments of the foot are on the plantar surface, since this is where loading of the arches produces tension. The plantar aponeurosis and the muscles of the first layer help to support the longitudinal arches by preventing flattening of the whole foot.

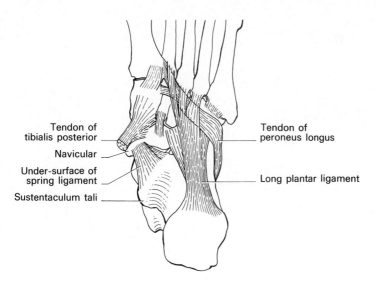

Tendon of tibialis posterior

Navicular

Under-surface of spring ligament

Sustentaculum tali

Tendon of peroneus longus

Long plantar ligament

**Fig. 7.23** The deep ligaments of the sole of the foot.

most powerful on the plantar surface of the foot (Fig. 7.22, p.163). Two of the strongest are the *long* and *short plantar ligaments*, which, respectively, pass from the calcaneus to the cuboid and metatarsal bases and from the calcaneus to the cuboid alone (Fig. 7.23, p. 163). There are also interosseous ligaments between other bones. The plantar aponeurosis and the muscles of the first layer help by spanning across the bases of the 'pillars' of the arches and so stopping the arch being flattened by the body weight. Holding up the arches from above are the tibialis anterior and posterior and peroneus longus muscles. The interossei help maintain the transverse arch by bunching the metatarsals together.

The arches, particularly the medial arch, are most important in the take-off stage of walking. When the foot is plantar-flexed by the calf muscles, the form of the foot ensures that the main weight is taken on the heads of the metatarsals, in particular the first (which is why it is so thick).

This is an extremely simplified account of the mechanism of the foot and it must be admitted that the functions of the arches are imperfectly understood; hence the different accounts that are given in different textbooks.

# Chapter 8
# The Thorax

The thoracic wall comprises the 12 thoracic vertebrae, the 12 pairs of ribs and the sternum. It is more or less circular in cross-section in a baby but in the normal adult it is flattened anteroposteriorly. It is important that you do not think of the thorax as a self-contained entity. Remember: (i) that it is largely covered by upper limb muscles such as latissimus dorsi and pectoralis major, which must be divided or retracted in thoracic operations; (ii) that the thoracic spinal nerves (which form the intercostal and subcostal nerves) supply the abdominal wall as well as the thoracic wall; and (iii) that the ribs enclose a number of abdominal viscera, such as the liver and spleen, as well as the heart and lungs.

## THE THORACIC VERTEBRAE

The parts of a typical thoracic vertebrae are shown in Fig. 8.1. Notice particularly the downwardly projecting spine, the facets on the transverse processes for the tubercles of the ribs, the demifacets on the body for the heads of the ribs and the slight asymmetry of the body. The left side of the body is slightly flattened in vertebrae from the middle of the series since in this region the descending thoracic aorta is directly in contact with the bodies to the left of the midline (see Fig. 8.6). More will be said about the vertebrae and the joints between them in Chapter 14.

## THE RIBS

A typical rib is shown in Fig. 8.2. The articular part of the tubercle articulates with its own transverse process (i.e. the 6th rib with the 6th vertebrae) and the demifacets on the head articulate with the demifacets on its own vertebra and the vertebra *above* (i.e. the 6th rib with the 6th and 5th vertebrae) as well as with the intervertebral disc. The angle marks the lateral edge of the erector spinae muscle (see Fig. 14.11) and, since this muscle is widest in the sacral and lumbar regions and tapers up to the neck, the angles of the ribs become more and more medial as one ascends until on the 1st rib the angle and tubercle correspond.

The anterior end of a typical rib is attached by means of its *costal cartilage* to the sternum. This is true of the upper seven ribs (*true ribs*) but the 8th to the 10th costal cartilages unite only with the costal cartilage above, while the 11th and 12th ribs are floating (Fig. 8.3). Ribs 8–12 are called *false ribs*. The 1st rib is flattened and very obliquely placed so that in an X-ray it is seen almost in face view (Fig. 8.4). It articulates with the body of the lst thoracic vertebra only. Its important relationships are described on p. 330. The 11th and 12th ribs articulate by their heads with the 11th and 12th vertebrae only; they have no tubercles.

## THE STERNUM

This consists of the *manubrium sterni*, the *body*, which consists of four segments

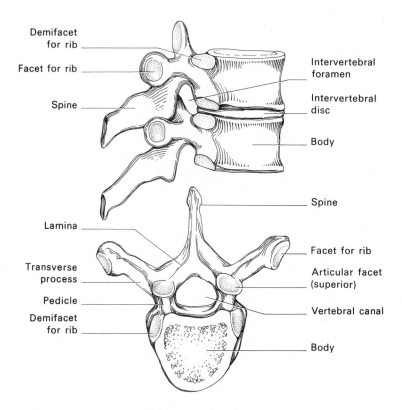

Demifacet for rib

Facet for rib

Spine

Intervertebral foramen

Intervertebral disc

Body

Spine

Lamina

Transverse process

Pedicle

Demifacet for rib

Facet for rib

Articular facet (superior)

Vertebral canal

Body

**Fig. 8.1** Thoracic vertebrae. Notice how two demifacets become one complete facet when the vertebrae are articulated.

or *sternebrae*, and the *xiphisternum* (xiphoid process). There is a secondary cartilaginous joint between the manubrium and the body and, as there is a slight angulation between the two, the region of the joint is called the *sternal angle* (of Louis). It is here that the 2nd costal cartilage joins the sternum and this is the starting place when counting the ribs to use them as landmarks, for example, for locating the apex beat (see below). The sternum is developed from bilateral sternal bars that fuse in the midline, and it is not uncommon to find a localized failure of fusion, which takes the form of a hole in the centre of the sternum, or a bifurcate xiphisternum, which forms a marked hollow in the chest. The sternum is often used to obtain a specimen of red bone marrow (sternal puncture) in the investigation of blood diseases.

## X-RAY APPEARANCES OF THE SKELETON OF THE CHEST

The sternum cannot be seen clearly in an anteroposterior (AP) X-ray since it has only a thin shell of compact bone on the outside and is superimposed on the shadow of the vertebrae. Since the ribs descend from the vertebrae and ascend to the sternum they show a criss-cross pattern which may be difficult to interpret at first. With increasing age the costal cartilages become calcified and are then visible.

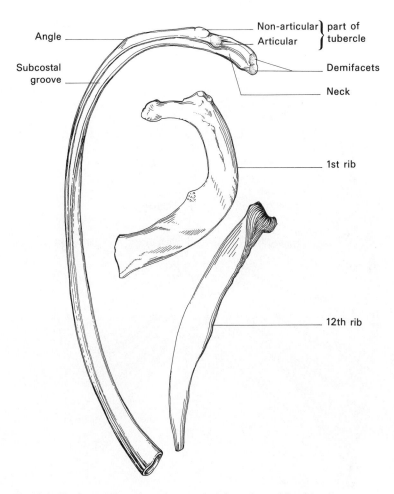

**Fig. 8.2** The first and last ribs and a typical rib from the middle of the thorax.

They may fracture when cardiac massage (p. 171) is being performed. The 1st thoracic vertebrae may be distinguished by its upwardly projecting transverse processes, as opposed to the downward direction of the transverse processes of the 7th cervical vertebra (Fig. 8.4).

## THE INTERCOSTAL SPACES

There are three layers of muscle in both the thoracic and the abdominal wall since both have a similar development (p. 196). They are the *external intercostal* (*oblique* in the abdomen), the *internal intercostal* (*oblique*) and the *transversus thoracis* (*abdominis*) layers. The vessels and nerves run in the plane between the internal and transversus layers (Fig. 8.5).

The external intercostals run downwards and forwards, being replaced towards the front by the *anterior intercostal membrane*. The internal oblique fibres run downwards and backwards, being replaced towards the back by the *posterior*

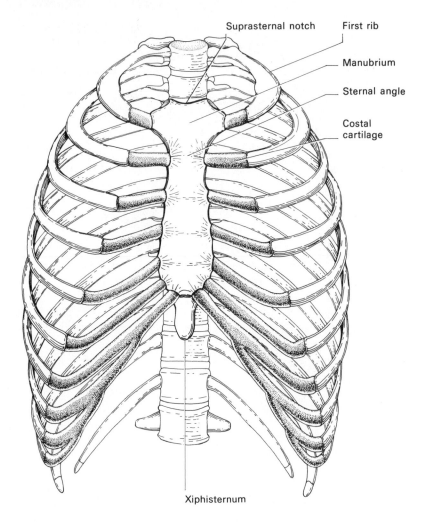

**Fig. 8.3** The skeleton of the thorax. Note that the sternal angle marks the level of the 2nd costal cartilage. You may like to use this diagram to practise marking out the lungs and pleura.

*intercostal membrane*. The transversus thoracis layer is incomplete, its most important part being anteriorly (*sternocostalis*), where it separates the internal thoracic vessels from the pleura.

The vessels and nerve lie near the upper part of each intercostal space, partly in the subcostal groove, in the order nerve, artery, vein from below upwards.

## THE DIAPHRAGM

This forms the floor of the thoracic cavity and will only be described briefly here (see Chapter 9). It is attached mainly to the lower six ribs, and the muscle fibres rise almost vertically before flattening out to be inserted into the central tendon.

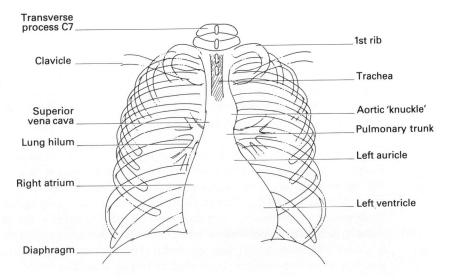

**Fig. 8.4**  Tracing of a chest X-ray. Notice that the vertebrae C7 and T1 can be identified by their converging transverse processes.

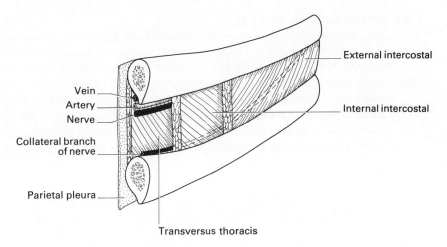

**Fig. 8.5**  Contents of an intercostal space.

The right dome or cupola of the diaphragm is higher than the left, owing to the presence of the liver below it.

### Nerves and vessels of the thoracic wall

These lie in the intercostal spaces but, whereas the vessels stay in the spaces, the lower nerves, anteriorly, continue their downwardly curved course to enter the abdomen. The first to do so is the 7th while the 12th thoracic along with the 1st lumbar supplies the region just above the pubis (see Fig. 1.10).

The nerves are the anterior rami of the thoracic nerves. The 12th thoracic nerve is below the 12th rib and is therefore called the *subcostal nerve*. Each gives a white ramus communicans to the sympathetic trunk ganglia and each receives one or more grey rami (see Chapter 15). It then gives a *collateral branch*, which runs along the lower part of the corresponding intercostal space, and continues to the front, where it gives an *anterior cutaneous branch*. Laterally, there is a *lateral cutaneous branch*, and there are also muscular branches to the intercostal muscles and (for the lower six nerves) the abdominal muscles (see Fig. 1.9).

The arteries anastomose with arteries of the upper limb such as the subscapular artery. Posteriorly, the *posterior intercostal arteries* are branches of the thoracic aorta. Each gives a *spinal branch* (something to be borne in mind when a segment of aorta has to be resected) and then runs forward in the intercostal space. It gives branches that follow branches of the intercostal nerves and ends by anastomosing with the *anterior intercostal arteries*, which come from the internal thoracic or the musculophrenic artery. The *internal thoracic* comes from the subclavian and descends just lateral to the edge of the sternum, separated from the pleura by sternocostalis. It ends by dividing into the *musculophrenic* and *superior epigastric* arteries. The former follows the attachment of the diaphragm to the ribs and the latter enters the rectus sheath (Chapter 9).

## THE THORACIC CAVITY

This contains principally the two pleural cavities with, between them, the mediastinum, as shown in Fig. 8.6. This is a simplified diagram but shows some important features. Note that the aorta lies in contact with the vertebral bodies and the oesophagus is anterior to it, just as the gut (fore-, mid- or hind-) is ventral to the dorsal aorta throughout the whole length of the embryo (see Fig. 1.8). The posterior border of the lungs is rounded because it is shaped by the ribs, but the anterior border is sharp because it fits in between the heart and the chest. From the front, therefore, part of the heart is directly in contact with the chest wall but part

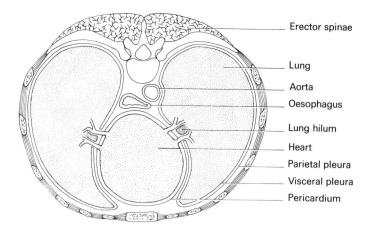

Fig. 8.6 Cross-section through the chest.

of it is overlaid by lung tissue. The ventricles of the heart may be compressed between the sternum and the vertebrae — external cardiac massage.

The mediastinum forms a complete septum across the chest, from front to back, so that a change in pressure in one pleural cavity may deflect the mediastinum to one or other side; the clinician can recognize this by feeling for the position of the trachea at the root of the neck. A horizontal line through the manubriosternal joint will pass through the lower border of the 4th thoracic vertebra. This line subdivides the mediastinum into *superior* and *inferior* parts. The *inferior mediastinum is further subdivided into the anterior mediastinum* (unimportant), the *middle mediastinum*, containing the heart, the lower parts of the great vessels and the bifurcation of the trachea, and the *posterior mediastinum*, containing the descending aorta, the oesophagus and various smaller structures.

## PLEURA AND LUNGS

The pleural cavity is a potential space, containing a little fluid, between the visceral and parietal pleurae. It is often said that there is a 'negative pressure' in the pleural cavity. There is no such thing as a negative pressure. What is meant is that the pressure is slightly below atmospheric, i.e. perhaps 755 mmHg on a fine day.

### The parietal pleura

This is a thin membrane that lines the thoracic cavity on each side of the mediastinum. It is therefore subdivided into *costal, cervical, mediastinal* and *diaphragmatic* portions. The cervical pleura is that part that rises above the level of the anterior end of the 1st rib into the root of the neck, extending for two fingerbreadths above the medial third of the clavicle. It is covered by a *suprapleural membrane* (Sibsen's fascia).

### The visceral pleura

This covers the outer surface of the lung and is continuous with the parietal pleura at the hilum, forming a sleeve around the bronchi, pulmonary vessels and other structures entering or leaving the lung. The sleeve is too big for the contained structures. If you pinch together the cuffs of a jacket below your wrist, you will understand how the lower part of the 'sleeve' at the hilum forms the *pulmonary ligament* (Fig. 8.7).

The visceral pleura is developed from the splanchnopleure (p. 25) so its nerve supply is from the autonomic system. The parietal pleura is from the somatopleure (p. 25) so it is supplied by the intercostal and phrenic nerves, the latter supplying the mediastinal pleura and the central part of the diaphragmatic pleura as well as the peritoneum lining the central part of its under-surface (see p. 208).

### Pneumothorax

As has already been mentioned, pressure in the (potential) pleural cavity is slightly below atmospheric whereas that in the lungs and air passages is more or less atmospheric (depending on the stage of respiration). Thus the lungs are held distended by the pressure inside them. If, however, the chest wall is penetrated or the visceral pleura is punctured, air enters the pleural cavity so that its pressure

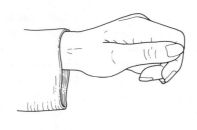

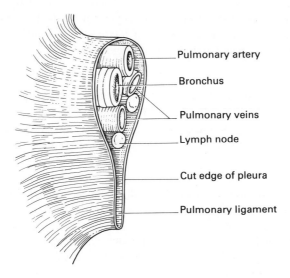

Pulmonary artery

Bronchus

Pulmonary veins

Lymph node

Cut edge of pleura

Pulmonary ligament

**Fig. 8.7** Major structures at the hilum of the lung seen from *behind*. The pulmonary ligament is a fold of pleura.

becomes the same as that inside the lungs. The lungs contain a great deal of elastic tissue and they therefore collapse.

### Surface markings of the pleural cavities

Think of the even numbers from 2 to 12. The outline of the left pleural cavity can be marked out as shown in Fig. 8.8. From the cervical pleura the anterior border of the pleural cavity passes behind the sternoclavicular joint and then passes medially, reaching the midline at the level of the 2nd costal cartilage. From here, the two pleural cavities are in contact down as far as the 4th cartilage. There is then an indentation for the heart which reaches the lateral border of the sternum at the 6th cartilage. Thence, the line crosses the 8th costal cartilage in the midclavicular line and the 10th in the midaxillary line and crosses the 12th rib at the back. On the right side the markings are similar but there is no cardiac indentation. The lung markings correspond to those of the pleura above, but are two ribs or costal cartilages higher in the lower part of the thorax. Thus the lung reaches down as far as the 6th costal cartilage or rib in the midclavicular line, the 8th in the midaxillary line, and the 10th rib posteriorly. The lung does not, therefore, fill the pleural

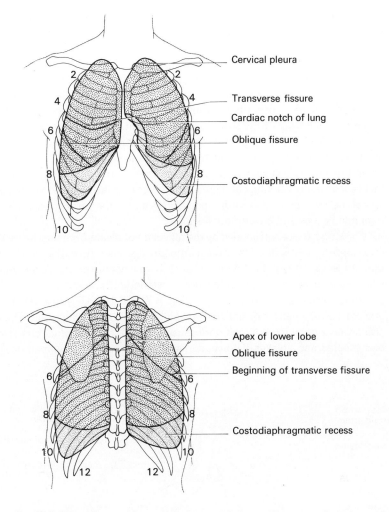

Cervical pleura

Transverse fissure

Cardiac notch of lung

Oblique fissure

Costodiaphragmatic recess

Apex of lower lobe

Oblique fissure

Beginning of transverse fissure

Costodiaphragmatic recess

**Fig. 8.8** The surface markings of the lungs and pleura. Note particularly the apices of the upper and lower lobes.

cavity. For example, between the 8th and 10th ribs in the midaxillary line there is a part of the pleural cavity unfilled by lung. This is the *costodiaphragmatic recess*. Here the costal and diaphragmatic pleura come into contact although during a deep breath the lower border of the lung moves downwards and separates the two layers. Radiologists call this region the *costophrenic angle*, and fluid in the pleural cavity will cause an opacity which obliterates the angle.

### The lungs

The right lung usually has three lobes and the left only two, but do not try to side a lung by counting its lobes as there may be variations. You can tell the side quite easily by looking at the general shape (see Fig. 8.6) and mentally trying to fit it into your own chest. The left lung has a deep cardiac impression on its medial surface, which produces a *cardiac notch* on the anterior border. There is also an arched groove produced by the aorta. The right lung has a deep hollow on its diaphragmatic surface produced by the underlying liver. Other smaller impressions are produced by mediastinal structures (Figs 8.9 & 8.10; see p. 191).

The lower lobe of the lungs, on both right and left sides, is the portion below the *oblique fissure*. This starts at the level of the 4th thoracic spine, follows the line of the medial border of the scapula when the arm is abducted, and reaches the lower border of the lung, in front, at the 6th rib. In the right lung there is an additional *transverse fissure*, which marks off a small *middle lobe*. This passes forward from the oblique fissure, following the line of the 4th rib. There is no middle lobe in the left lung but there is often a tongue-like projection of the upper lobe at the lower end of the cardiac notch. This part of the lung corresponds to the right middle lobe and is called the *lingula*.

The lung hila are surrounded by the reflection of pleura from the mediastinum on to the lung; below, this forms the *pulmonary ligament*. As will be seen later, the uppermost structure in the hilum is usually the pulmonary artery (Artery Above), while the main bronchus is situated more posteriorly (Bronchus Behind). In front and below are the pulmonary veins (Figs 8.7 & 8.10). Closely related to the bronchi are the small *bronchial arteries*, which supply them with blood. There are also *bronchopulmonary lymph nodes*, usually black because of carbon that has been inhaled, and a plexus of nerves derived from the vagus and sympathetic trunk

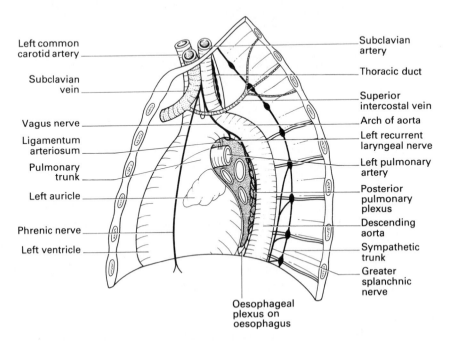

Left common carotid artery

Subclavian vein

Vagus nerve

Ligamentum arteriosum

Pulmonary trunk

Left auricle

Phrenic nerve

Left ventricle

Subclavian artery

Thoracic duct

Superior intercostal vein

Arch of aorta

Left recurrent laryngeal nerve

Left pulmonary artery

Posterior pulmonary plexus

Descending aorta

Sympathetic trunk

Greater splanchnic nerve

Oesophageal plexus on oesophagus

**Fig. 8.9** Structures visible on the left side of the mediastinum (i.e. the medial relations of the left lung). They are of course covered by mediastinal pleura.

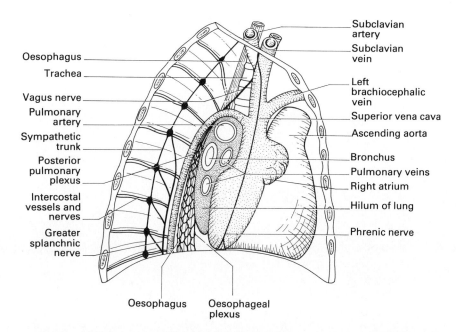

Subclavian artery

Subclavian vein

Oesophagus

Trachea

Left brachiocephalic vein

Vagus nerve

Superior vena cava

Pulmonary artery

Ascending aorta

Sympathetic trunk

Bronchus

Posterior pulmonary plexus

Pulmonary veins

Right atrium

Intercostal vessels and nerves

Hilum of lung

Greater splanchnic nerve

Phrenic nerve

Oesophagus    Oesophageal plexus

**Fig. 8.10** Structures visible on the right side of the mediastinum (i.e. the medial relations of the right lung).

(p. 190). These provide the nerve supply for the smooth muscle in the arteries and bronchi, and carry afferent fibres from the bronchi and lung which are important in the Hering–Breuer and other reflexes. The Hering–Breuer is probably less important in man than in animals.

### The bronchi

The trachea (described on p. 333) divides into the right and left bronchi at the level of the manubriosternal joint. Since the trachea deviates slightly to the right in the lower part of its course, the right bronchus is more in line with the trachea than the left and, since it is also wider than the left, inhaled foreign bodies are more likely to be found in the right lung than the left. The right main bronchus, as it approaches the hilum of the lung, gives off an *upper lobe bronchus* and then divides into a *middle* and a *lower lobe bronchus*. On the left side, the middle lobe bronchus is replaced by the *lingular bronchus*. Each lobar bronchus divides again into *segmental bronchi*, each supplying one *bronchopulmonary segment* of the lung. The arrangement of these segments and their supplying bronchi is shown in Figs 8.11 & 8.12. The segmental bronchi and their segments are named and numbered as follows:

**1** Apical
**2** Posterior
(1 & 2 from a common apico-posterior stem on the left side)
**3** Anterior

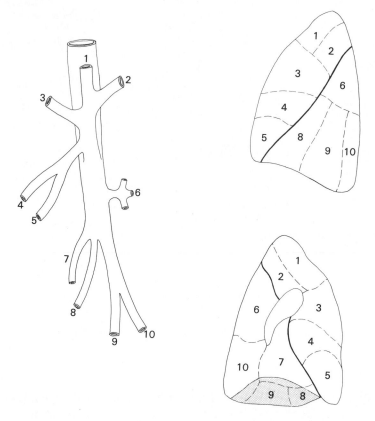

**Fig. 8.11** The bronchopulmonary segments of the left lung, lateral and medial surfaces. The segmental bronchi, seen on the left of the diagram, are viewed from the *lateral* side.

**4 & 5**   Lateral and medial middle lobe (superior and inferior lingular on left side)
  **6**   Superior (apical).
  **7**   Medial basal (cardiac on left)
  **8**   Anterior basal
(7 & 8 often by a common stem on left)
  **9**   Lateral basal
  **10**   Posterior basal
Each bronchus is accompanied by a branch of the pulmonary artery, which branches with it. Even in a histological section you can find a small artery related to each bronchial subdivision. The tributaries of the pulmonary veins, however, are *intersegmental* in position so that each vein drains more than one segment and each segment drains into more than one vein. A knowledge of the segments is important in localizing areas of lung disease or inhaled foreign bodies in X-rays, and also in planning postural drainage. If you wish to drain sputum from the lung bases, for example, the patient needs to be tilted head down whereas the upper lobe drains better in the sitting position. Note that both upper and lower lobes have

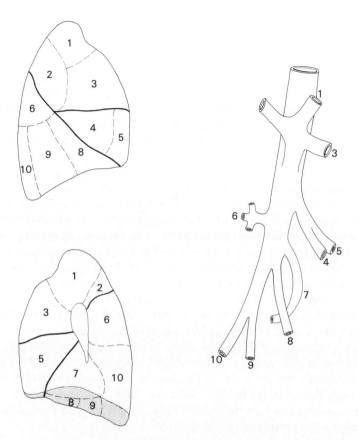

**Fig. 8.12** The bronchopulmonary segments of the right lung, lateral and medial surfaces. The segmental bronchi, seen on the right of the diagram, are viewed from the *lateral* side.

an apical segment (see Figs 8.11 & 8.12). That of the lower lobe, reaching up as far as the spine of T4, is the most dependent part of the lung when lying down and is a region where intrapulmonary fluid may collect in some conditions.

## RESPIRATION

There are two mechanisms of breathing, thoracic and abdominal, carried out by means of the ribs and chest muscles and by the diaphragm respectively. Neither is essential: in the baby the ribs are horizontal and breathing is entirely abdominal while in the later stages of pregnancy the diaphragm cannot descend and breathing is thoracic.

### Thoracic breathing

The ribs are elevated during inspiration in two ways. Firstly, the anterior ends of the ribs are raised in the so-called *pump-handle* action and, since they are normally below the posterior ends, this will increase the anteroposterior diameter of the chest. Secondly, the most lateral parts of the ribs are raised in the so-called *bucket-handle* action. Since the centre of most of the ribs is normally below the anterior and posterior ends, this will increase the transverse diameter. In the case of the false ribs, the main movement consists of an opening out anteriorly, in a caliper-like action, although the other types of movement can also occur.

In quiet respiration, the rib movements are not very marked; in fact the first ribs and the manubrium form a more or less fixed unit with the other ribs moving below it. The muscles involved in quiet respiration are mainly the scalene muscles and the external intercostals. In forced inspiration, other muscles that are attached to the ribs come into play, for example sternomastoid, the pectoral muscles and latissimus dorsi.

### Abdominal breathing

The diaphragm will be described in Chapter 9. Here it will be enough to remind you that it is composed of skeletal muscle and a central tendon. The peripheral muscle fibres are more or less vertical and take origin from the lower six ribs. When the muscle contracts the diaphragm descends as a whole, retaining its shape and thus increasing the vertical diameter of the thorax. While doing this, however, it will be compressing the abdominal contents and its vertical movement must finally come to a stop. The central tendon is then pressed over the upper surface of the liver and acts as an *origin* for the muscle fibres, which therefore now pull on the lower six ribs to elevate them in the final stages of inspiration.

If the diaphragm is composed of skeletal muscle, how do you manage to keep breathing even when asleep or anaesthetized? The answer is that in these conditions (and under normal awake conditions) the contraction is controlled by the respiratory centres in the medulla and pons. These can be overridden, when required, by impulses from the cerebral cortex but the final common pathway in both cases is the phrenic nerve (C3, 4 and 5).

### Expiration

The above descriptions have all referred only to inspiration. This is because expiration is normally a passive process, produced by the elastic recoil of the lungs and the tissues of the chest wall. Forced expiration, however, such as occurs in playing wind instruments or in coughing, is the result of muscular activity, principally those muscles that will compress the abdomen and thorax such as the muscles of the abdominal wall and latissimus dorsi (p. 51).

## THE HEART

Students studying anatomy are often misled by diagrams in textbooks of physiology in which there are two atria on top with two ventricles below them. To convert a 'physiologist's heart' into an 'anatomist's heart', two manoeuvres are necessary (Fig. 8.13). The heart is first laid on its left side and then rotated in a clockwise direction. In this way it can be understood that the right border of the heart is formed by the right atrium (Figs 8.4 & 8.14), with the superior and inferior venae cavae entering it above and below (at the levels of the 3rd and 6th costal cartilages respectively). The anterior surface of the ventricles consists of about two-thirds right ventricle and one-third left ventricle. This proportion is reversed on the

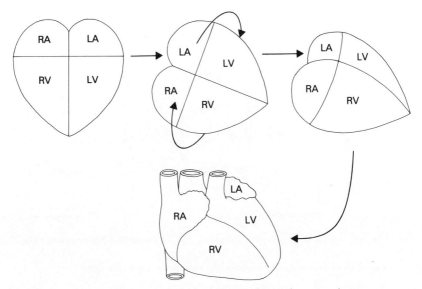

**Fig. 8.13** The heart does not resemble diagrammatic hearts; however, the true arrangements of the chambers may be remembered by lying a diagrammatic heart on its side and rotating it slightly.

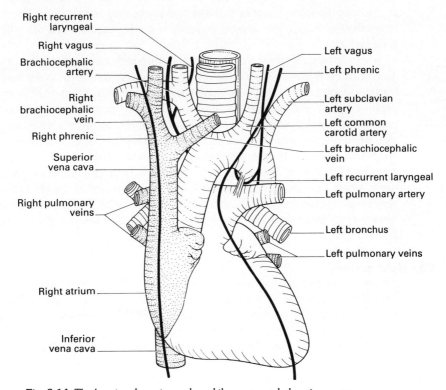

**Fig. 8.14** The heart and great vessels and the vagus and phrenic nerves.

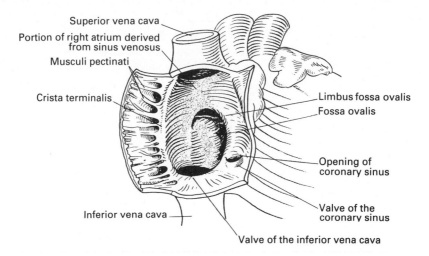

**Fig. 8.15** The interior of the adult right atrium, seen from the front.

diaphragmatic surface, i.e. one-third right ventricle and two-thirds left ventricle. The posterior surface, or base of the heart, consists almost entirely of the left atrium, only its auricle being visible from the front. Figure 8.13 also helps to explain why the septa do not lie in the sagittal plane; their right surfaces also face forwards (Fig. 8.15). The heart is enclosed in the pericardium and this will be described later.

It should be understood that the heart is only a rather specialized part of the general vascular system, as will be seen when its development is (very briefly) considered later. An artery or a vein consists of a *tunica intima*, whose most important part is the endothelium, a *tunica media*, composed largely of smooth muscle, and a *tunica adventitia*, which consists mainly of connective tissue, nerves and, in the larger arteries, vasa vasorum. Similarly the heart is lined by the *endocardium*, which is covered by endothelium, outside this is a *myocardium*, composed of cardiac muscle, and the outermost layer consists of connective tissue, which includes a good deal of fat, and this is covered by the *epicardium*, which is also called the visceral layer of serous pericardium. It is in the connective tissue and fat that nerves and the 'vasa vasorum' of the heart, i.e. the coronary arteries, run.

### The right atrium

This develops from two sources: the original embryonic right atrium and the right horn of the sinus venosus. In the adult, these two components are separated by the *crista terminalis*. The smooth area behind it, into which the two venae cavae open, is the sinus venosus portion, while the true atrium is represented by the part lined by muscular ridges, called the *musculi pectinati* (Fig. 8.15). The left horn of the embryonic sinus venosus develops into the *coronary sinus*, whose opening is near the inferior vena cava. Neither of the caval orifices has a functioning valve. The so-called *valve of the inferior vena cava*, as well as that of the coronary sinus, is an embryological remnant (the right venous valve). The other interesting feature of the right atrium is the interatrial septum. This develops from a rather thick, annular *septum secundum* and, to its left, a thin and flexible *septum primum*, which overlaps it. In the embryo, it is essential that these two components of the septum remain unfused so that oxygenated blood from the placenta can pass from right atrium to left atrium through the slit-like *foramen ovale* between the two septa. At birth, the pressure rises on the left side of the heart and falls on the right so the septum primum is pressed against the septum secundum to close the opening. In the adult the floor of the *fossa ovalis* represents the septum primum and the *limbus fossa ovalis* is the septum secundum. An opening may still be present (*probe patency*) but, as long as the septa overlap, pressure in the left atrium keeps the valve-like opening closed and the condition will be symptomless. If, however, the septa do not overlap, the resulting atrial septal defect (ASD) will allow blood to pass from left to right after birth, since the pressure is higher on the left side of the heart. The same is true of a ventricular septal defect. Such defects are not, *per se*, the cause of a 'blue baby'. Cyanosis will only occur if there are other defects present, such as a pulmonary stenosis, that will raise the pressure on the right side of the heart and give rise to a right–left shunt.

### The left atrium

This also develops from two sources. The embryonic left atrium has a single pulmonary vein opening into it, this vein dividing into left and right branches to (or, rather, from) the two lungs. The vein becomes absorbed into and forms part of the wall of the atrium as far as its branching point and then as far as the first branching of the left and right veins so that eventually there are four veins opening into the atrium.

### The right ventricle

This has a thinner wall than the left ventricle since the blood pressure is less in the pulmonary circulation than in the systemic (25 mmHg compared with 120 mmHg systolic). This difference, however, develops only after birth since the fetal circulation is very different from that of the adult. The *tricuspid* valve between the right atrium and ventricle has, of course, three cusps arranged in three planes, *septal* (parallel to the septum), *anterior* (in the coronal plane) and *posterior* (parallel to the diaphragm). The cusps are not entirely separate from each other but are joined together near their bases by thin *commissures*. To the edges and outer surfaces of the cusps are attached the *chordae tendineae*, which, in turn, are attached to the *papillary muscles*. Their function is not to open the valves but to prevent the valves blowing back into the atrium during ventricular systole. Other muscular ridges in the wall of the ventricle are called *trabeculae carneae* ('meaty ridges', as in chilli con carne). The exit from the right ventricle to the pulmonary trunk is guarded by the *pulmonary valve*. This has three cusps but these are semilunar in form and therefore very different from the tricuspid valve.

### The left ventricle

This has a thick wall, necessary to produce the high systolic pressure, but otherwise is very similar to the right ventricle. The *mitral* (or bicuspid) valve, however, has only two cusps, an *anteromedial* and a *posterolateral*. The antero-medial cusp is interesting since its chordae tendineae are attached only to the edge so that both surfaces are smooth. This is because blood flows over both surfaces of the valve: over the anterior surface *upwards* into the aorta and over the posterior surface *downwards* from the atrium into the ventricle.

The opening from the left ventricle into the aorta is guarded by the tricuspid *aortic valve*. As in the pulmonary valve, the cusps are semilunar in form. In the Aorta, one valve is Anterior and, in the Pulmonary trunk, one valve is Posterior. Disregard the official terminology (*Nomina Anatomica*), which seems, wrongly, to think otherwise.

## The conducting system

The contraction of the heart is generated at the *sinu-atrial node*, situated near the orifice of the superior vena cava (the pacemaker). The action potential travels via the atrial muscle to reach the *atrioventricular node*, which is near the tricuspid valve. From this, the specialized conducting cells known as Purkinje fibres form a distinct bundle (the *bundle of His*), which lies in the interventricular septum and divides into right and left branches. These run on the corresponding sides of the septum before breaking up into branches which ascend from the apex of the heart to the walls of the ventricles.

## The blood supply of the heart

This is by means of the left and right *coronary arteries*, so called because they form a circle (corona) around the atrioventricular groove (cf. coronet). These arteries and their branches do anastomose with each other but the anastomoses are not sufficiently large to maintain a collateral circulation if a major branch is occluded (see Chapter 1). They are therefore functional, if not anatomical, *end arteries*. Thus the sudden blockage of a major branch may be a cause of sudden death, but a more slowly developing blockage of smaller branches may give time for a collateral circulation to develop and may be compatible with a long life.

The *left coronary artery* (Fig. 8.16) arises from the aorta just above the left posterior cusp of the aortic valve and runs around the heart in the atrioventricular groove. It gives a large *anterior interventricular branch*, which supplies the septum as well as the adjacent ventricular wall; the main trunk anastomoses with the right coronary artery at the back of the heart. Other branches supply both atria and ventricles. The *right coronary artery* arises from the aorta above the anterior cusp of the aortic valve, runs down in the atrioventricular groove and then turns round to the posterior surface of the heart. It gives a *posterior interventricular branch*, which runs on the diaphragmatic surface of the heart. A *marginal branch* passes down the side of the right ventricle. Remember that the interventricular septum contains the bundle of His and its branches, so that interventricular arteries are of particular importance (especially the inferior one). The pattern that has been described is the usual one but variations are common. For example, the left coronary artery may be large and supply both anterior and posterior inter-ventricular branches, or the right coronary artery may be large and provide the

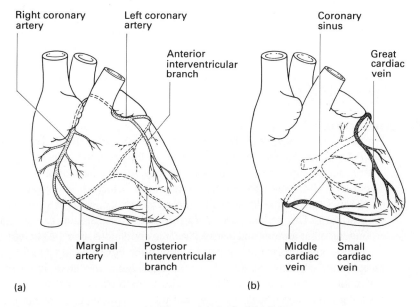

**Fig. 8.16** Arteries (coronary) and veins (cardiac) of the heart.

posterior interventricular branch and also supply a large part of the left ventricle. Obviously, therefore, the effects of the occlusion of one of the major branches of the coronary arteries will depend upon the arterial pattern.

Coronary artery occlusion may be treated by a bypass operation, in which one or more lengths of great saphenous vein, reversed so that valves do not obstruct the flow, are used to connect the ascending aorta or the internal thoracic artery to the affected vessel beyond the occlusion.

The veins of the heart follow, more or less, the pattern of the arteries but unfortunately the names are different. The *great cardiac vein* follows the anterior interventricular artery upwards to the atrioventricular groove, turns around the left border of the heart and ends by entering the left end of the coronary sinus (Fig. 8.16). This sinus lies posteriorly and opens into the right atrium. The *small cardiac vein* follows the marginal artery up the right ventricle, turns around the right border of the heart and enters the right end of the coronary sinus. Other small veins, draining the atria and the ventricles, enter the sinus and there are also some small veins, *venae cordis minimae*, that enter the right atrium directly.

The coronary circulation is unusual in that the arterial tree fills mainly during diastole, as a result of the elastic recoil of the aortic wall, since during systole the blood is squeezed out of the arterial bed by the muscle contraction.

### Nerve supply of the heart

The heart is, of course, supplied by the autonomic nervous system. There are sympathetic fibres from both the cervical and the thoracic sympathetic ganglia and parasympathetic fibres from the cervical part of the vagus. These nerves form the *cardiac plexus* (superficial and deep parts), which consists of both nerve fibres and ganglion cells (parasympathetic). The node is supplied with both types of fibre and the heart rate is controlled by them. Vagal stimulation slows the heart while sympathetic stimulation speeds it up. In heart transplantation, the posterior parts of the atria are left *in situ* when the heart is removed so that the recipient's sinu-atrial node (and its nerve supply) and the openings of the pulmonary veins are still in working order.

## The pericardium

The outermost covering of the heart is the *fibrous pericardium*. This takes the form of a simple fibrous covering that encloses the heart and fuses with the outer coats of the proximal parts of the great vessels and with the central tendon of the diaphragm. It encloses the *serous pericardium*, which, like the other serous membranes (pleura, peritoneum and tunica vaginalis), consists of a parietal and a visceral layer. The parietal layer lines the fibrous pericardium and is reflected around the great arteries and veins to become continuous with the visceral layer.

The serous pericardium can best be understood by comparing it to the pleura, except that, whereas the lung has only one hilum, the heart has two. The embryonic heart tube is originally suspended from the dorsal pericardial wall by a mesentery, the *dorsal mesocardium*, but this soon disappears so that the tube penetrates the pericardium at its two ends, the arterial end and the venous end. Figure 8.17 shows the arterial and venous hila of the heart, with the gap between them that is caused by the disappearance of the dorsal mesentery. This gap is the *transverse sinus* of the pericardium. The S-shaped form taken by the embryonic heart leads to a blind-ended pericardial space behind the heart and below the venous hilum. This is the *oblique sinus*. The arterial end of the embryonic heart tube becomes divided by a spiral septum into the aorta and the pulmonary trunk; hence these two vessels are enclosed in a common tube of pericardium. The venous hilum is more complicated as it contains the four pulmonary veins as well as the superior and inferior venae cavae. The pericardial sleeve that surrounds these vessels is moulded closely around them to produce the shape shown in the third part of Fig. 8.17. Thus a finger placed in the transverse sinus passes in front of the superior vena cava and behind the aorta and pulmonary trunk, while the oblique sinus separates the back of the left atrium from the oesophagus.

## THE SUPERIOR MEDIASTINUM

This contains so many important structures that it is best to consider it in stages. The most posterior structure, closely related to the vertebrae, is the oesophagus, with the thoracic duct running up its left side (see Fig. 8.9). In front of its upper part is the trachea, which inclines slightly to the right and bifurcates at the level of the manubriosternal joint. Because of the inclination, the right bronchus is more in line with the trachea than the left and it is also larger (Fig. 8.18). Between the trachea and the oesophagus on the left side is the *left recurrent laryngeal nerve*, which comes from the vagus nerve and hooks under the *ligamentum arteriosum* (see Fig. 8.9). The right vagus crosses the trachea obliquely (see Figs 8.10 & 8.20) and both vagi then break up behind the roots of the lungs to form the *posterior pulmonary plexuses* and then, lower down, the *oesophageal plexus*. The *arch of*

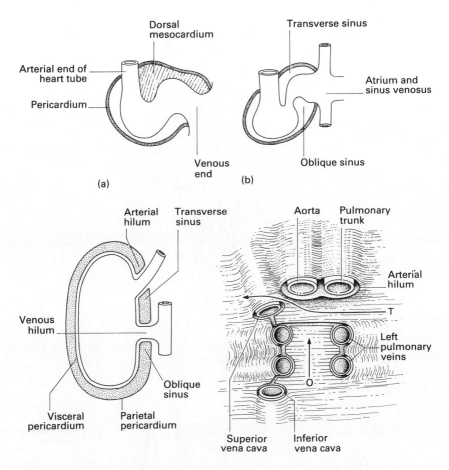

**Fig. 8.17** The sinuses of the serous pericardium (described in the text). The upper two diagrams show the embryonic heart tube. T = transverse sinus, O = oblique sinus.

*the aorta* arches over the root of the left lung and the *azygos vein* (shown, unlabelled, in Fig. 8.10) over the root of the right.

In front of the tracheal bifurcation is the *pulmonary trunk*, dividing into left and right pulmonary arteries. This has the appearance of a T with a sloping crosspiece (Fig. 8.19). The beginning of the left pulmonary artery is connected to the under-surface of the aorta by the *ligamentum arteriosum*, a remnant of the fetal ductus arteriosus that short-circuited the functionless lungs by diverting most of the right ventricular outflow into the aorta. Not shown in Fig. 8.19 are the *tracheobronchial lymph nodes* (see below) and the superficial and deep cardiac plexuses. The former plexus lies to the right of the ligamentum arteriosum and the latter is related to the tracheal bifurcation. They are formed by various branches from the sympathetic trunk and the vagi.

Figure 8.19 shows how the arch of the aorta and the ascending aorta are related to the structures described above. Remember that the arch passes

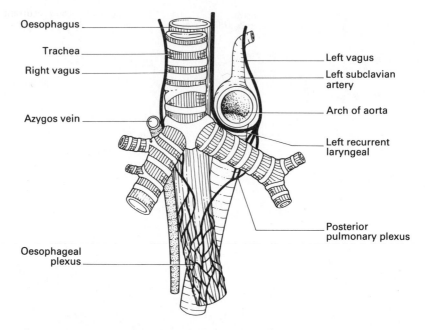

**Fig. 8.18** Figs 8.18, 8.19 and 8.20 should be studied together. This diagram shows the trachea, oesophagus and aorta, and the two vagus nerves.

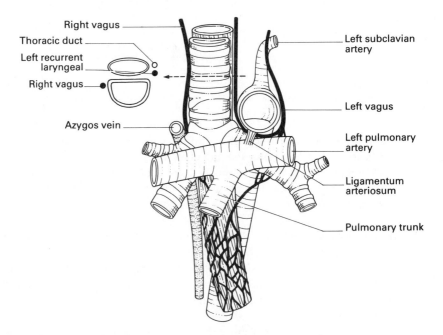

**Fig. 8.19** The pulmonary trunk has been added to Fig. 8.18.

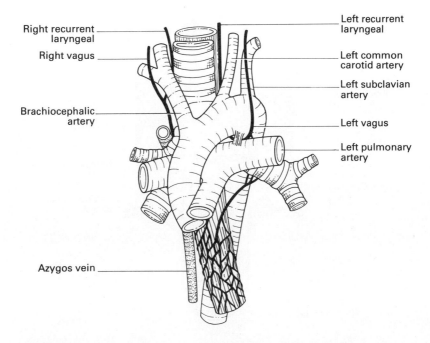

Right recurrent laryngeal

Right vagus

Brachiocephalic artery

Azygos vein

Left recurrent laryngeal

Left common carotid artery

Left subclavian artery

Left vagus

Left pulmonary artery

**Fig. 8.20** The remainder of the aorta has been added to Fig. 8.19. The pulmonary veins are not shown.

*backwards* as well as to the left so that in an anteroposterior X-ray it appears in an almost end-on view, as the *aortic knuckle*. The junction between the ascending aorta and the arch is at the level of the lower border of T4. Thus the whole arch is in the superior mediastinum. The major branches of the arch spiral around the trachea and oesophagus (Fig. 8.20), a reminder of the way in which the embryonic aortic arches, from which they develop, embrace the embryonic pharynx (see Fig. 8.21). The *left phrenic nerve* crosses the arch of the aorta in front of the vagus (see Fig. 8.14). The *left superior intercostal vein* crosses the arch from back to front, over the vagus and under the phrenic (see Fig. 8.9), relationships similar to those of the azygos vein on the right side (see Fig. 8.10), which is embryologically equivalent to it. When faced with a question about the relationships of the arch of the aorta, do not forget the most obvious structure, the left lung.

By comparing Figs 8.4 and 8.14 you will be able to remember the structures that form the left and right borders of the heart in a chest X-ray.

The remaining structures in the superior mediastinum are the *great veins* (see Fig. 8.14) and the *thymus*. In the embryo the venous system is, at first, symmetrical but two large cross-connections drain most of the blood across the midline to the right. These cross-channels are the embryonic equivalents of the *left brachiocephalic vein* and the *left renal vein*. Hence both superior and inferior venae cavae are on the right and open into the right atrium. Each *brachiocephalic vein* is formed by the junction of the corresponding *internal jugular* and *subclavian veins*; the left brachiocephalic crosses the midline just above the arch of the aorta. Since

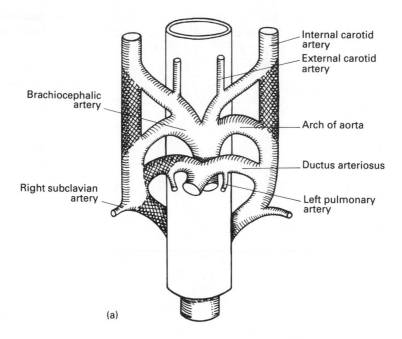

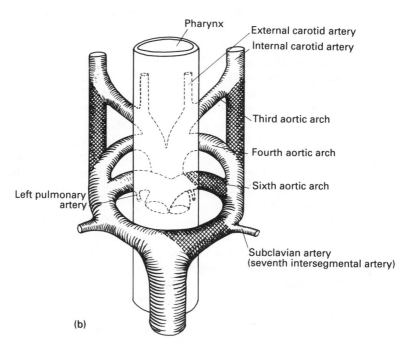

**Fig. 8.21** The relation of the aortic arches to the pharynx, (a) viewed from in front and (b) viewed from behind. The cross-hatched vessels will later disappear. Diagram (b) will be helpful in understanding the congenital malformations that affect these vessels, particularly 'vascular rings'.

the arch is entirely behind the manubrium sterni, the left brachiocephalic vein is only just below the jugular (suprasternal) notch and is actually above it in children. The brachiocephalic veins unite to form the *superior vena cava*, which lies just to the right of the ascending aorta before opening into the right atrium. It is separated from the aorta and pulmonary trunk by the transverse sinus of the pericardium (p. 184). The only other tributary of the superior vena cava is the azygos vein. The brachiocephalic veins receive a number of tributaries, including the *left superior intercostal vein* (into the left brachiocephalic), the *inferior thyroid veins* (p. 327), which come down from the neck in front of the trachea, the *vertebral veins* and the *internal thoracic veins*.

### Development of the great arteries

The great arteries are derived from the 3rd, 4th and 6th aortic arches of the embryo (you can forget about arches 1 and 2 and the 5th arch never exists). Figure 8.21 shows the embryonic aortic arch system, with the parts that will later disappear shown as cross-hatched areas. it can be seen that the oesophagus and its offshoot, the trachea, are encircled but the ring is broken by the disappearance of part of the right dorsal aorta. Persistence of this will give rise to a *vascular ring*, in which both oesophagus and trachea may be compressed. One of the most common anomalies (occurring in about 2% of the population) is the occurrence of the right subclavian artery as the last branch of the aortic arch (*arteria lusoria*). It is due to the persistence of the lower segment of the right dorsal aorta (cross-hatched in Fig. 8.21) and the loss of the right 4th arch. The anomalous artery passes *behind* the oesophagus and may cause difficulty in swallowing (*dysphagia*). The right recurrent laryngeal nerve enters the larynx directly since the part of the 4th arch that it normally recurs around has disappeared.

The ductus arteriosus may remain patent after birth; the direction of blood flow in it will be reversed because the pressure is now higher in the aorta than in the pulmonary trunk.

*Coarctation of the aorta* is a constriction usually occurring just distal to the origin of the left subclavian artery. A collateral circulation is established by, *inter alia*, anastomoses between the arteries of the upper limb with the intercostal arteries and also via the spinal arteries.

### The thymus gland

This important component of the lymphatic system lies behind the manubrium sterni but may extend up into the neck and down to about the 4th costal cartilage in the anterior mediastinum. It is moulded around the great vessels and trachea but you have probably not been able to recognize it in the dissecting room since in adult life it is gradually replaced by fat. It reaches its largest size just before puberty but, relative to the adjacent structures, it appears at its largest about the time of birth.

### The nerves in the thorax

These have been mentioned here and there in this chapter but it will be helpful now to deal with them systematically.

### Left vagus

This enters the thorax between the left common carotid and subclavian arteries, having been related to the common carotid in the carotid sheath (p. 316). The carotid leads it down to the arch of the aorta, which it crosses before giving off its recurrent laryngeal branch. This hooks around the ligamentum arteriosum and ascends to the larynx in the groove between the trachea and oesophagus (but see p. 333). *Note*: it is more correct to say that it 'recurs' around the ligamentum than around the aorta. The nerve is the nerve of the embryonic 6th aortic arch. The 6th arch on the left forms the proximal part of the pulmonary artery and the ductus arteriosus; the aorta is derived from the 4th aortic arch (Fig. 8.21). The left recurrent nerve therefore loops around the ligamentum on the left but, since the distal part of the 6th arch disappears on the right, the right recurrent nerve is able to slip up as far as the 4th arch, i.e. the right subclavian artery.

After giving off the recurrent laryngeal nerve, the vagus passes behind the hilum of the left lung and breaks up to form the *posterior pulmonary plexus*, along with branches from the 2nd–5th thoracic sympathetic ganglia. Small offshoots pass to the front of the hilum to form the *anterior pulmonary plexus*. The vagal fibres then pass on to the oesophagus to form, with the right vagus, the *oesophageal plexus*. These nerves are reduced to one or two branches on the front of the lower oesophagus and one or two on the back. The former are derived chiefly from the left vagus and the latter from the right. In the embryo, the stomach lies in the sagittal plane with left and right surfaces supplied by the corresponding vagi; but, with the rotation of the gut, the stomach 'falls over' to the right so that its nerve supply is as described above.

### Right vagus

The right vagus enters the thorax by crossing in front of the right subclavian artery and descends medial to the brachiocephalic artery. It then crosses the trachea obliquely from front to back, is crossed by the azygos vein and reaches the region of the posterior pulmonary plexus. From here down its course is similar to that of the left vagus. As it crosses the right subclavian artery it gives off the right recurrent laryngeal nerve, which hooks under the artery and ascends behind the common carotid artery to reach the larynx.

### Left phrenic

This enters the thorax by leaving the anterior surface of the scalenus anterior muscle (p. 318) and descending alongside the thoracic portion of the subclavian artery. It then crosses the arch of the aorta in front of the vagus and lies on the side of the pulmonary trunk and the left ventricle before entering the diaphragm.

### Right phrenic

In a coloured illustration, this nerve is usually shown in yellow, which makes a nice contrast to its background, which is blue all the way through the thorax. It lies on, from above downward, the right brachiocephalic vein, the superior vena cava, the right atrium and, for a very short distance, the inferior vena cava.

Both phrenic nerves arise from C3, **4** and 5 (mainly 4 but branches from C3 and 5: 3, 4, 5, keeps the diaphragm alive!). Remember that they are not only motor (p. 208) but also sensory to the central part of the diaphragm, so that irritation of the overlying pleura or peritoneum (for example by gall bladder disease on the right or conditions of the spleen on the left) may give referred pain in the 4th dermatome, which covers the tip of the shoulder.

### Other nerves in the thorax

The sympathetic system will be described in detail in Chapter 15. Here it will be enough to mention the small nerves that cross the arch of the aorta between the vagus and phrenic and others that pass behind the aorta. These are derived from cardiac branches of the vagi and sympathetic trunks in the neck and from the 2nd–5th thoracic sympathetic ganglia. They form the *superficial* and *deep cardiac plexuses*. In the lower part of the thorax are the *greater, lesser* and *lowest splanchnic nerves*. These are derived respectively from the 5th–9th, 10th and 11th and the 12th sympathetic ganglia and they pass through the diaphragm to the abdomen (see Figs 8.9 & 8.10, and p. 295).

## Medial relations of the lungs

The structures which are mentioned above form the medial relations of the lungs, being separated from them by the mediastinal pleura. They are shown in Figs 8.9 and 8.10. Some of them make deep grooves on the lungs. The left lung is indented by the left ventricle of the heart, the arch of the aorta, the subclavian artery and the left brachiocephalic vein, and perhaps the lower part of the oesophagus. The right lung carries impressions for the right atrium, the subclavian artery, the brachio-cephalic vein, the superior vena cava, the azygos vein and the oesophagus.

## THE POSTERIOR MEDIASTINUM

This contains the oesophagus, the descending aorta, the azygos venous system and the thoracic duct, and it may be regarded as a duct leading from the neck and superior mediastinum to the abdomen so that all the structures mentioned above (in addition to the inferior vena cava) have to pass through the diaphragm.

## Oesophagus

This begins in the neck at the level of the 6th cervical vertebra, where the pharynx ends. It passes through the diaphragm at the level of T10 and, after a short intra-abdominal course, it enters the stomach. Like the rest of the gut, it has an inner circular and an outer longitudinal coat but in its upper quarter the muscle is skeletal and this is gradually replaced by smooth muscle until in the lower half all the muscle is smooth. Its lining epithelium is of the 'wear and tear' type, i.e. stratified squamous, but this is suddenly replaced by columnar epithelium at the cardiac orifice. While oesophageal epithelium is excellent at resisting the effects of friction produced by swallowing crisps and toast, it is highly susceptible to the effects of acid and gastric enzymes, hence the importance of the (non-existent) cardiac 'sphincter', which will be described in Chapter 10 (p. 222).

The oesophagus follows the curve of the vertebral column for most of its length but moves forwards and to the left in its lower reaches so that it has a 'mesentery' formed from the parietal pleura. It lies to the right of the thoracic aorta for much of its course and is later anterior (ventral) and to its left. Between the oesophagus and the vertebrae, for at least part of its course, are the azygos and hemi-azygos veins, with their interconnections, the thoracic duct and the right intercostal arteries. In the superior mediastinum the trachea is in front and, slightly to the right and below this, an important relationship is the oblique sinus of the pericardium, which separates it from the left atrium (p. 184). Other relationships are shown in Figs 8.18–8.20.

The oesophagus has four slight constrictions (Fig. 8.22): (i) where it begins, in the form of the cricopharyngeus muscle; (ii) where it is crossed by the arch of the aorta; (iii) where it is crossed by the left bronchus (left because the trachea veers to the right); and (iv) where it pierces the diaphragm. The vertebral levels of these are of academic interest only. The clinician needs to know where they are when he or

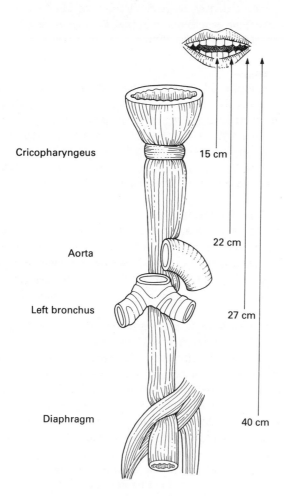

**Fig. 8.22** The four structures that constrict the oesophagus and their distances from the incisor teeth. Cricopharyngeus is extremely diagrammatic.

she is passing a tube or instrument; hence their distance from the incisor teeth is more important: respectively 15 cm, 22 cm, 27 cm and 40 cm.

## The descending aorta

This begins at the level of the lower border of T4 and at first lies to the left of the midline and then moves medially to pierce the diaphragm in the midline at the level of T12. The bodies of vertebrae T4–8 are therefore slightly flattened on their left side. Its branches are mostly small, supplying structures in the posterior mediastinum, but it also gives off *posterior intercostal arteries*. Where the aorta lies to the left of the midline, the right posterior intercostals have to cross the bodies of the vertebrae behind the oesophagus to reach the right intercostal spaces. Individual intercostal arteries are not important as far as their supply to the chest wall is concerned but their posterior divisions have spinal branches to supply the spinal cord (p. 290). Unfortunately, these are not always large and the longitudinal anastomoses on the cord are not always good so it must be borne in mind that there is usually a particularly large spinal branch in the lower thoracic or upper lumbar regions (*artery of Adamkiewicz*) that must be preserved at all costs.

## The azygos system of veins

Since the azygos and hemi-azygos veins are developed from a complex network of veins in the embryo, they are subject to great variation and the following account only gives the usual arrangement. The azygos vein begins by the union of the right subcostal vein and one or more veins coming up from the lumbar region. It passes through the aortic opening alongside the thoracic duct and behind the aorta, ascends on the vertebral bodies to T4 and then arches over the root of the lung to enter the superior vena cava (see Fig. 8.10). It receives the *right posterior intercostal veins* either directly or via the *right superior intercostal vein*, which drains the upper spaces.

The hemi-azygos vein begins on the left in a manner similar to that of the azygos and receives the *lower* posterior intercostal veins. The accessory hemi-azygos drains the *middle* posterior intercostal veins. These two venous channels cross the midline, usually at about the 7th and 8th vertebral levels, and enter the azygos. The upper posterior intercostals form the *left superior intercostal vein*, which has already been mentioned as a relation of the arch of the aorta.

## The thoracic duct

The thoracic duct drains lymph from the whole body except for the right side of the head and neck, the right upper limb, the right side of the thorax and the upper surface of the liver. These areas drain into right *jugular, subclavian* and *broncho-mediastinal* trunks, which enter the veins on the right side of the neck.

The thoracic duct begins in the *cisterna chyli* (p. 23), which lies in front of the bodies of the upper two lumbar vertebrae and receives lymph from the abdomen and lower limbs. The duct passes through the aortic opening in the diaphragm and ascends on the vertebral bodies until, at the level of T4, it crosses the midline and comes to lie on the left side of the oesophagus (see Fig. 8.9). It ends in the neck by entering the venous system at the junction of the left subclavian and internal jugular veins.

## LYMPHATIC DRAINAGE OF THE THORAX

The principal groups of lymph nodes in the thorax are the *internal thoracic (parasternal)* and *intercostal* nodes, clustering around the corresponding vessels, *diaphragmatic* nodes, *brachiocephalic* nodes around the corresponding veins, *posterior mediastinal* and, perhaps the most important, *tracheobronchial* (Fig. 8.23).

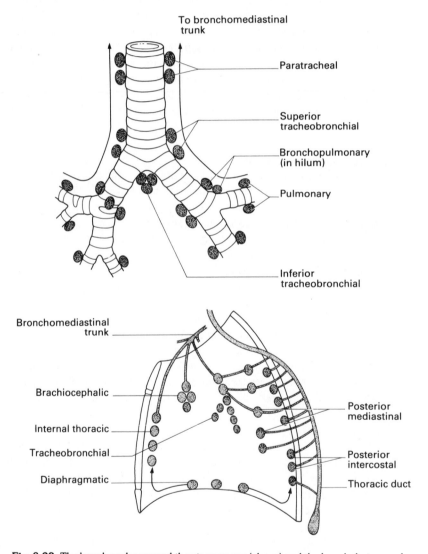

**Fig. 8.23** The lymph nodes around the air passages (above) and the lymph drainage of the thorax.

The superficial tissues of the chest wall drain into the axillary and internal thoracic nodes (see p. 65 for the lymph drainage of the breast). The deeper tissues drain to the internal thoracic, intercostal and diaphragmatic nodes and thence, directly or indirectly, into the thoracic duct.

The heart drains by left and right collecting trunks into the inferior tracheo-bronchial and brachiocephalic nodes respectively. The lungs have a plexus of vessels under the visceral pleura that drain around the periphery of the lungs into the bronchopulmonary nodes. The deep lymphatics follow the air passages to the various groups of tracheobronchial nodes. There are, at least in the normal lung, no communications between the superficial and deep systems since there are no lymphatics in the alveoli of the lung so that the deep lymphatics do not reach the surface.

Efferents from the brachiocephalic, tracheobronchial and posterior mediastinal nodes unite to form right and left bronchomediastinal trunks. These are joined by vessels from the internal thoracic, intercostal and diaphragmatic nodes.

# Chapter 9
# Abdominal Wall, External Genitalia, Diaphragm

Before commencing a detailed study of the abdomen, it is worth considering a few general points. It is helpful to remember that the abdomen and thorax are part of the same entity, the trunk, and that the serous cavities (*pleural* and *peritoneal cavities* and the *pericardium*, as well as the *tunica vaginalis* in the male) formed one continuous *coelomic duct* in the embryo. They are thus very similar in having a parietal layer, which is part of the chest or abdominal wall and is supplied by the spinal nerves, and a visceral layer which is part of the lung or the gut and is supplied by autonomic system of nerves (see Fig. 1.8). Again, the chest wall and abdominal wall are each formed by the migration of cells from the mesodermal somites TI–LI into the embryonic body wall so that the muscles are similar in the two regions (p. 167) and the lower thoracic nerves supply the abdominal wall.

Both chest and abdominal cavities are variable in size, but in quite a different way. The chest can be *increased* in volume as a result of the action of the respiratory muscles (p. 177), to produce inspiration, but expiration is largely a passive process of elastic recoil. The abdominal cavity can be *decreased* in volume by the constricting action of the abdominal muscles and when these relax its volume increases again. The diaphragm is an active partition between the two; when it contracts, the chest volume increases and the abdominal volume decreases. When the abdominal muscles contract and the diaphragm relaxes, the chest volume diminishes and the result is a forced expiration. Of course, these changes also produce changes in pressure so that, if the abdominal muscles contract at the same time as the diaphragm, the intra-abdominal pressure will rise steeply. Thus any expulsive effort from the abdomen, such as vomiting or defaecation, is preceded by a deep inspiration.

Finally, remember that the 'chest' (i.e. the ribcage) encloses the upper abdominal contents as well as the lungs and heart. The liver, under the right dome of the diaphragm, rises to the level of the 5th intercostal space, as does the fundus of the stomach under the left dome. Wounds that penetrate the lower intercostal spaces may therefore involve both thoracic and abdominal viscera.

## THE ANTERIOR ABDOMINAL WALL

If you place your hands on either side of your abdomen with the finger tips just touching in the midline, the palms of your hands lie mainly over the oblique muscles and your fingers over their aponeuroses, so that the anterior part of the abdominal wall is mostly aponeurotic except for the vertical rectus abdominis muscle. Since the abdomen needs to be distensible, there is no true deep fascia over the abdomen but the deepest part of the superficial fascia is somewhat thickened to form the *membranous layer* of superficial fascia or *Scarpa's fascia*.

The oblique muscles, *external* and *internal*, and *transversus abdominis* form three distinct layers, similar to those in the intercostal spaces (p. 167) and, as in the intercostal spaces, the principal nerves and vessels lie between the internal oblique and transversus muscles. Internal to the muscles is the *transversalis fascia* (do not confuse this with the transversus muscle). This is a rather dense layer of fascia, especially in the lower abdomen, deep to which lies a layer of extraperitoneal fat, followed by the peritoneum itself.

## Membranous layer of superficial fascia (Scarpa's fascia)

This deep layer of superficial fascia is particularly dense in the lower abdomen. It crosses the inguinal ligament and is attached to the fascia lata of the thigh below the ligament. It is prolonged over the penis, as far as the base of the glans, to form the *fascia penis*. It covers the scrotum, where it contains the smooth muscle fibres of the *dartos* muscle. It may best be pictured by thinking of a pair of wet bathing trunks which cling to the penis and scrotum, are sewn to the thighs below and parallel to the inguinal ligament and have a small hole through which the glans protrudes (Fig. 9.1). Between the thighs the fascia is prolonged into the perineum as *Colles' fascia* (p. 271). In the female the fascia blends with the fascia of the labia majora.

## External oblique

For the most part, the fibres of this muscle run downwards and forwards from their posterior attachments. The muscle arises from the *outer* surfaces of the lower eight ribs. Serratus anterior arises from the upper eight so the two muscles interdigitate on the middle four. The muscle fibres posteriorly pass vertically down to the iliac crest so that there is a posterior free border. The attachment to the ilium continues forward as far as the anterior superior spine, where it becomes

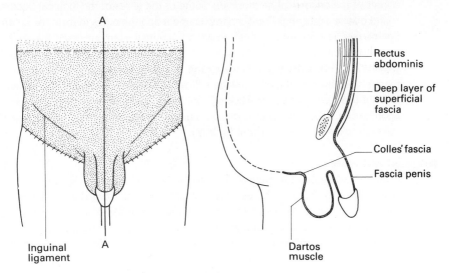

**Fig. 9.1** The membranous layer of superficial fascia can be likened to a pair of bathing trunks sewn to the thighs and clinging to the penis and scrotum (except for the glans).

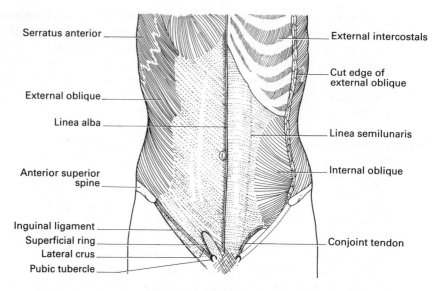

**Fig. 9.2** The anterior abdominal wall. The left external oblique muscle has been removed.

aponeurotic, and the aponeurosis jumps from here to the pubic tubercle, bridging across the muscles, nerves and vessels that are entering the thigh (see Fig. 6.8). The free lower border of the aponeurosis is thickened and rolled inwards to form the *inguinal ligament*. (Remember that it is not a ligament and is not the cord-like structure that is so often depicted in diagrams and seen in models of the pelvis. In a similar way, the vocal ligament (p. 362) is not a cord but is the thickened *upper* border of the cricovocal membrane.) Some of the fibres of the inguinal ligament round off the acute angle between the ligament and the pubis to form the *lacunar ligament*. The aponeurotic fibres above the inguinal ligament reach the midline, where they intermingle with the fibres from the other side to form the *linea alba* (Fig. 9.2). This is attached above to the xiphoid process but above the costal margin the fibres of external oblique still reach the midline to form the anterior layer of the rectus sheath (see below). In the inguinal region there is a gap in the aponeurosis, called the *superficial inguinal ring*, which will be described in detail in the section on the inguinal canal (p. 202).

### Internal oblique

Over a good deal of the abdominal wall, the fibres of this muscle run roughly at right angles to those of the external oblique. It arises from the thoracolumbar fascia, the dense fascia that covers the back muscles (p. 239). The fascia leads the origin down to the iliac crest and thence, via the anterior superior spine, to the lateral half of the inguinal ligament. The lowermost fibres arch downwards over the spermatic cord or round ligament to be inserted into the *pecten pubis* (*pectineal line*) along with the aponeurotic fibres of transversus. The two form the *conjoint tendon* (*falx inguinalis*) but again it must be stressed that this is just the lower part of an extensive pair of aponeuroses. The higher fibres of the aponeurosis of internal oblique reach the midline in front of the rectus muscle and higher still they form anterior and posterior layers, which will be discussed later. The uppermost fibres are inserted into the ribs and costal cartilages, which form the costal margin.

### Transversus abdominis

This arises, like the internal oblique, from the thoracolumbar fascia and the iliac crest but from only the lateral one-third of the inguinal ligament. It also rises from the *inner* surfaces of the lower six ribs, interdigitating with the origin of the diaphragm. Most of the fibres of transversus pass more or less horizontally towards the midline, taking part in the formation of the rectus sheath. The fibres from the inguinal ligament arch over to join the conjoint tendon.

It is helpful to remember that, of the three muscles already mentioned, external oblique lies superficial to the lower ribs, internal oblique is attached to the costal margin and transversus lies deep to the lower ribs (Fig. 9.3).

### Rectus abdominis and its sheath

Rectus extends from the 5th, 6th and 7th costal cartilages to the pubic crest and tubercle and the front of the symphysis pubis. The vertical muscle fibres are interrupted by (usually) three *tendinous intersections* in the upper part of the muscle, which are adherent to the anterior rectus sheath. The lateral border of each rectus muscle can often be recognized in a thin person's abdomen as a curved line, the *linea semilunaris*. It crosses the costal margin in the transpyloric plane (p. 211) at the tip of the 9th costal cartilage and, on the right side, this is the surface marking of the fundus of the gall bladder.

The muscle is enclosed in the *rectus sheath*, which is derived from the aponeuroses of the other abdominal muscles. Throughout the greater part of the

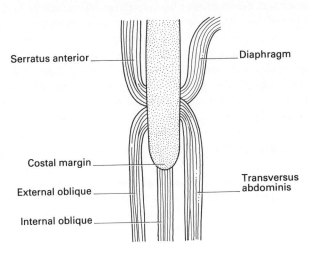

**Fig.9.3** Section through the costal margin. External oblique is outside the ribs, transversus is inside the ribs and internal oblique is attached to the costal margin.

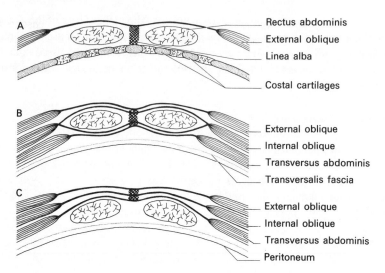

A

Rectus abdominis

External oblique

Linea alba

Costal cartilages

B

External oblique

Internal oblique

Transversus abdominis

Transversalis fascia

C

External oblique

Internal oblique

Transversus abdominis

Peritoneum

**Fig. 9.4** Cross-sections through the rectus and its sheath; A, above the costal margin; B, above the umbilicus; and C, above the pubic symphysis.

length of the muscle there are one and a half aponeuroses in front and one and a half behind. The aponeurosis of internal oblique splits, part going in front of the muscle and part behind (Fig. 9.4), so that the anterior sheath is composed of the external oblique aponeurosis and the anterior layer of internal oblique. The posterior sheath is composed of the posterior layer of internal oblique and transversus. However, in the uppermost part of the abdomen, *muscular* fibres of transversus almost reach the midline; otherwise the sheath is entirely aponeurotic.

Above the costal margin, only the external oblique aponeurosis is present (Fig. 9.3) so that this forms the anterior rectus sheath. Posteriorly the muscle lies directly on the chest wall. About halfway between the umbilicus and the pubis, the one and a half layers of the posterior rectus sheath are usually described as coming to an abrupt stop, their free lower borders being known as the *arcuate line*. Below this level, all three aponeuroses pass in front of rectus so that, posteriorly, the muscle lies directly on the transversalis fascia. In fact, however, there is usually no clearly demarcated arcuate line, as the aponeurotic fibres transfer from the posterior sheath to the anterior sheath gradually so that the posterior sheath becomes thinner and thinner until, some distance above the pubis, all the aponeurotic fibres pass anterior to rectus. Behind the muscle but within the rectus sheath are the *superior* and *inferior epigastric arteries*, which will be described below.

### Nerve supply and actions of the abdominal muscles

As has already been mentioned, the lower six thoracic and the lst lumbar nerves run between transversus abdominis and internal oblique. These nerves supply the abdominal muscles and the skin of the abdomen (see Fig. 1.10).

The abdominal muscles have three roles: to produce movement, to protect the viscera and to increase intra-abdominal pressure. Rectus abdominis is a flexor of the lumbar spine and contracts strongly on sitting up from a lying position. It can also fix the ribcage so that muscles acting on the neck have a fixed base to work from; rectus therefore contracts on lifting the head from the pillow. Similarly, it can fix the pelvis so that flexors of the hip joint such as iliacus have a fixed base; thus it contracts if the legs are raised from the bed. It can also tilt the pelvis by elevating the pubis. The oblique muscles help in flexion but also produce other movements. The *right* external oblique can pull the right ribcage forwards, so rotating the trunk to the left, acting in conjunction with the *left* internal oblique. When all the muscles on one side contract, they produce side flexion. Transversus, owing to its transverse fibres, can pull in and flatten the abdominal wall.

All muscles, particularly rectus, contract involuntarily when expecting a blow to the abdomen and thus protect the viscera. Acting together, they also increase the intra-abdomir.al pressure. If this is combined with contraction of the diaphragm, there is a large increase in pressure, which is used for defaecation, parturition and other expulsive efforts. If the diaphragm is relaxed, the abdominal muscles will produce a forced expiration as in coughing. It is a pity, therefore, that the pain from abdominal incisions may inhibit patients from coughing up sputum and unless they are encouraged to cough there may be chest complications after surgery.

## Blood vessels

Between transversus and internal oblique lie branches of the lower two intercostal arteries, the subcostal arteries and the lumbar arteries. Within the rectus sheath and behind the muscle are the epigastric arteries. The *superior epigastric* is one of the two terminal branches of the internal thoracic (the other being the *musculophrenic*) and the *inferior* is one of the two branches of the external iliac (the other being the *deep circumflex iliac*). The epigastric arteries pass downwards and upwards, respectively, and anastomose freely.

## The umbilicus

The umbilicus is a scar, that is, it is an area of dense fibrous tissue that plugs the gap in the abdominal wall that was present in the fetus to allow the passage of the umbilical vessels (two arteries, one vein). Remnants of these vessels are present in the adult, as are remnants of the allantois and, sometimes, the vitello-intestinal duct which represents the remains of the stem of the yolk sac (see Fig. 1.8). These are shown diagrammatically in Fig. 9.5. The two *umbilical arteries* are represented by the *medial umbilical ligaments*. The arteries originally extended from the aorta to the umbilicus but the proximal part of this system enlarged to form the common iliac arteries and their branches while the ligament, after birth, forms a fibrous band from the superior vesical artery to the umbilicus. The allantois originally extended from the developing urogenital sinus into the umbilical cord. After making a contribution to the bladder, the remnants of the allantois form a fibrous cord from the apex of the bladder to the umbilicus (the *urachus*). Uncommonly, part or all of it may remain patent to form cysts or even a complete fistula (rare). The *umbilical vein* originally passed from the umbilical cord to the under-surface of the liver and thence the blood flowed through the ductus venosus to the inferior vena cava. After birth, it is represented by the *ligamentum teres*,

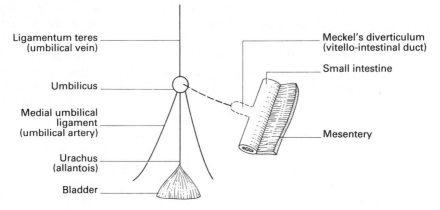

**Fig. 9.5** The four interesting embryological remnants at the umbilicus.

which lies in the free edge of the falciform ligament (neither of these are, in fact, ligaments!). It is accompanied by some small veins that unite the portal vein with the veins of the anterior abdominal wall (p. 231).

Finally, remnants of the *vitello-intestinal duct* may be present. This originally passed from the distal part of the ileum to the umbilicus and it is the proximal part that usually persists. This takes the form of a *Meckel's diverticulum*, a blind-ended sac that is classically 2 inches long, 2 feet from the ileocaecal junction and present in 2% of cases. It may be joined to the umbilicus by a fibrous cord or, rarely, there may be a complete fistula. A fibrous cord may lead to intestinal obstruction by acting as an axis round which volvulus (p. 233) may occur.

The umbilicus is therefore a hotbed of embryology.

### The inguinal canal

This is a passage through the abdominal wall that lets the spermatic cord (or round ligament) out without letting anything else out. Any gap in the wall of the abdomen (including the diaphragm, pelvic muscles or posterior abdominal wall) is a potential weak spot through which abdominal contents may be forced when the intra-abdominal pressure is raised. The inguinal canal is, therefore, not a true canal but a series of three openings which are staggered so that no two openings lie opposite each other. The openings are in the *external oblique aponeurosis*, the combined *internal oblique and transversus* muscles, and the *transversalis fascia*. They are shown in Fig. 9.6.

The opening in the external oblique aponeurosis is the *superficial inguinal ring*, which is not a ring and cannot be seen easily because the aponeurosis that forms the edges of the ring is continuous with a layer of fascia that surrounds the spermatic cord, known as the *external spermatic fascia*. The opening is an elongated triangle, the medial and lateral sides of which are the *crura*. The lateral crus is attached to the pubic tubercle and the medial crus crosses the midline and is attached to the front of the pubic symphysis. The main part of the opening is thus above and medial to the pubic tubercle, and the spermatic cord emerges from it.

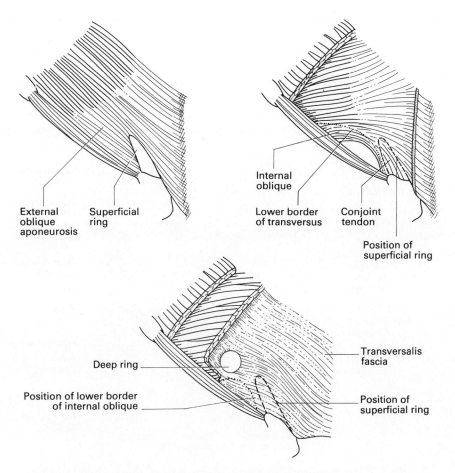

**Fig. 9.6** The three layers and three holes of the inguinal canal.

The second opening is below the arch formed by internal oblique and transversus as they unite and pass over the spermatic cord to form the conjoint tendon. From these two muscles a layer of fascia containing muscle fibres is prolonged around the spermatic cord, deep to the external spermatic fascia. These are the *cremasteric fascia* and the *cremaster muscle*. Cremaster is supplied by the genital branch of the genitofemoral nerve and is striated muscle but is not under voluntary control (compare this with the ciliaris muscle in the eye, which is smooth muscle but is under voluntary control). It may be made to contract and thus to elevate the testis by stimulation of the inside of the thigh. This may best be seen after a hot bath, when relaxation of the smooth muscle of *dartos* (see above) allows the scrotum to become slack.

The third opening, the *deep inguinal ring*, is in the transversalis fascia, but again cannot be seen easily because the fascia is continuous with the innermost covering of the cord, the *internal spermatic fascia*. The surface marking of the deep ring is important; it lies just above the inguinal ligament *halfway between the anterior superior spine and the pubic tubercle.* You will remember that the external iliac/femoral artery passes down under the inguinal ligament *halfway between the anterior superior spine and the midline*, so the ring is lateral to the artery and to its branch, the inferior epigastric (Fig. 9.7). More about this later.

Within all three layers of the spermatic cord lie the ductus deferens and the testicular vessels. The three openings in the abdominal wall thus form an oblique passage through which these structures can pass (Fig. 9.8) and, moreover, the superficial and deep rings receive an extra protection. Thus, the superficial ring is

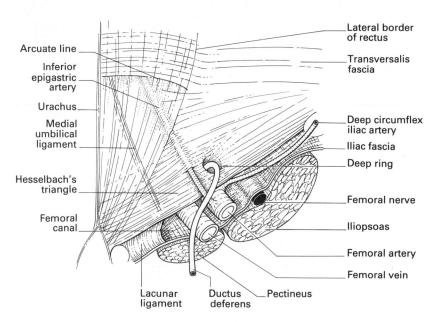

**Fig. 9.7** The deep surface of the anterior abdominal wall. Compare with Fig. 6.8.

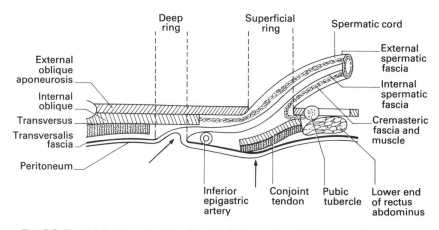

**Fig. 9.8** Roughly horizontal section through the inguinal canal. The two arrows show the position of indirect and direct inguinal hernia.

reinforced posteriorly by the conjoint tendon and the deep ring is reinforced anteriorly by the muscular fibres of origin of internal oblique from the lateral half of the inguinal ligament. The deep ring itself is quite a strong structure since the transversalis fascia in the vicinity is reinforced by bands of denser connective tissue that run below the ring, like supports. Thus, when contraction of the muscles raises the intra-abdominal pressure, the three rings are pressed closer to each other and the arch of internal oblique and the conjoint tendon flattens and compresses the cord.

### The femoral canal

The femoral canal has been mentioned already as being enclosed within the femoral sheath. The sheath itself is a fascial sleeve that encloses the femoral artery and vein, as well as the canal, and is produced by an outpouching of the innermost lining of the abdominal wall, i.e. the transversalis fascia above the inguinal ligament, and the fascia covering iliacus and psoas below. It is thus similar to the axillary sheath, which is an outpouching of the prevertebral fascia (Chapter 17).

The canal lies between the femoral vein and the lacunar ligament so that it provides a space into which the vein can expand when the venous return from the leg is increased. The upper opening of the femoral canal is called the *femoral ring* and its boundaries, therefore, are medially the lacunar ligament, laterally the vein, anteriorly the inguinal ligament and posteriorly pectineus and its fascia. The canal normally contains a certain amount of fat and some lymphatics and lymph nodes. Note particularly that the opening of the canal is below and lateral to the pubic tubercle, an important diagnostic feature in hernia.

## THE SPERMATIC CORD AND TESTIS

A little embryology helps to explain some of the features of the external genitalia. Both testis and ovary develop on the medial surface of the mesonephros within the abdomen. Each is embedded in the upper end of a thick gelatinous cord of mesenchyme, the *gubernaculum*, which extends downwards through the region of the future inguinal canal to the *genital swelling* (the future scrotum or labium majus). A diverticulum of peritoneum, the *processus vaginalis*, passes down into the developing scrotum and when the testis descends down the track prepared for it by the swelling of the gubernaculum (it is not actively pulled down by the gubernaculum) it lies behind the processus. The testis drags down behind it its duct, the *ductus deferens*, developed from the mesonephric duct, as well as its blood vessels and lymphatics. The coverings of the cord differentiate around these structures in the mesenchyme of the gubernaculum. Thus the gubernaculum ends up as a mere remnant at the bottom of the scrotum and the ductus passes through the inguinal canal. In the female, however, the ovary descends only as far as the pelvis (although the processus vaginalis passes through the inguinal canal). Thus, the inguinal canal in the female is occupied only by remnants of the gubernaculum in the form of the *round ligament* (Fig. 9.9). In both sexes, the processus normally pinches itself off from the peritoneal cavity and in the male forms the *tunica vaginalis* of the testis.

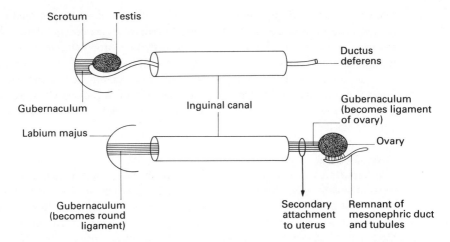

**Fig. 9.9** The fate of the gubernaculum and the mesonephric duct in the male and the female.

The spermatic cord therefore contains the *ductus deferens*, the *testicular artery* (and some other smaller arteries), the *pampiniform plexus of veins, lymphatics* and *autonomic nerves*. It is covered by the internal spermatic fascia, the cremasteric fascia and the external spermatic fascia, all of which have been described already. The ductus deferens is a thick-walled tube extending from the lower end of the epididymis (see below) through the inguinal canal into the abdomen, eventually to enter the prostatic urethra. It can easily be distinguished from the other contents of the cord by the sense of touch since it feels quite hard. Traditionally the ductus has been known as the *vas* (deferens) and this name has tended to stick. Whoever heard of a ductectomy clinic?

The testicular artery arises from the aorta within the abdomen and the veins and lymphatics end in the same region, the result of the intra-abdominal development of the testis. The testicular vein is formed near the deep ring by the joining together of the *pampiniform plexus* of veins. These form a dense plexus within the spermatic cord and act as a countercurrent heat exchanger so that heat from the body carried by the testicular artery is transferred to the veins. A similar heat-retaining system is found in the flippers of marine mammals. The lymphatics from the testis and epididymis end in the lateral and pre-aortic lymph nodes; remember that the lymph drainage of the scrotum is into the superficial inguinal lymph nodes so that enlargement of these nodes accompanying a testicular tumour suggests that the disease has spread to the scrotum.

### Cryptorchidism

This is the condition in which the testis cannot be palpated. The testis normally descends into the scrotum during the 8th or 9th month *in utero* but the descent is not as spectacular as you might imagine from your knowledge of adult anatomy. The testis is never far from the deep inguinal ring and this is where the undescended testis is often found. It may also lie in the inguinal canal but is difficult to

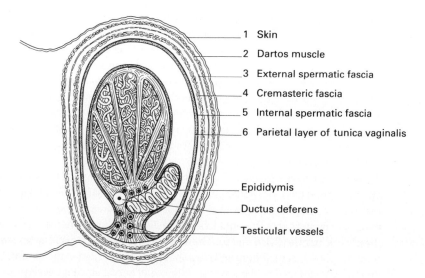

1  Skin
2  Dartos muscle
3  External spermatic fascia
4  Cremasteric fascia
5  Internal spermatic fascia
6  Parietal layer of tunica vaginalis

Epididymis
Ductus deferens
Testicular vessels

**Fig. 9.10** Horizontal section through the testis and scrotum.

feel here as it is usually softer than the normal testis. A powerful cremasteric reflex may be able to retract a normal testis into the inguinal region and this condition needs no treatment. Finally, the testis may descend into a position outside the scrotum (*ectopic testis*) and may be found, for example, somewhere in the perineum.

### The testis and epididymis

The testis consists of a mass of seminiferous tubules which open into a knot of tubules near the upper pole, the *rete testis*. From here, *ductuli efferentes* pass to the upper end of the epididymis. The testis is covered by the thick fibrous *tunica albuginea* and is invested by the serous coat of the tunica vaginalis as though it had been pushed into the potential cavity of the tunica from behind (Fig. 9.10). Outside this there are the three layers that also form a covering for the spermatic cord, and finally by the superficial fascia, the dartos muscle and the scrotal skin. The epididymis consists of one, long, highly convoluted tubule, about 7 m long. The coils are arranged to form a *head* (caput), *body* and *tail* (cauda). The tubule finally leaves the cauda epididymis and becomes the ductus deferens. The epididymis is partly covered by the visceral layer of the tunica vaginalis (Fig. 9.10). Various remnants of the mesonephric and paramesonephric ducts and tubules may be found around the testis and epididymis. The most important of these are the *appendices* of the testis and of the epididymis, both of which are at the upper end and both of which may be the site of cysts or of torsion. The female genitalia will be described in Chapter 13.

## THE DIAPHRAGM

The diaphragm is composed of ordinary striated muscle but is under dual control. It is thus possible voluntarily to alter the rate and depth of respiration by nerve impulses from the motor cortex but when asleep the muscle still contracts rhythmically, controlled by impulses from the respiratory centres in the pons and medulla. It takes origin from the inner surfaces of the lower six ribs (see Fig. 9.3); from two *crura* that are attached to the upper two or three lumbar vertebrae (three on the right, two on the left because, it is said, the right crus needs to be stronger to pull down the liver — this does not sound convincing); from the xiphoid process; and from the *arcuate ligaments* (lumbocostal arches), which will be described below. From these attachments the muscle fibres ascend almost vertically at first and are then inserted into the *central tendon*. As can be seen in a chest X-ray (see Fig. 8.4), the right dome of the diaphragm, which covers the liver, is higher than the left. The upper surface of the diaphragm is largely covered by parietal pleura and the central tendon is adherent to the fibrous pericardium. The under-surface is covered by parietal peritoneum except for the bare area of the liver, where it is directly in contact with the right lobe of the liver. The motor nerve supply is from the phrenic nerves (C3, **4** and 5). The sensory supply (which also carries sensation from the pleura, pericardium and peritoneum) is also carried in the phrenic but the periphery is supplied by the lower six intercostal nerves since it develops from the chest wall. The actions of the diaphragm have already been discussed (pp. 178 and 201).

### The arcuate ligaments (lumbocostal arches)

These are not really ligaments, but merely fibrous thickenings in the fascia covering psoas major and quadratus lumborum (Fig. 9.11). The *lateral arcuate ligament* arches over quadratus lumborum, being attached to the transverse process of L1 and to the 12th rib; the *medial arcuate ligament* arches over psoas major and is attached to the transverse process and to the body of L1. The *median*

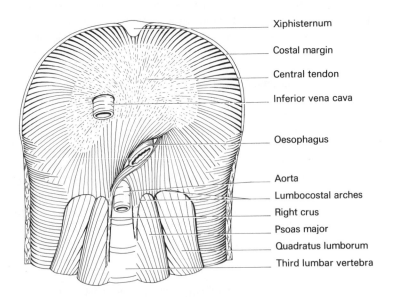

Xiphisternum

Costal margin

Central tendon

Inferior vena cava

Oesophagus

Aorta

Lumbocostal arches

Right crus

Psoas major

Quadratus lumborum

Third lumbar vertebra

**Fig. 9.11** The under-surface of the diaphragm.

*arcuate ligament* is a little different since it arches over from one crus to the other, crossing the aorta *en route*. The 'aortic opening' is therefore not really an opening at all since the aorta is actually behind the diaphragm.

The sympathetic trunk enters the abdomen behind the medial arcuate ligament and the greater and lesser splanchnic nerves pierce the crura.

### The openings in the diaphragm

There are *three* major openings, for the *inferior vena cava*, the *oesophagus* and the *aorta*. They lie respectively at the levels of T8, T10 and T12. The opening for the inferior vena cava is in the central tendon luckily; if there were muscle fibres in the vicinity, they would compress the rather thin-walled vein. The oesophageal opening is a little to the left of the midline but, surprisingly, is surrounded by fibres of the *right crus*, which crosses the midline (Fig. 9.11). The oesophageal opening also transmits the oesophageal branches of the left gastric vessels (a portal–systemic shunt, p. 230) and the anterior and posterior gastric nerves (see vagus nerve in Chapter 8). The aortic opening is in the midline between the two crura and behind the median arcuate ligament. It also transmits the thoracic duct.

## HERNIA

A hernia is the protrusion of some of the abdominal contents through a weak spot in the abdominal wall. Anything which increases intra-abdominal pressure may predispose to the condition, so that it may occur as a complication of a chronic cough, an enlarged prostate causing difficulty in passing urine, or even a large intra-abdominal tumour. The commonest hernias are *inguinal* (especially in males), *femoral* (especially in females), *umbilical* and *diaphragmatic*. The contents of a hernia are usually intraperitoneal structures, such as omentum, and small or large intestine, and they are enclosed in a diverticulum of parietal peritoneum called the *hernial sac*. Occasionally extraperitoneal structures are involved, for example, part of the bladder may be found in a femoral hernia.

### Inguinal hernia

There are two types, *indirect* and *direct*. In the former, the sac passes through the deep ring and down the spermatic cord, being covered by all its three layers. It usually enters the scrotum if it is large, because such hernias are probably all the result of the persistence of a complete processus vaginalis. The neck of the sac is thus lateral to the inferior epigastric artery. *Direct hernias* push straight through the abdominal wall opposite the superficial ring, so that the conjoint tendon is usually stretched out over the hernia. The neck of the sac is somewhere within *Hesselbach's triangle*, a triangle formed by the inferior epigastric artery, the lateral border of rectus and the inguinal ligament (see Fig. 9.7).

### Femoral hernia

The hernial sac passes through the femoral canal into the thigh. The neck of the sac will therefore have the lacunar ligament medial to it and the femoral vein lateral. The former may be divided if necessary to free the hernial sac but remember the possible course of the obturator artery (p. 265). Since the fascia lata is attached to the inguinal ligament, the hernia will lie deep to the fascia but may then bulge forwards through the saphenous opening. The neck of a femoral hernial sac is below and lateral to the pubic tubercle and that of an inguinal hernia above and medial.

## Diaphragmatic hernia

Although this may be a congenital hernia passing through an opening caused by a persistence of the embryonic coelomic duct (usually near the upper pole of the left kidney), the commonest type is through the oesophageal opening, which may be greatly enlarged. This is known as a *hiatus hernia* and may be of various types. The main difficulty is that the hernia may interfere with the mechanism that normally prevents reflux of stomach contents into the oesophagus.

## Umbilical hernia

This, too, may be of a congenital type, resulting either from failure of the midgut to return to the abdominal cavity during the 2nd month of development (*exomphalos*) or from a small protrusion of gut through the umbilicus at birth. The latter is easily treated without surgery. In the adult, the hernia occurs through a gap in the fibrous tissue in the region of the umbilical scar.

# LYMPH DRAINAGE OF THE ABDOMINAL WALL

## Superficial tissues

Anteriorly, there is a watershed approximately at the level of the umbilicus. The lower part of the abdominal wall drains downwards towards the superficial inguinal nodes and the upper part drains upwards towards the pectoral and subscapular nodes. Posteriorly, the back above the iliac crests drains to the subscapular nodes in the axilla.

## Deep tissues

From the upper part of the abdominal wall, lymphatics drain into the internal thoracic (parasternal) nodes, while the lower part drains to lymph nodes around the inferior epigastric and external iliac arteries. The posterior abdominal wall (see Chapter 12) drains to the lateral aortic nodes.

# Chapter 10
# The Upper Abdomen

## REGIONS OF THE ABDOMEN

The abdomen is divided by a noughts and crosses grid into nine regions which are useful for descriptive purposes. These are shown in Fig. 10.1. The *transpyloric plane* is officially halfway between the jugular notch of the sternum and the pubic symphysis, but a better guide to it is the level at which the lateral border of rectus abdominis crosses the costal margin. It does not necessarily indicate the level of the pylorus since that is a mobile region. For example, in a tall thin person who has just breathed in, who is standing up and who has recently eaten a large bowl of porridge, the pylorus would be a very long way indeed below the transpyloric plane. The *transtubercular plane* is level with the tubercles on the iliac crest. In terms of vertebral levels, the *transpyloric, subcostal* (lowest part of the costal margin), *umbilical* (at least in the young person) and *transtubercular* planes lie respectively at the levels of the *lower* borders of L1, 2, 3 and 4.

Clinically, these terms are often modified. The iliac region is usually called the *iliac fossa* (abbreviated to LIF and RIF), the hypogastrium is often called the *suprapubic region* and the back of the lumbar region in the angle between the 12th rib and the midline is called the *renal angle*

## PERITONEAL CAVITY

As in the case of the pleural cavity, the parietal layer develops as part of the abdominal wall and is supplied by the thoracic and 1st lumbar nerves while the visceral layer is part of the gut and is therefore supplied by the autonomic system (p. 25). Pain from the parietal layer is like pain from the skin. It is sharp and stabbing and well localized. Pain from the visceral layer is aching, poorly localized and usually referred to another region of the abdomen. Thus the initial pain in appendicitis is usually a diffuse pain referred to the umbilical region but later, when the infection has spread to the parietal peritoneum, it changes its character and the patient points to the RIF as the site of the pain.

The original gut tube in the embryo is subdivided into *foregut, midgut* and *hindgut*, each with its own artery, respectively the *coeliac, superior mesenteric* and *inferior mesenteric*. In adult terms, the foregut extends down as far as the second part of the duodenum, just beyond the entrance of the common bile and pancreatic ducts. From here to about two-thirds of the way along the transverse colon is midgut, while the hindgut extends to halfway down the anal canal. Remember the hindgut particularly, as it will help you to avoid one of the commonest sources of error (p. 226).

The whole of the embryonic gut has a midline dorsal mesentery, attached down the line of the dorsal aorta. The foregut also has a ventral mesentery, which connects it to the future diaphragm and the anterior abdominal wall. As the gut

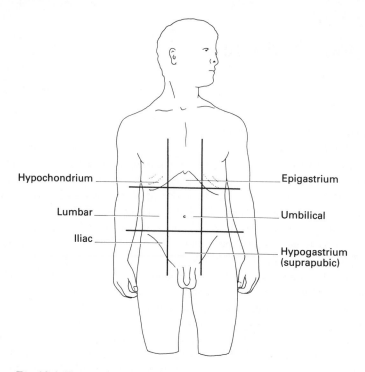

Hypochondrium

Epigastrium

Lumbar

Umbilical

Iliac

Hypogastrium
(suprapubic)

**Fig. 10.1** The noughts and crosses grid divides the abdomen into nine regions.

lengthens, however, it becomes increasingly convoluted and, where a part of the mesentery lies close to the parietal peritoneum, the two usually fuse so that the corresponding part of the gut becomes retroperitoneal. All this embryology may seem academic but it is essential to the understanding of adult anatomy and of variations in normal anatomy that may be met clinically.

The peritoneal cavity is subdivided into a main cavity (sometimes called the *greater sac*) and a diverticulum thereof, the *omental bursa* (*lesser sac*). The omental bursa may best be understood by imagining a rubber hot-water bottle facing to the right with its opening behind the lesser omentum and its main part lying behind the liver and stomach (Fig. 10.2) and extending between the layers of the greater omentum. There is only one way into the omental bursa without cutting something, and that is through the neck of the bottle, which is called the *epiploic foramen* or the *foramen of Winslow*. The guide to this foramen is the free border of the lesser omentum, which extends from the lesser curvature of the stomach and the first 2 cm or so of the duodenum to the liver.

The vertical extent of the lesser sac may best be understood by looking at a section along the line A–A in Fig. 10.2. This is shown in Fig. 10.3, from which you will see that the lesser omentum extends from the upper border (lesser curvature) of the stomach to the liver and that the upper recess of the omental bursa is behind both the lesser omentum and the part of the liver that is called the *caudate lobe* (p. 227). If you were to push your finger into the epiploic foramen, then you

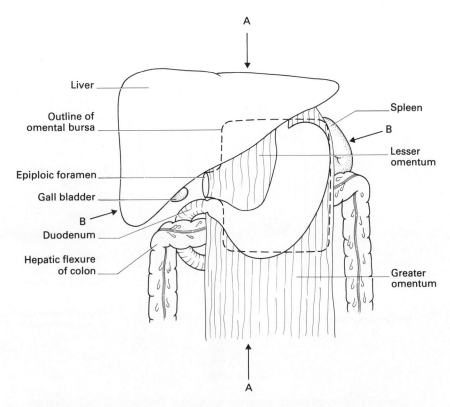

**Fig. 10.2** An 'exploded view' of the lesser sac: a rubber hot-water bottle with its neck facing to the right.

could push it upwards behind the caudate lobe until it would be stopped by the reflection of peritoneum from the diaphragm on to the liver. Downwards, you could pass behind the stomach and for a varying distance between the four layers of the greater omentum. In life, the greater omentum is a huge 'apron' of peritoneum, fat and blood vessels that hangs down from the lower border (*greater curvature*) of the stomach. If you throw it upwards over the costal margin, you will see the transverse colon apparently embedded in its posterior surface. Figure 10.3 shows that the greater omentum is actually a double layer of peritoneum that covers the front and back of the stomach, hangs down from the greater curvature and is then folded back on itself at the lower border. The transverse colon seems to be between the posterior two layers but, as will be seen later, it originally had a mesentery of its own, called the *transverse mesocolon*. This name, in the adult, is given to the 'mesentery' formed by the fusion between the transverse mesocolon and the posterior two layers of the greater omentum (indicated by a dotted line in Fig. 10.3). It attaches the transverse colon to the pancreas. Above the attachment of the transverse mesocolon, the posterior wall of the lesser sac is formed by the 'stomach bed', shown in Fig. 10.9.

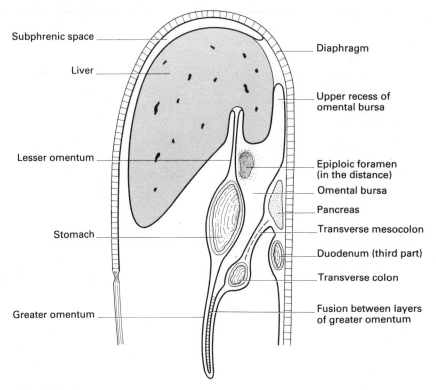

Subphrenic space

Liver

Lesser omentum

Stomach

Greater omentum

Diaphragm

Upper recess of
omental bursa

Epiploic foramen
(in the distance)

Omental bursa

Pancreas

Transverse mesocolon

Duodenum (third part)

Transverse colon

Fusion between layers
of greater omentum

**Fig. 10.3** A vertical section along A–A in Fig. 10.2. The fissure for the ligamentum venosum in the liver, to which the lesser omentum is attached, is drawn very wide for clarity.

How far to the left could you push your finger? This is shown in Fig. 10.4(a). The right free border of the lesser omentum can be seen with the epiploic foramen behind it and the three tubes, which will be described later, between its layers. The left border of the omental bursa can now be seen to consist of the *lienorenal ligament* from the kidney to the spleen, and the *gastrosplenic ligament* from the stomach to the spleen. What if the section were taken a little lower, below the level of the spleen? This is shown in Fig. 10.4(b), which is similar to Fig. 10.4(a) except that the spleen is missing. The double layer of peritoneum folded back on itself is the *left edge of the greater omentum*, which you have already seen to be composed of four layers of peritoneum. Figure 10.4(c) shows another section, similar to the first two but *above* the level of the spleen. Again, a double layer of peritoneum passes from the stomach to the posterior abdominal wall, but this section is above the kidney. This fold is therefore called the *gastrophrenic ligament*. You will now appreciate that the gastrosplenic, lienorenal and gastrosplenic ligaments, along with the greater omentum, are all parts of the same folded, double layer of peritoneum, which is complicated by the presence of the spleen. It is, in fact, the dorsal mesentery of the embryonic foregut, often called, in the region of the stomach, the *dorsal mesogastrium*.

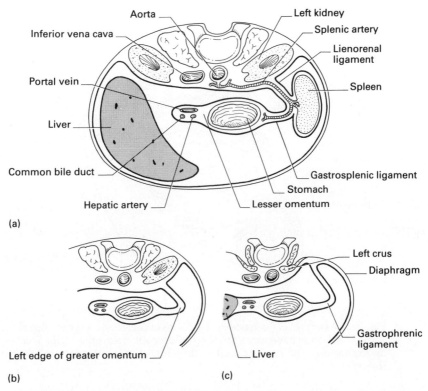

**Fig. 10.4**  (a) A section along the line B–B in Fig. 10.2. (b) A similar section a little below the spleen. (c) A section a little above the spleen.

The embryonic foregut also has a ventral mesentery, which has a free border at the termination of the foregut. When the hepatic diverticulum grows out, it passes into the ventral mesogastrium and its stem, which will later become the common bile duct, occupies the free border. Later, blood vessels that will become the hepatic artery and the portal vein also come to lie in the free border. As the liver increases in size, it becomes so big that the ventral mesogastrium, instead of passing from the stomach to the diaphragm and anterior abdominal wall, now appears to pass from stomach to liver and from liver to diaphragm and anterior abdominal wall. Thus the ventral mesogastrium becomes the *lesser omentum* and the *falciform, triangular* and *coronary ligaments*. These will be described later.

Turning now to the dorsal mesentery, this was originally attached to the dorsal body wall in the midline. When the midgut is returned to the abdomen, the stomach 'falls over' to the right so that instead of having left and right surfaces it comes to have anterior and posterior surfaces. This is why the left vagus becomes the anterior gastric nerve(s) and the right the posterior (Fig. 10.5). The dorsal mesogastrium correspondingly is thrown over to the left and the spleen begins to develop within its mesoderm. Above and below the spleen, however, it passes directly back to the dorsal body wall. The dorsal mesentery then becomes partly fused with the dorsal body wall, as explained at the beginning of this chapter

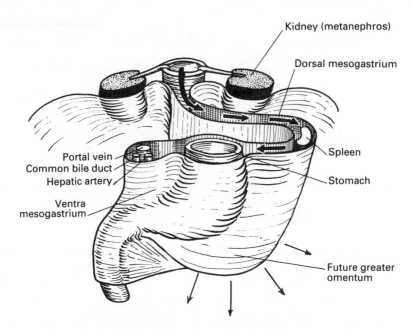

**Fig. 10.5** As a result of the ingrowth of the bursa omentalis and of the rotation of the stomach, the dorsal mesentery is thrown over to the left. It also grows in the direction of the small arrows. The larger arrows show the course of arteries that pass to the spleen and stomach.

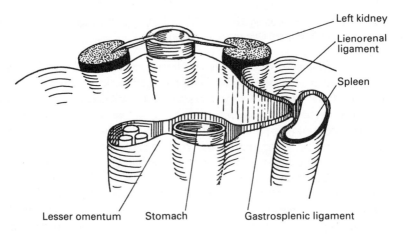

**Fig. 10.6** A similar view to Fig. 10.5, after fusion of the dorsal mesentery with the dorsal body wall.

(Fig. 10.6). The attachment of the dorsal mesentery to the midline becomes transferred to the region of the kidney. As the spleen grows in size, the left border of the lesser sac becomes defined, as described above. The dorsal mesentery, where it is attached to the greater curvature of the stomach, enlarges downwards to become the greater omentum. Behind it, the dorsal mesentery of the transverse colon becomes horizontally orientated and, again, fusion of mesenteries occurs, as shown in Fig. 10.3. The transverse colon thus appears to be embedded in the back of the greater omentum.

The pancreas develops from two outgrowths at the end of the foregut and, although one of these is at first ventral, both parts of the pancreas end up in the dorsal mesentery. When the stomach falls over to the right, the duodenum and pancreas likewise are pushed to the right and fusion of their dorsal mesentery with the dorsal body wall occurs. Thus both duodenum and pancreas become retroperitoneal, except for the end of the tail of the pancreas, which lies in that part of the dorsal mesogastrium that will become the lienorenal ligament. Uncommonly, the ventral pancreas fails to migrate and fuses with the dorsal pancreas to surround the duodenum (*annular pancreas*).

The above account of the embryology of the foregut is very (an academic embryologist would use the word grossly) simplified but it is well worth while studying it in detail as this is the only way to understand properly the peritoneal arrangements in the upper abdomen. The viscera and their peritoneal attachments will now be described in detail.

## RELATIONSHIPS OF VISCERA IN THE UPPER ABDOMEN

First, it is important to understand the relationships between the structures in the upper abdomen since these are important surgically, radiologically and, not least, to examination candidates. The best way to appreciate these relationships is to reconstruct the viscera diagrammatically, starting with the midline and the transpyloric plane. The latter is at the level of the lower border of the lst lumbar vertebra and although, as explained above, it does not necessarily indicate the level of the pylorus it will be convenient to place the pylorus here in the diagram, a little to the right of the midline (Fig. 10.7). That done, the duodenum can be added. The first part of its first part has the edge of the lesser omentum attached to it and so is relatively mobile, but the rest of the duodenum is retroperitoneal. The second part descends vertically downwards to the level of L3, the opening into it of the common bile duct marking the end of the embryonic foregut. The third part crosses the midline horizontally while the fourth part ascends to the left of the midline to the level of L2. Having drawn the duodenum, it is easy to insert the pancreas, whose head is enclosed within the C-shape of the duodenum.

The duodenum and pancreas are guides to the position of the kidneys. The second part of the duodenum overlies the hilum of the right kidney and, therefore, the renal pelvis and the renal vessels. The tail of the pancreas crosses the hilar region of the left kidney before insinuating itself between the layers of the lienorenal ligament (see above). The suprarenal glands can now be added.

All the structures mentioned above are related directly or indirectly to the posterior abdominal wall, of which the aorta and inferior vena cava are the principal features. The aorta lies in the midline in contact with the vertebral bodies and the inferior vena cava is to its right (Fig. 10.8). The beginning of the transverse colon crosses the right kidney and the second part of the duodenum before it becomes suspended from the pancreas by the transverse mesocolon. The transverse colon becomes retroperitoneal again on the left side of the left kidney. Above

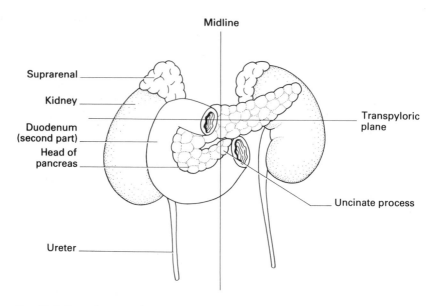

**Fig. 10.7** This diagram should be studied together with Figs 10.8, 10.9 and 10.10. It shows the structures on the posterior abdominal wall, all of them retroperitoneal.

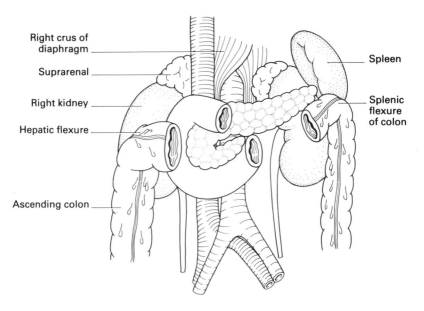

**Fig. 10.8** Other structures on the posterior abdominal wall have been added to Fig. 10.7, along with the colon and spleen.

the kidneys are the suprarenal glands, the right one being tucked behind the inferior vena cava. The spleen may also be added. It sits on the upper lateral part of the left kidney and the tail of the pancreas reaches its hilum via the lienorenal ligament. The spleen, however, is enclosed in peritoneum so does not come into *direct* contact with the renal surface. All these relationships may be appreciated in Fig. 10.11, which shows the anterior relations of the kidneys.

The three anterior (gut) branches of the aorta may now be added (Fig. 10.9). The foregut artery, the coeliac, arises just above the upper border of the pancreas and immediately breaks up into three branches. The small *left gastric artery* climbs towards the oesophagus, where it turns forwards and inserts itself between the layers of the lesser omentum, giving off some *oesophageal branches*. The *splenic artery* is characterized by its tortuous course along the upper border of the pancreas, into the lienorenal ligament and finally into the hilum of the spleen. The *hepatic artery* passes to the right and then turns upwards into the right free border of the lesser omentum and thence to the liver. As it turns upwards it gives off the *right gastric artery*, to the lesser curvature of the stomach, and the *gastroduodenal artery*. The latter runs behind the first part of the duodenum and when it reaches the lower border thereof it divides into the *right gastro-epiploic artery*, which follows the greater curvature of the stomach, and the *superior pancreaticoduodenal artery*. The latter runs down between pancreas and the duodenum, supplying both and anastomosing with the *inferior pancreaticoduodenal* from the superior mesenteric. There is thus, not surprisingly, an anastomosis here between the foregut and midgut arteries.

The superior mesenteric artery arises behind the pancreas so its origin cannot be seen, but it emerges at the lower border of the pancreas and crosses the third

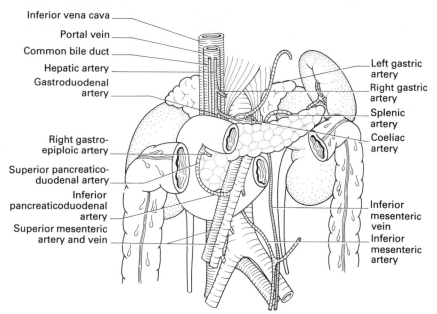

**Fig. 10.9** Blood vessels have been added to Fig. 10.8.

part of the duodenum. This artery and the aorta thus form nutcrackers that enclose the left renal vein (not visible in Fig. 10.9) and the duodenum (see Figs 10.12 & 1.13). The pancreas sends an *uncinate process* behind the artery (uncus means hook; look at the shape of the uncinate gyrus in the brain).

The *superior mesenteric vein* lies to the right of the artery and it, too, passes behind the pancreas. It is joined by the *splenic vein*, which is below the corresponding artery and therefore also behind the pancreas. The union of the two veins forms the *portal vein*, which is not only behind the pancreas but also behind the first part of the duodenum. It emerges, travelling upwards, into the free border of the lesser omentum, lying behind the hepatic artery, and finally reaches the liver. In doing so, it is anterior to the inferior vena cava but since it is lying between the layers of the free edge of the lesser omentum it is separated from the vena cava by the epiploic foramen.

It has already been mentioned that the free border contains three tubes. The only one that has not so far been mentioned is the *common bile duct*. This leaves the liver and gall bladder region and travels down in the lesser omentum, lying to the right of the hepatic artery and in front of the portal vein. It runs behind the first part of the duodenum and the head of the pancreas and opens into the second part of the duodenum, in company with the main pancreatic duct.

The origin of the *inferior mesenteric artery* is hidden by the third part of the duodenum but it can be seen emerging below this and beginning to give off branches for the supply of the hindgut. Also emerging from behind the duodenum are the *testicular* or *ovarian arteries*, but these are not shown in the diagram (see Fig. 12.1). The approximate outline of the lower part of the liver is shown as a broken line in Fig. 10.10. The two *inferior phrenic arteries* are also shown, the

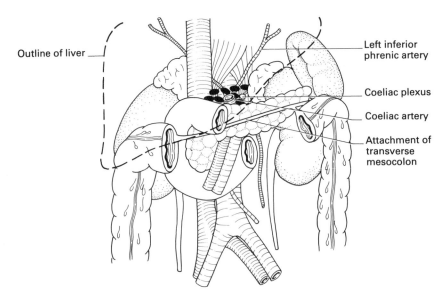

Outline of liver

Left inferior phrenic artery

Coeliac plexus

Coeliac artery

Attachment of transverse mesocolon

**Fig. 10.10** The phrenic arteries, the coeliac plexus and the outline of the liver have been added to Fig. 10.9, but some arteries have been omitted for clarity.

right one being a posterior relation of the inferior vena cava. Clustered around the coeliac artery are the *coeliac ganglia* and plexuses but the sympathetic trunks cannot be seen since they lie behind the aorta along the vertebral bodies. Also indicated is the attachment of the transverse mesocolon to the front of the pancreas. Finally, remember that the *inferior mesenteric vein* leaves the corresponding artery as it is followed upwards. It runs alongside the fourth part of the duodenum in a fold of peritoneum (the *paraduodenal fold*) and then behind the pancreas to join the splenic vein or the portal vein.

The area above the attachment of the transverse mesocolon and to the left of the three tubes in the free border of the lesser omentum is the posterior wall of the bursa omentalis. It is also known as the *stomach bed* (Fig. 10.9) so the posterior relations of the stomach (apart from the bursa omentalis) are the pancreas, the upper part of the left kidney and suprarenal, the coeliac artery and two of its branches, the coeliac ganglia and the spleen.

If you have followed the 'build-up' of the structures shown in Figs 10.7–10.10, you will now find it easy to understand the anterior relationships of the kidneys (Fig. 10.11) and the 'nutcracker' at the origin of the superior mesenteric artery (Fig. 10.12).

The anatomy of all these structures will now be described in a little more detail. It will involve a certain amount of repetition but this will aid the learning process.

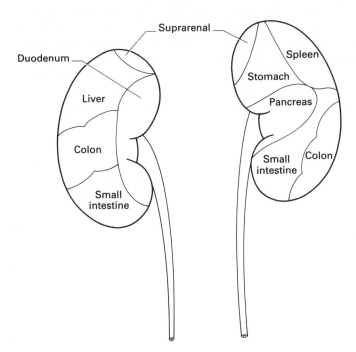

**Fig. 10.11** Anterior relations of the kidneys. Study in conjunction with the previous four figures.

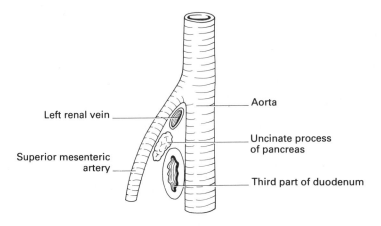

**Fig. 10.12** The aorta and the superior mesenteric artery form nutcrackers that enclose three important structures.

## OESOPHAGUS

The abdominal part of the oesophagus is not much more than 1 cm in length. After piercing the right crus of the diaphragm (p. 208), it is partly covered with peritoneum and is accompanied by the *anterior* and *posterior gastric nerves* (from the vagi) and *oesophageal branches of the left gastric vessels*. The corresponding veins are important since they belong to the portal system and they anastomose with oesophageal tributaries of the azygos veins in the thorax (p. 193). The lining epithelium changes from stratified squamous ('wear and tear') epithelium to columnar mucus-secreting ('acid-resistant') epithelium at the cardiac orifice of the stomach. Reflux of stomach contents into the oesophagus has deleterious effects on the oesophageal epithelium but it is still not entirely clear how such reflux is prevented. There is no true cardiac sphincter but a combination of the circular

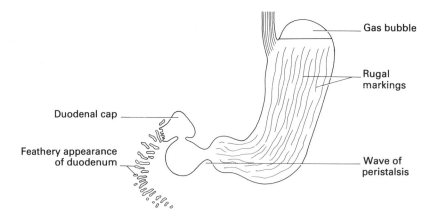

**Fig. 10.13** The appearance of a barium meal.

smooth muscle fibres, the muscle of the right crus of the diaphragm, the acute angle between the left border of the oesophagus and the fundus of the stomach (the cardiac notch) and the corresponding folds of mucous membrane inside all seem to combine to produce a functional sphincter.

## THE STOMACH

The fundus rises to the level of the 5th intercostal space; hence the name 'heartburn' given to pain from this region. The right border, the lesser curvature, usually has a notch called the *angular incisure* (incisura angularis) about two-thirds of the way down and beyond this is the pyloric part of the stomach. The pyloric sphincter is a marked thickening of the circular smooth muscle coat. The mucous membrane is thrown into large folds, called *rugae*, which produce a characteristic appearance when the first mouthful of barium is swallowed during a barium meal (Fig. 10.13).

## THE OMENTA

### The lesser omentum

This is a double layer of peritonuem extending from the lesser curvature of the stomach to the liver, where it is attached in the *fissure for the ligamentum venosum*. The right free border of the lesser omentum extends a little beyond the stomach and connects the first part of the duodenum to the *porta hepatis* in the liver (see below). It contains the three tubes, the portal vein behind, the hepatic artery in front and to the left and the common bile duct in front and to the right. Behind the free border is the epiploic foramen (of Winslow) and it can be seen from Fig. 10.4 that a finger in the foramen lies between two large veins, the portal vein and the inferior vena cava. The finger, with a thumb outside the foramen, can compress both vessels that supply blood to the liver. The right and left gastric arteries run within the lesser omentum along the lesser curvature.

### The greater omentum

This is a large, fat-impregnated and highly vascular quadruple layer of peritoneum that hangs down from the greater curvature, covering over the coils of small intestine below and having the transverse colon embedded in it posteriorly (p. 213). The blood supply of the greater omentum is derived from the *gastro-epiploic arteries*, right and left, which supply both the stomach and the epiploon (omentum). The omentum serves as a storehouse for fat but its main function is to go to trouble spots; hence Rutherford Morison's name for it, the *abdominal policeman*. When peritoneum is inflamed, it becomes 'sticky' and attaches itself to adjacent layers of peritoneum, forming 'adhesions'. You will certainly have met old adhesions, and mistaken them for various ligaments and mesenteries, in the dissecting room. Thus, if, for instance, the gall bladder is inflamed, the omentum will stick on to all adjacent structures and eventually seal off the danger area so that what might have led to a general peritonitis will become localized to form an abscess. The omentum is therefore often sewn around, for example, sites of intestinal anastomosis, to localize any possible leakage.

## DUODENUM

It is named from the Latin *duodecim* (12) because it is said to be 12 finger-breadths in length. The four parts of the duodenum have been described earlier in this chapter. The beginning of the first part is slightly more mobile than the rest because the greater and lesser omenta are attached to it, but the rest of the duodenum is retroperitoneal, along with the greater part of the pancreas, with which it is closely associated. Remember that the C-shape of the duodenum, as shown in Fig. 10.7, is not quite accurate since the first part, as well as passing towards the right, also turns backwards around the prominence of the vertebral bodies, so that from the front you are almost looking down the lumen. Thus, in a barium meal, the first part of the duodenum is seen as a roughly triangular *duodenal cap*, perched on top of the pyloric end of the stomach (Fig. 10.13). On the surface, the junction between the pylorus and the duodenum is marked at operation by a vein, the *prepyloric vein of Mayo*, but you will not have seen this in the dissecting room. The duodenal papilla, in the second part of the duodenum, is described below.

The duodenum is the first part of the small intestine, whose main function is absorption so its internal surface area is relatively enormous. With the naked eye one can see the *circular folds* in the mucous membrane, which increase the surface area considerably and produce the feathery appearance seen after a barium meal. With the light microscope one can see the *villi*, which give the mucous membrane a velvety appearance, and with the electron microscope one can see the *microvilli*, which enormously increase the surface area of individual cells.

The end of the fourth part of the duodenum turns forwards at the duodeno-jejunal flexure and acquires a mesentery to become the *jejunum*. There are various peritoneal folds around this region but the only one you need to remember at this stage is the *paraduodenal fold*, which encloses the inferior mesenteric vein.

Embryologically speaking, this description has transgressed a boundary, that between the foregut and midgut. The end of the foregut in the embryo is marked by the hepatic outgrowth, which, in the adult, is represented by the biliary system. Its original site of outgrowth is marked by the point of entry of the common bile duct into the second part of the duodenum. This, then, is where there is anastomosis between a branch of the artery of the foregut (superior pancreaticoduodenal) and a branch of the artery of the midgut (inferior pancreaticoduodenal). It is now obviously time to describe the biliary system.

## THE LIVER

The liver is the largest organ in the abdomen and is even bigger (relatively speaking) in the embryo since it doubles as a haemopoietic organ during its development. Its lower border corresponds approximately to the right costal margin (although, nearer the midline, it crosses the epigastrium below the margin). It is not normally palpable but when enlarged it can be felt, moving downwards with a deep inspiration, if it is hard as well as enlarged. A *soft* enlarged liver is not palpable: it has been pointed out that in a butcher's shop you would not be able to feel a piece of fresh liver through a piece of steak.

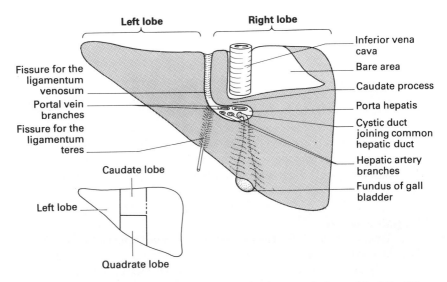

**Fig. 10.14** The posterior surface of the liver may be drawn on the basis of the letter 'H'.

The upper surface of the liver lies beneath the diaphragm and rises approximately to the level of the xiphisternal joint. The inferior and posterior surfaces are continuous and show a number of landmarks so that, when the liver is seen from behind, its division into lobes can clearly be seen. These are shown in Fig. 10.14. The liver is divided into right and left lobes; the right lobe is the larger because it includes the *caudate* and *quadrate lobes*.

### The porta hepatis and the fissure for the ligamentum venosum

The porta hepatis is the gateway to (and from) the liver. It thus contains the *portal vein*, the *hepatic artery* and the *hepatic ducts* (described below). You will remember that the two vessels and the common bile duct are found in the free border of the lesser omentum. The lesser omentum itself is attached to the liver in the fissure for the ligamentum venosum, its free border thus enclosing the porta hepatis (Fig. 10.15a). The *ligamentum venosum* is a fibrous remnant lying in the depths of its fissure and joining the left branch of the portal vein to the inferior vena cava. You should think of this in conjunction with the *ligamentum teres*, which runs from the umbilicus, in the edge of the falciform ligament, on to the undersurface of the liver and joins the left branch of the portal vein opposite the ligamentum venosum. There is thus a complete 'ligamentous' bridge between the umbilicus and the inferior vena cava. The ligamentum teres is the remains of the *umbilical vein* (carrying oxygenated blood) and the ligamentum venosum is the remains of the *ductus venosus*. In the fetus, therefore, oxygenated blood from the placenta can pass directly into the inferior vena cava, most of it bypassing the liver completely. These abdominal 'ligaments' are confusing. There are peritoneal folds such as the gastrosplenic and lienorenal ligaments, embryological remnants such as the ligamentum venosum and the round ligament of the uterus, and other structures such as the inguinal and arcuate ligaments. None of these bears any resemblance to a real ligament such as the iliofemoral.

## The inferior vena cava (hepatic portion)

The inferior vena cava, before passing through the central tendon of the diaphragm, is deeply embedded in the liver substance, so deeply that in museum specimens a segment of vena cava is often left within the groove in the liver. This is a pity as it conceals the very large *hepatic veins* that open into this segment of the vena cava. Do not confuse the hepatic and the portal veins. The *hepatic veins are nowhere near the hepatic artery*. The inferior vena cava is, of course, a retroperitoneal structure so that it is related to the bare area of the liver (Fig. 10.15a).

## The gall bladder

This is a reservoir for bile and also concentrates it and adds mucus. It lies in a groove in the liver and is covered on its free surface by peritoneum. Its fundus projects a little below the inferior margin of the liver; the most convenient surface marking of the fundus is the point where the linea semilunaris (lateral border of the rectus muscle) crosses the costal margin. Remember that the gall bladder drains upwards. Its fundus is correctly named since it means the base or bottom of an organ (cf. 'fundamental'), unlike the fundus of the uterus or of the stomach.

The gall bladder leads into a *cystic duct*, which, near the porta hepatis, joins the *common hepatic duct* (formed by the union of right and left *hepatic ducts*) to

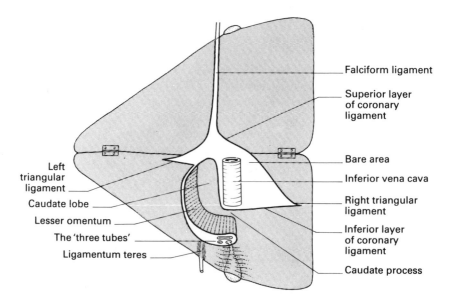

**Fig. 10.15** (a) The reflections of peritoneum from the liver on to the posterior abdominal wall. The liver is depicted as though viewed from behind and split in the coronal plane, with its upper surface turned up to show the diaphragmatic surface. The shaded area is covered by peritoneum.

form the *common bile duct*. The further course of this has been described already (p. 220). The union of these ducts is subject to variation and it is not uncommon to find the cystic duct and the common hepatic duct together in the free border of the lesser omentum. The blood supply to the gall bladder is by means of the *cystic artery*, which usually (but not always) rises from the right branch of the hepatic artery.

## The lobes of the liver

The boundaries of the caudate and quadrate lobes are shown in Fig. 10.14. The two lobes are often confused. Remember that the caudate lobe is the one with a tail (like the caudate nucleus), the tail being the caudate process. These terms are useful for description but they do not correspond to the true right and left lobes of the liver, which are those parts supplied by the right and left branches of the hepatic artery and portal vein and drained by the right and left hepatic ducts. The true left lobe is, therefore, the left lobe as shown in Fig. 10.14 plus the quadrate and most of the caudate lobes. The division is roughly marked by a line through the gall bladder and the inferior vena cava.

## The peritoneal relations of the liver

Figure 10.15(a) shows a liver that has been removed from a body and Fig. 10.15(b) shows the body from which it has been removed. Thus the pattern of cut edges of the peritoneum on the liver is a mirror-image of the pattern of the body. The lesser omentum has been cut along the lesser curvature of the stomach and so remains attached to the liver. The liver is almost completely covered by perito-neum except for the *bare area*, in which the inferior vena cava is embedded. From the upper surface of the liver the peritoneum is reflected to the diaphragm and the anterior abdominal wall to form the *falciform ligament*. If the left leaf of this

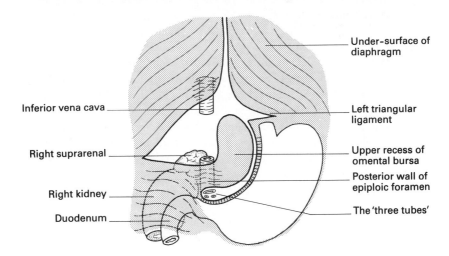

**Fig. 10.15** (b) The posterior abdominal wall and diaphragm from which the liver shown in (a) has been removed.

ligament is followed backwards, it folds back on itself to form the *left triangular ligament*. This contains no major blood vessels and may be divided if the left lobe of the liver has to be retracted to expose, for example, the oesophagus. This leaf of peritoneum, then, forms the anterior layer of the lesser omentum, which is folded back on itself at the free border to form the posterior layer. From the peritoneum-covered caudate lobe and caudate process, the peritoneum is reflected on to the diaphragm above and to the right, to define one side of the *bare area*. This area is bounded above and below by other peritoneal reflections from the liver to the diaphragm. These two reflections, shown in Fig. 10.15, are called the superior and the inferior layers of the coronary ligament, and they join to form the *right triangular ligament*. Note the position of the epiploic foramen in Fig. 10.15(b). Three of its boundaries can be seen: the lesser omentum with its three tubes in front, the inferior vena cava covered with peritoneum behind and the caudate process above.

### The subphrenic spaces

Intraperitoneal infection in the upper abdomen may arise from such lesions as perforated gastric or duodenal ulcers and gall bladder disease. It sometimes gives rise to localized collections of pus (abscesses) and these are particularly likely to be found in four peritoneal spaces, the condition being known as a *subphrenic abscess*. Only two of the spaces are, in fact, directly subphrenic, the other two being subhepatic. The *left* and *right subphrenic spaces* are between the liver and the diaphragm (see Fig. 10.3) and they are separated by the vertical barrier of the falciform ligament. The *right subhepatic space* is just outside the epiploic foramen, with the liver and gall bladder, duodenum and right kidney in immediate relationship. Clinically, this is often called *Rutherford Morison's pouch*. The *left subhepatic space* is the omental bursa itself. There are also some extraperitoneal spaces, such as the bare area of the liver, where pus sometimes collects, but these are less important.

## THE PANCREAS

The position and relations have already been described. The head, neck, body and part of the tail are all retroperitoneal as a result of the fusion of the embryonic dorsal mesentery with the posterior abdominal wall, but the last piece of the tail lies in an unfused part of the mesentery, i.e. the lienorenal ligament, which carries it to the hilum of the spleen along with the splenic vessels. The main part of the splenic artery undulates along the upper border of the pancreas and supplies it with blood, particularly by means of a large artery, the *arteria pancreatica magna*.

The pancreas is, of course, an endocrine and an exocrine gland. The endocrine part is represented by the islets of Langerhans while its exocrine part secretes pancreatic juice by means of one main duct and often an accessory duct that opens proximal to the main duct. The main duct runs the length of the pancreas, rather towards its posterior surface, and receives a number of tributaries. It joins the common bile duct, which has passed behind the first part of the duodenum and the head of the pancreas, and the resulting combined duct is dilated to form the *hepatopancreatic ampulla* (the *ampulla of Vater*). This opens on to the surface of the duodenal papilla halfway down the left side of the second part of the duodenum. The terminal parts of the two ducts, and the ampulla, are surrounded by a smooth muscle sphincter, the *sphincter of Oddi*. Sometimes the two ducts open separately.

## THE SPLEEN

Before the days of metrication, the spleen was said to measure $1 \times 3 \times 5$ inches, weigh 7 ounces and lie opposite ribs 9 to 11 (compare with the surface markings of the pleura, p. 172). It does not, of course, lie directly against the ribs because the diaphragm and the costodiaphragmatic recess of the pleura intervene. It is also related to the tail of the pancreas, the left kidney, the colon and the stomach (see Figs 10.8–10.10). An important feature is the presence of several notches on its anterior border, which not only help you to orientate the spleen in an oral examination but help the clinician to identify an enlarged spleen by palpation since the notches persist even when the spleen is huge. The normal spleen is not palpable.

The spleen is a 'floating' structure, i.e. it can be pulled forwards into the wound area during an operation, since it has two mesenteries, the gastrosplenic and lienorenal ligaments.

## THE BLOOD SUPPLY OF THE UPPER ABDOMINAL VISCERA

### Arteries

Many of the blood vessels involved have been mentioned already but it may be helpful to bring them all together here as an aid to revision. The coeliac artery, or axis, arises from the aorta soon after the latter enters the abdomen (Fig. 10.16) and immediately breaks up into three branches, the *left gastric, splenic* and *hepatic*. The left gastric ascends behind the peritoneum to the oesophagus and then inserts itself between the layers of the lesser omentum and supplies the lesser curvature of the stomach, anastomising with the right gastric artery. It also gives *oesophageal branches*, the accompanying veins of which are important (see below).

The splenic artery follows a tortuous course along the upper border of the pancreas, which it supplies, and then reaches the spleen via the lienorenal ligament. It gives off pancreatic branches and the *short gastric* and *left gastro-epiploic arteries*, which reach the stomach via the gastrosplenic ligament.

The hepatic (common hepatic) artery travels to the right behind the peritoneum and then inserts itself between the layers of the free border of the lesser omentum. This conveys it to the porta hepatis, where it divides into right and left branches, the right branch usually giving rise to the *cystic artery*. The main trunk gives off the *right gastric artery*, which anastomoses with the left gastric along the lesser curvature, and the *gastroduodenal artery*. The latter passes behind the duodenum and, at its lower border, divides into the *right gastro-epiploic* and the *superior pancreaticoduodenal* arteries. The former anastomoses with the left gastro-epiploic on the greater curvature and the latter supplies the duodenum and

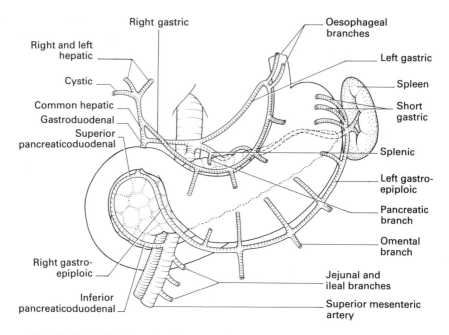

**Fig. 10.16** The blood supply of the stomach.

pancreas and ends by anastomosing with the *inferior pancreaticoduodenal branch of the superior mesenteric*, thus establishing an anastomosis between the fore- and midgut arteries.

### Veins

*Never forget that the venous drainage of the gut and its accessories is by means of the portal system.* A portal vessel is one that has capillaries at each end so portal veins are the only veins that have branches; all others have tributaries. There are other portal systems in the body, for example, the *hypophyseal portal system*, which begins in the capillaries of the hypothalamus and ends in the capillaries of the adenohypophysis.

The (hepatic) portal vein is formed by the junction of the *splenic* and *superior mesenteric veins* behind the head of the pancreas. It ascends behind the first part of the duodenum and enters the free edge of the lesser omentum before dividing into left and right branches in the porta hepatis. The veins corresponding to the branches of the coeliac and superior mesenteric arteries drain into the portal vein or one of its tributaries. The inferior mesenteric vein ascends alongside the fourth part of the duodenum and usually ends up in the splenic vein, but sometimes in the portal vein itself.

There are no valves in the portal system so that obstruction, for example by disease of the liver, causes a rise in pressure throughout the system (*portal hypertension*). In order to escape, the blood passes through any available anastomoses between the portal and systemic system and the anastomotic veins concerned become dilated and may bleed. These portal–systemic anastomoses are therefore very important. One group which you will be familiar with already are the veins that accompany the oesophageal branches of the left gastric artery. These unite the left gastric vein with tributaries of the azygos system in the thorax. *Oesophageal varices* form when the portal blood pressure is increased and these may rupture and cause *haematemesis* (vomiting of blood). Some small *para-umbilical veins* (not to be confused with the fetal umbilical vein) accompany the ligamentum teres and unite the portal vein with the veins of the anterior abdominal wall. The enlarged veins can be seen around the umbilicus, where they form a *caput Medusae* (Medusa was a Greek woman of hideous appearance with snakes in place of hair). This appearance is spectacular but rare. Another portal–systemic anastomosis occurs in the anal canal and will be described below. Finally, there are numerous anastomoses between small veins of the portal system and veins of the diaphragm or posterior abdominal wall, for example in the region of the bare area of the liver.

# Chapter 11
# Midgut and Hindgut Derivatives

Part of the embryonic midgut has been described in Chapter 10, namely from the middle of the second part of the duodenum down to the duodenojejunal flexure. This chapter will therefore continue with the jejunum and ileum, the colon, the rectum and part of the anal canal. The junction between the midgut and the hindgut is about two-thirds of the way along the transverse colon and here there is an anastomosis between branches of the superior and the inferior mesenteric arteries.

## THE SMALL INTESTINE

There is no sharp junction between the jejunum and the ileum, merely a gradual change in a number of features. Common to both, and to the duodenum, is the very large internal surface area (p. 224) to help in absorption. However, the circular folds of mucous membrane gradually diminish in size along the small gut so that the jejunum feels fuller than the ileum when held between finger and thumb. It is also pinker in colour and has less complex vascular arcades. Distinguish between *ileum*, part of the small intestine, and *ilium*, part of the hip bone. One of the characteristic features of the small intestine is the presence of aggregated lymph follicles in the mucous membrane. These are large enough to be seen with the naked eye in the ileum, where they are known as *Peyer's patches*.

The 6 m or so of jejunum and ileum is suspended from the posterior abdominal wall by *the mesentery*. This is attached along a line from the duodenojejunal flexure to the end of the ileum so that its attachment is about 15 cm long. The other border of the mesentery must obviously be about 6 m long so that the whole mesentery resembles a very full skirt with perhaps a 55 cm waist and a hem of 3 m (Fig. 11.1). The gut is thus extremely convoluted and when a loop is pulled out of an operation wound, the only way to tell which end is proximal and which is distal is by running the fingers down each side of the mesentery to straighten it out and thus one can determine the direction of the root of the mesentery. You will also appreciate how easy it is for a loop of small intestine and its segment of mesentery to become twisted (*volvulus*) so that the vessels in the mesentery and the gut itself are obstructed (see also the vitello-intestinal duct below).

The superior mesenteric artery and vein, having crossed the third part of the duodenum, enter the root of the mesentery and curve down towards the right iliac region, giving off *jejunal* and *ileal branches* from the left side. The branches divide and reunite to form a series of arches and arcades from which secondary arcades may arise and, in the ileum, there may be tertiary arcades as well until finally vessels run straight to the gut wall. Thus one or more of the main branches may be ligatured, or may become obstructed, without affecting the viability of the gut. The

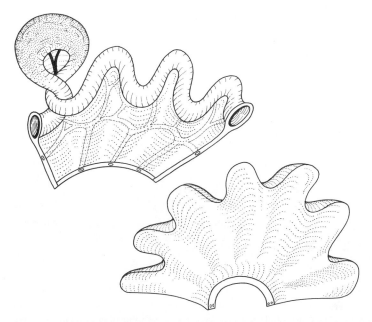

**Fig. 11.1** The mesentery of the small intestine is like a very full skirt with a short attached border and a long and convoluted free border. Part of the small intestine has become twisted and gangrenous (a volvulus).

anastomoses between the first two or three jejunal branches, however, are not good and cannot be depended upon to form a collateral circulation.

In the early embryo, the midgut is continuous with the yolk sac. Later, the neck of the yolk sac constricts to form a *vitello-intestinal duct* (see Fig. 1.8). The duct normally disappears but it may persist in whole or in part. Most commonly the proximal part persists, giving rise to a *Meckel's diverticulum* (p. 202).

## THE CAECUM

The ileum ends in the right iliac region by joining the *caecum*, which is the blind-ended sacculation at the beginning of the colon. Caecum means blind-ended, for example in the foramen caecum in the skull or in the tongue. The ileocaecal junction takes the form of a pair of lips that project into the caecum and act as a form of valve. The caecum in the fetus is a roughly conical sac with the appendix as its apex. This appearance is occasionally seen in the adult but normally one side of the caecum grows faster than the other so that the appendix is attached asymmetrically.

## THE APPENDIX

This is about the size of a moderately large earthworm (but not as long) and, being a fairly useless piece of equipment, frequently becomes inflamed and has to be removed. Because of this, its blood supply is important (Fig. 11.2). The superior mesenteric artery ends by becoming the *ileocolic artery*, which gives off *ileal, colic*

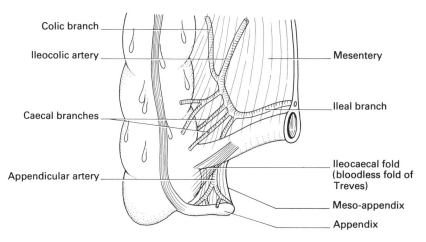

**Fig. 11.2** The blood supply of the ileocaecal region.

and *caecal* branches and the *appendicular artery*. This runs behind the terminal part of the ileum and then into the free border of the mesentery of the appendix, the *meso-appendix*, before running alongside the appendix itself. In appendicitis, therefore, it may be involved in the inflammatory process and become thrombosed.

Removal of the appendix may not always be as easy as one might think. It may be quite difficult to find as its position is so variable. It may lie below the caecum and pass over the pelvic brim to lie near the ovary in the female; it may also lie in front of or behind the terminal ileum but its commonest position is tucked into a diverticulum or recess of the peritoneal cavity that lies behind the caecum, the *retrocaecal position*. The best guide to it is, or are, the *taeniae coli*. These are three strips of longitudinal muscle on the surface of the colon and they converge to meet at the base of the appendix.

## THE COLON

The colon may be recognized by its size, by its three taeniae coli, which cause it to be sacculated (*haustrations*), and by its *appendices epiploicae*, which are numerous fatty tags scattered over its surface. Furthermore, some parts of the colon lack a mesentery and are retroperitoneal. In the embryo, the whole colon was suspended from a dorsal mesentery, but, after the return of the gut to the abdomen in the 2nd month, the mesentery in the region of the ascending and descending colon and the rectum fuses with the posterior abdominal wall (Fig. 11.3). Thus these parts of the intestine, along with their blood vessels and lymphatics, are behind the parietal peritoneum. The transverse colon and the sigmoid (pelvic) colon both retain their mesenteries: the *transverse* and *sigmoid mesocolons*.

The ascending colon becomes the transverse colon at the *hepatic flexure*, while the transverse becomes the descending colon at the *splenic flexure* (the higher of the two). The transverse colon is not often transverse; it normally hangs down in a loop, which may be as low as the pelvis. The sigmoid colon leads down

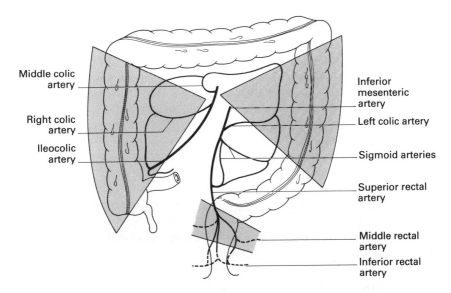

Middle colic artery
Right colic artery
Ileocolic artery
Inferior mesenteric artery
Left colic artery
Sigmoid arteries
Superior rectal artery
Middle rectal artery
Inferior rectal artery

**Fig. 11.3** The blood supply of the colon. The shaded areas indicate the fusion of the embryonic dorsal mesentery with the dorsal body wall so that the ascending and descending colon and the rectum do not have mesenteries.

into the pelvis to become the *rectum* at the level of the 3rd sacral vertebra. Its mesentery is bent back on itself in the shape of an upside-down 'V'.

## THE RECTUM

Rectum means straight, as in rectus abdominis or rectangle, but the rectum by no means lives up to its name. Besides some slight lateral curves to either side, the rectum, as a whole, follows the curve of the sacrum before bending back in front of the tip of the coccyx to become the *anal canal*. The taeniae coli, when they reach the rectum, spread out to form two wide bands in front and behind. The rectum also lacks haustrations and appendices epiploicae. Its upper third is covered by peritoneum in front and at the sides, its middle third has a peritoneal covering anteriorly only, while its lower third is below the level of the peritoneum. But beware of dividing the rectum into thirds (see 'blood supply' below).

The mucous membrane is folded into the lumen to form a number (usually three) of horizontal *crescentic folds* (*the valves of Houston*). The middle fold is the largest and below it the rectum is dilated to form the *ampulla*. The relations of the rectum will be described in Chapter 13.

## THE ANAL CANAL

This is about 4 cm long and passes downwards and backwards. It is surrounded by a complex sphincter consisting of both smooth and striated muscle, but since this is intimately related to the levator ani muscle it, too, will be discussed in Chapter 13. The lining of the anal canal is quite complicated. About halfway down, the mucous membrane is ringed by a series of small cup-like folds, the *anal valves*, from which extend upwards a series of vertical folds, the *anal columns*. The anal valves enclose small depressions, the *anal sinuses*, into which open anal glands. These extend into the submucosa. The epithelium in the upper half is mostly 'gut' epithelium, i.e. columnar, but it contains patches of other types, such as stratified squamous. It is plum-coloured. Below the anal valves the epithelium is a whitish non-keratinized, stratified squamous; this region is called the *pecten*. Lower down again is skin. The boundaries between these zones are not clear-cut. The anal canal develops from the end of the hindgut (endoderm) and an invagination of ectoderm, the *proctodaeum*. This dual development has an important bearing on the anatomy of the anal canal and will be discussed later.

## THE BLOOD SUPPLY OF THE LARGE INTESTINE

The midgut of the embryo terminates about two-thirds of the way along the transverse colon and is supplied by the superior mesenteric artery. The terminal part of this is the *ileocolic artery*, which is shown in detail in Fig. 11.2, supplying the ileum, caecum, appendix and colon. Other branches of the superior mesenteric are the *right colic* and the *middle colic* (Fig. 11.3). The right colic is retroperitoneal, since the embryonic mesentery of the ascending colon fused with the posterior abdominal wall. The middle colic arises high up from the main trunk and reaches the middle of the transverse colon by travelling through the transverse mesocolon. It is important to remember the presence of this artery since one of the ways the surgeon can get into the omental bursa is by incising the mesocolon.

When the various arteries reach the colon (and this is true also of the colic branches of the inferior mesenteric artery), each divides into two branches that travel proximally and distally respectively, to anastomose with a branch of another vessel. There is thus established a continuous chain of anastomoses along the colon (Fig. 11.3), sometimes known as the *marginal artery of Drummond*. A good collateral circulation can thus be established if one or more of the colic arteries are obstructed. The final distally running branch of the middle colic artery anastomoses about two-thirds of the way along the transverse colon with the first branch of the *inferior mesenteric artery*, the artery of the hindgut.

The inferior mesenteric artery gives off the *left colic artery*, which runs retroperitoneally before dividing into two branches. There follow several *sigmoid branches* and the artery finally descends into the pelvis in the sigmoid mesocolon and changes its name to the *superior rectal artery*. A warning has been given already about undue enthusiasm for dividing the rectum into thirds. This is legitimate for the peritoneal relationships of the rectum but one of the commonest errors made by medical students and by postgraduate candidates is to believe that the superior, middle and inferior rectal arteries supply the upper, middle and lower thirds of the rectum. If you remember that the inferior mesenteric and its termination, the superior rectal, supply the *hindgut* and that the hindgut ends at about the level of the anal valves you will not make this mistake. Thus the superior rectal supplies the *whole* of the rectum and the upper half of the anal canal, while the inferior rectal supplies only the lower half of the canal. The middle rectal

(Chapter 13) is small, supplies only the muscle coats, and anastomoses only slightly with the other two arteries.

When it reaches the rectum, the superior rectal artery first divides into two branches, which run on either side, and then the right branch divides into two again. The branches descend to the level of the anal valves, where they anastomose with branches of the inferior rectal artery. They are accompanied by similar tributaries of the superior rectal vein, which drain, of course, into the portal system. The tributaries of the inferior rectal vein drain into the systemic system (Chapter 13) so the anal canal is a site of a portal–systemic anastomosis. Distension of the veins gives rise to the condition known as (internal) haemorrhoids and although this may occasionally be due to obstruction in the portal system it is usually due to other causes. Because of the arrangement of the main vessels mentioned above, the main dilated veins are found at 3, 7 and 11 o'clock when the anal canal is viewed with the patient lying on his back with his legs flexed. This is the *lithotomy position* (Greek 'lithos', meaning a stone) and is the position in which the patient was placed for removal of bladder stones in pre-anaesthetic days. See Pepys' Diary.

It is worth remembering the embryology of the anal canal for many reasons. If you remember that the upper half is gut whereas the lower half is ectoderm, you will appreciate the following anatomical points:

**1**   The upper half is lined mostly by columnar epithelium, the lower half with stratified squamous.

**2**   The upper half is supplied by the autonomic nervous system and the lower half by spinal nerves. Thus when haemorrhoids are injected at their upper end the injection is painless but lesions in the lower part of the canal are extremely painful.

**3**   The upper half drains into the portal venous system, the lower half into the systemic.

**4**   The lymphatic drainage of the upper half is upwards into the abdomen while that of the lower half is to the superficial inguinal nodes.

# Chapter 12
# The Structures of the
# Posterior Abdominal Wall

The posterior abdominal wall consists of the vertical ridge produced by the vertebral bodies, the *psoas, quadratus lumborum* and *transversus abdominis* muscles (Fig. 12.1) and, in its uppermost part, the *posterior part of the diaphragm* along with its crura (p. 207). The greater part of the posterior abdominal wall is covered by parietal peritoneum, which also covers the retroperitoneal organs such as the duodenum and the kidneys, these lying on the vertebrae and muscles. There is a good deal of extraperitoneal connective tissue and fat, particularly in the region of the kidneys.

## THE MUSCLES

### Psoas major

The 15 origins of psoas major have been mentioned in Chapter 6 (five transverse processes, five intervertebral discs and five fibrous arches over the sides of the bodies), as has its insertion. It fills in the angle formed by the sides of the body and

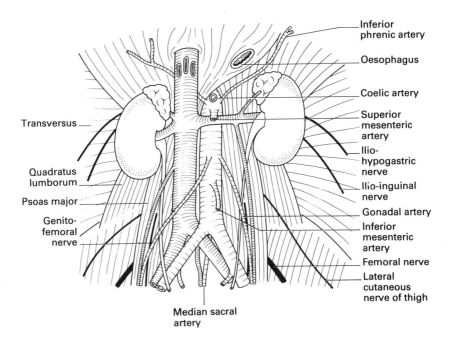

Fig. 12.1 The major structures of the posterior abdominal wall. This diagram will seem less complicated when you have read the text.

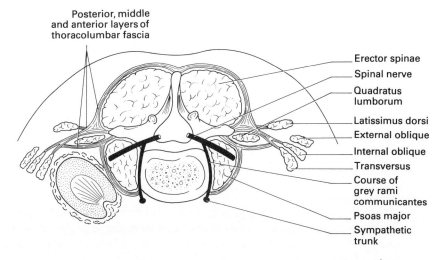

Posterior, middle
and anterior layers of
thoracolumbar fascia

Erector spinae

Spinal nerve

Quadratus
lumborum

Latissimus dorsi

External oblique

Internal oblique

Transversus

Course of
grey rami
communicantes

Psoas major

Sympathetic
trunk

**Fig. 12.2** The thoracolumbar fascia, which encloses erector spinae, quadratus lumborum and psoas major.

the transverse processes and it therefore covers over the intervertebral foramina (Fig. 12.2). Thus, as soon as the lumbar spinal nerves leave the vertebral canal, they enter psoas major and the branches of the lumbar plexus emerge from the surface of this muscle. *Psoas minor* is occasionally seen as a small muscle belly and long tendon lying on the surface of psoas. Psoas major is intimately related to iliacus, which rises from the concave inner surface of the ilium. The two muscles are covered by a rather dense layer of *iliac fascia* so that the lumbar plexus is behind the fascia although the iliac vessels are in front of it. This explains the formation of the femoral sheath (p. 205).

### Quadratus lumborum

This is a quadrate muscle lying lateral to psoas and running between the iliac crest and the 12th rib (Fig. 12.1). It is a side flexor of the trunk and is supplied by the adjacent lumbar nerves.

The uppermost parts of quadratus and of psoas are spanned by the diaphragm, which crosses them by the lateral and medial *lumbocostal arches* (*arcuate ligaments*) respectively.

### Transversus thoracis

Posteriorly, this muscle arises from the thoracolumbar fascia and thus lies lateral to quadratus. Its other attachments have been described in Chapter 9.

## THE AORTA

The abdominal aorta lies mostly in the midline. It enters the abdomen under the median arcuate ligament at the level of T12 and ends at the level of L4, slightly to the left of the midline, by dividing into the *common iliac arteries*. From an embryological point of view, these two arteries are, in fact, side branches of the aorta since they were, at one time, the origins of the umbilical artery. The main continuation of the aorta is therefore represented by a small midline branch, the *median sacral artery* (Fig. 12.1).

The other branches of the abdominal aorta may be subdivided into three groups of three. These are: (i) *ventral*, to the gut, (ii) *lateral*, to three glands; and (iii) *branches to the body wall*.

## Ventral branches

These are, of course, the fore-, mid- and hindgut arteries: the *coeliac, superior mesenteric* and *inferior mesenteric* arteries.

## Lateral branches

These supply the suprarenal glands, the kidneys and the gonads. The largest of these are the *renal arteries*, since they take about a quarter of the whole cardiac output. They arise just below the superior mesenteric artery. The right renal artery passes behind the inferior vena cava. The reason for this will be discussed later. Both renal arteries give small branches to the suprarenals and to the renal pelvis.

The suprarenal branch of the aorta is also called the *middle suprarenal artery*. The *gonadal arteries*, ovarian or testicular, are small and arise from the aorta just below the renal arteries. They descend, lying on the anterior surface of psoas, to reach the ovary or to pass through the inguinal canal into the scrotum. The long course of the vesels is due to the descent of the gonads but these do not travel such a great distance as one might imagine because their apparent descent is largely due to the posterior abdominal wall growing away from them.

## Branches to the body wall

These are the *inferior phrenic*, the four *lumbar arteries* and the *median sacral*. The inferior phrenic arteries give some small branches to the suprarenals and then ramify on the under-surface of the diaphragm. The median sacral artery descends in the midline on the anterior surface of the sacrum. The lumbar arteries form a series that are equivalent to the posterior intercostal arteries. Each artery runs around the side of a vertebral body under the fibrous arches from which part of psoas arises. Each gives a posterior branch that goes through to the back and gives a spinal branch, while the anterior branch runs in the abdominal wall between transversus and the internal oblique muscle.

## THE INFERIOR VENA CAVA

This begins in front of the body of the 5th lumbar vertebra and ascends to the right of the aorta, to pierce the central tendon of the diaphragm at the level of T8. In the embryo there were right and left venae cavae which were interconnected in the region of the iliac veins and the renal veins. The main blood flow was then shifted to the right side and the left inferior vena cava disappeared (Fig. 12.3). This explains the asymmetric arrangement of the abdominal veins. Furthermore, the segment of inferior vena cava from the renal veins upwards developed from a rather ventrally lying longitudinal vessel, whereas the segment below the renal

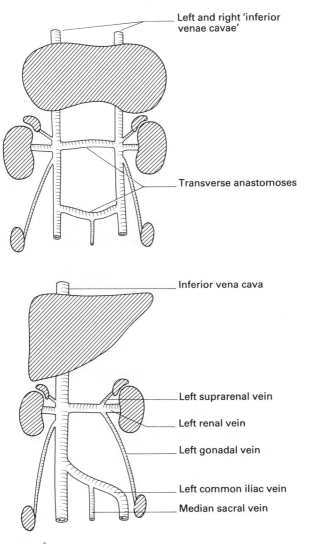

Left and right 'inferior venae cavae'

Transverse anastomoses

Inferior vena cava

Left suprarenal vein

Left renal vein

Left gonadal vein

Left common iliac vein

Median sacral vein

**Fig. 12.3** The development of the inferior vena cava. Above, early stage when the veins are symmetrical. Below, the two cross-connections enlarge to bring the blood flow over to the right side so that the right lobe of the liver becomes larger than the left and the inferior vena cava is on the right.

veins lay more dorsally. You will therefore find that, from the renal veins upwards, veins lie anterior to the corresponding arteries while, below this level, the arteries are in front of the veins.

Bearing this short excursion into embryology in mind, the tributaries of the inferior vena cava may be deduced from the scheme for the branches of the aorta. Firstly, the ventral branches of the aorta may be omitted; all the corresponding veins drain into the portal system. But this should remind you of the final tributaries, namely the three or more large *hepatic veins* that open into the inferior vena cava as it lies in the groove in the liver. The lateral tributaries from the three glands are similar to the arteries on the right but on the left the *suprarenal* and *gonadal veins* open into the left renal vein (Fig. 12.3). Of the tributaries from the body wall, the *inferior phrenic* and *lumbar veins* drain into the inferior vena cava but the *median sacral*, since it is in the midline, opens into the left common iliac.

## THE KIDNEYS AND URETERS

The anterior relations of the kidneys have been shown already in Fig. 10.11; the posterior relations are relatively simple (Figs 12.1 & 12.4), bearing in mind that the right kidney is lower than the left.

The kidneys lie on psoas major, quadratus lumborum and the origin of transversus abdominis, from medial to lateral. The upper part of the kidney lies directly on the diaphragm, outside which is the costodiaphragmatic recess of the pleura and the lower ribs, the 12th on the right and the 11th and 12th on the left. The subcostal, iliohypogastric and ilio-inguinal nerves, which will be described later, are also posterior relations.

The kidneys are surrounded by a thick layer of fat that helps to retain them in position. This, in turn, is enclosed in a layer of fascia, called the *renal fascia*. This is a membranous condensation of extraperitoneal tissue that splits to enclose each kidney, but the anterior and posterior layers remain separate below the kidney.

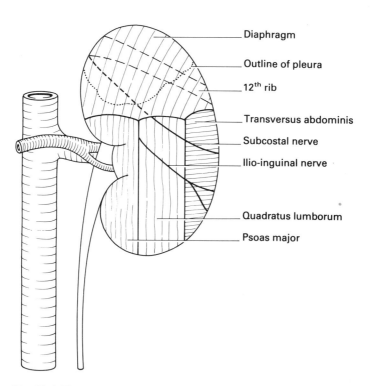

Diaphragm

Outline of pleura

12$^{th}$ rib

Transversus abdominis

Subcostal nerve

Ilio-inguinal nerve

Quadratus lumborum

Psoas major

**Fig. 12.4** The posterior relations of the kidney. Study in relation to Fig. 12.1.

This fascia (also called Gerota's fascia) must not be confused with the renal capsule, which directly surrounds the kidney.

### Blood supply of the kidney

The renal artery divides in the hilum, or perhaps before, usually into five *segmental arteries*. These supply corresponding segments of the kidney, as shown in Fig. 12.5. It is important to understand that each of these vessels is an end artery, i.e. there is no anastomosis between it and the branches of adjacent segmental arteries. Aberrant renal arteries are common and since these are remnants of the 'ladder' of mesonephric arteries they arise from the aorta and pass across to enter the kidney substance directly rather than through the hilum. These, too, are end arteries. Ligature of a segmental or of an aberrant artery will cause ischaemia and death of the corresponding segment of the kidney. The intrarenal veins, however, do anastomose.

### Internal appearance of kidneys

If a kidney is split open, the cortex and medulla can be recognized, as can the division of the kidney substance into lobes (Fig. 12.6). Each unit or lobe of cortex and medulla is separated by a deep prolongation of cortical substance called the *renal column of Bertin*. Each segmental artery gives rise to *lobar branches*, which, in turn, give *interlobar* arteries between the lobes. These give off *arcuate arteries*, which arch out like the spokes of an umbrella between cortex and medulla, where

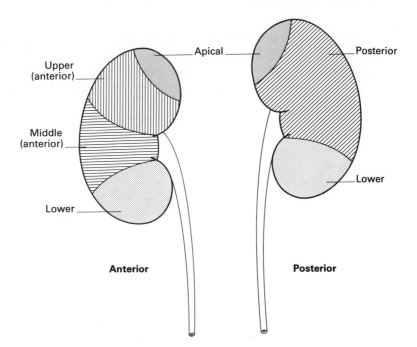

**Fig. 12.5** The usual (but not invariable) arrangement of the arterial segments of the kidney.

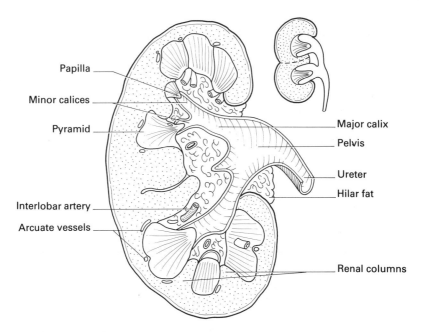

**Fig. 12.6** Section of the kidney. The small picture shows how the columns of Bertin represent the cortices of adjacent fused lobes.

they can be seen, along with their accompanying veins, with the naked eye. From these, *cortical radial arteries* radiate out towards the surface of the cortex. Some of them emerge on the surface and anastomose with the capsular vessels. The corresponding veins may be recognized on the surface of the kidney as *stellate veins*.

The cortical radial arteries give rise to the *afferent arterioles* of the glomeruli, and the glomerular *efferent arterioles* supply the peritubular plexus and, in the case of the juxtamedullary glomeruli, the medulla.

Each unit of medullary substance is called a *pyramid* and the free part is the *papilla*. This projects into one of the *minor calices* of the pelvis and these unite to form *major calices* and, finally, the renal pelvis itself. A calix (meaning a cup or chalice) should be distinguished from a calyx (a husk or covering as in glycocalyx). The cup-shaped outline of the normal calix can be seen clearly in an intravenous urogram.

The calices and pelvis are smooth and are lined by transitional epithelium, as are the ureter, the bladder and the upper reaches of the urethra. The ureter begins at the pelvi-ureteric junction and runs down on the surface of psoas fascia and behind the peritoneum. It then enters the pelvis, where its further course will be described in Chapter 13. It is about 25 cm in length and has three narrow regions: at the pelvi-ureteric junction, where it crosses the brim of the pelvis and as it enters the bladder.

At operation, when peritoneum is lifted from the posterior abdominal wall, the ureter comes away with the peritoneum. It can also be recognized by the waves of peristalsis that travel down it.

The relations of the upper parts of the ureters are shown in Figs 10.7–10.10. The lower parts are crossed by the gonadal vessels and by the vessels that supply the ascending and descending colon. The right ureter is crossed by the root of the mesentery and the terminal ileum. Since the ureter is transplanted along with the kidney, it is lucky that it has a very good blood supply from the renal artery, the aorta, the gonadal vessels and various pelvic vessels. In an intravenous urogram the ureters normally lie approximately opposite the tips of the lumbar transverse processes.

## Renal transplantation

Kidneys for transplantation, from a cadaver or from a living donor (usually a relative), are most commonly placed outside the peritoneum in the iliac region, the renal artery being anastomosed to the divided internal iliac artery and the renal vein to the external iliac vein. Other sites for the anastomoses may, however, be chosen, especially if aberrant renal arteries are present. The ureter is implanted into the bladder. Immunosuppressive drugs such as cyclosporin are used to discourage rejection.

## Developmental anomalies of the kidneys

In the embryo, the mesonephros is a large functioning organ, with glomeruli, tubules and a duct (the *mesonephric* or *Wolffian duct*) leading into the *urogenital sinus*. The *ureteric bud* is an outgrowth from the lower end of the mesonephric duct and it elongates in a cranial direction, picking up a *metanephrogenic cap* from the mesoderm. The tip of the ureteric bud divides over and over again to form the collecting ducts and (by coalescence of the first generations of divisions) the calices. The dilated end of each of the collecting ducts induces the differentiation of a glomerulus and its associated renal tubules from the metanephrogenic cap, and the distal tubule unites with the collecting duct. The lower end of the mesonephric duct, along with the opening into it of the ureteric bud (now the ureter), is taken into the urogenital sinus to form the trigone, and the ureteric orifice moves upwards and laterally to reach the upper corner of the trigone.

As the developing kidney continues to ascend it 'climbs the ladder' of the parallel series of mesonephric arteries until further progress is blocked by the suprarenal gland and the current mesonephric artery becomes the renal artery. At the same time the kidney rotates so that the hilum faces medially rather than forwards.

### Ectopia

The kidney may fail to ascend properly and is found low down in the abdomen, obtaining its blood supply from the lower aorta or the common iliac artery. It may also fail to rotate (*malrotation*) so that the hilum faces forwards and the kidney is flattened and deformed (Fig. 12.7b). In *crossed ectopia*, one kidney, usually the left, crosses the midline and comes to lie below the other, often fusing with it (Fig. 12.7a).

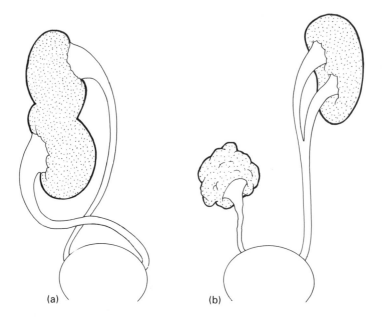

**Fig. 12.7** Some malformations of the kidneys and ureters. (a) Sigmoid kidney, one form of crossed ectopia. (b) A malrotated, unascended kidney on the left and a bifid system on the right.

### Horseshoe kidney

If the two kidneys fuse at their lower poles early in development, when they are close together, normal ascent is impossible as the isthmus cannot get past the inferior mesenteric artery (Fig. 12.8). This artery, as well as the two ureters, thus lies in front of the isthmus and ureteric obstruction may occur.

### Duplex systems

The ureteric bud may divide low down so that the kidney has two hila (*bifid pelvis*), the ureters uniting at any point between the pelvi-ureteric junction and the bladder (Fig. 12.7b). If there are two ureteric buds, there will be double pelves and ureters, the normally placed (*orthotopic*) ureter opening into the bladder somewhere near the normal position and the other (*ectopic*) ureter opening below it. The orthotopic ureter drains the lower part of the kidney (Fig. 12.9). Ectopic ureteric openings have the disadvantage that the normal valvular action at the orifice (p. 259) is disturbed and vesico-ureteric reflux may occur.

### Autosomal dominant polycystic disease

This is a relatively common condition occurring in about 1 in 1000 of the population. One or both kidneys are grossly enlarged and 'bubbly' in appearance due to the presence of an enormous number of cysts. These are local dilatations of tubules which are continuous with normal tubules outside the cyst. Any part of the nephron or the collecting duct may be involved.

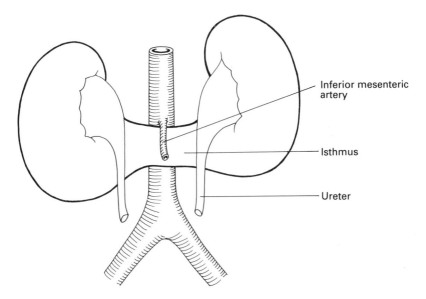

**Fig. 12.8** Diagram of a horseshoe kidney. Note that both the inferior mesenteric artery and the ureters pass in front of the isthmus.

## THE SUPRARENAL GLANDS

These are situated within the renal fascia at the upper poles of the kidneys. The left gland is more accessible to the surgeon than the right because the right gland is behind the bare area of the liver and its vein is very short and liable to tear away from the inferior vena cava. On sectioning the gland, the cortex and medulla can easily be distinguished. They are, like the cortex and medulla of the kidney, two quite different organs with a different development, a different blood supply and a different function. Although both parts of the suprarenal are endocrine glands, the cortex is stimulated to secrete by adrenocorticotrophic hormone (ACTH) whereas the medulla is stimulated by its nerve supply, which consists of preganglionic fibres from the coeliac plexus.

The gland is very vascular, and receives its blood supply by means of a number of branches from the renal and phrenic arteries and from the aorta (not just one branch from each, as is sometimes stated). The venous drainage is by a surprisingly large vein into the left renal vein (on the left) or the inferior vena cava (on the right), as shown in Fig. 12.3.

Before birth, the suprarenals are almost as large as the kidneys because of the presence of a '*fetal cortex*'. This disappears after birth, so that the gland becomes smaller but it is still relatively large in the child.

## THE LUMBAR PLEXUS

The lumbar plexus lies in the substance of psoas major and is formed from the anterior rami of L1, 2, 3 and 4. The sacral plexus (Chapter 13) is from L4, 5, S1, 2, 3 and 4. The fourth lumbar nerve is therefore sometimes called the *nervus furcalis* because it splits itself between the lumbar and the sacral plexuses. One

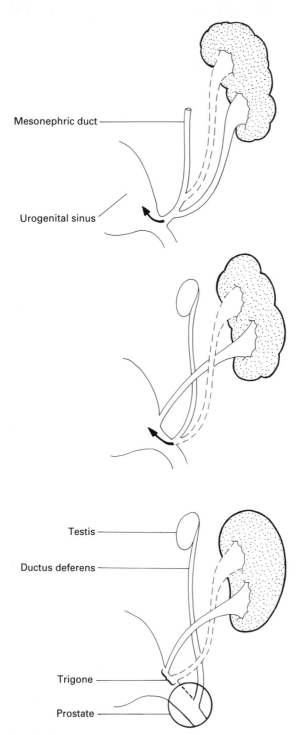

**Fig. 12.9** The development of a duplex system. The upper ureter is shown by dashed lines. Note that the orifice of the lower ureter is taken into the bladder first and so travels further up the bladder wall than the upper ureter.

representative of the thoracic nerves is seen on the posterior abdominal wall, although, of course, the lower six are found in the anterior abdominal wall (p. 200). The 12th thoracic nerve is not in an intercostal space but is below the last rib and is therefore called the *subcostal nerve*. It emerges from under cover of the lateral lumbocostal arch and so immediately finds itself on quadratus lumborum. It then slips in between transversus abdominis and internal oblique and is distributed like the other lower intercostal nerves to the abdominal wall. It has a collateral branch like the other intercostals, and so does the next nerve down, the lst lumbar nerve. The main trunk of this is the *iliohypogastric* nerve, which is the last of the series of nerves that supply the anterior abdominal wall and, like the main inter-costal nerves, has a lateral cutaneous branch. Its collateral, which, like the inter-costal collaterals, has no lateral cutaneous branch, is called the *ilio-inguinal* nerve; it runs between the layers of the abdominal wall but ends by emerging from the superficial inguinal ring and supplying skin on the medial side of the thigh and the scrotum or labium majus. It will thus be seen during operations for inguinal hernia.

The *genitofemoral nerve* (L1 and 2) emerges from the anterior surface of psoas major and runs down deep to the psoas fascia to supply the cremaster muscle in the male via its genital branch and a small area of skin below the inguinal ligament via its femoral branch. The *lateral cutaneous nerve of the thigh* (L2 and 3) emerges from the lateral border of psoas, sweeps around the iliac fossa and leaves the abdomen by passing under (or sometimes through) the inguinal ligament near its lateral attachment to the anterior superior iliac spine.

The two largest branches of the lumbar plexus both have the same root value (L2, 3 and 4), which you should remember. Remember also that, during develop-ment, the lower limb rotates medially so that the original dorsal surface faces forwards. The great saphenous vein is a rough guide to the boundary between dorsal and ventral surfaces. Therefore the femoral nerve, which supplies skin and muscles on the front of the thigh, is derived from the *dorsal* divisions of L2, 3 and 4 while the obturator nerve, which supplies muscles, and sometimes skin, on the medial side of the thigh is derived from *ventral* divisions. (NB These are dorsal divisions of the anterior rami, not of the spinal nerves themselves.)

The *femoral nerve* is large and emerges from the lateral border of psoas quite low down. It may give branches to psoas and iliacus although these usually come directly from the spinal nerves. Like all other branches of the lumbar plexus, it lies behind the fascia that covers psoas and iliacus and it is therefore outside the femoral sheath as it passes beneath the inguinal ligament.

The *obturator nerve* emerges from the medial border of psoas near the brim of the pelvis so that it lies behind the common iliac vessels. It then travels forwards and downwards, anterior to obturator internus (see Fig. 13.2), and leaves the pelvis by passing through the upper part of the obturator foramen.

## SOME RELATIONS OF PSOAS MAJOR

These have all been mentioned already but two points are worth repeating here. Firstly, all branches of the lumbar plexus emerge from the lateral border of psoas except for the *genitofemoral nerve*, which emerges on the front, and the *obturator nerve*, which emerges on the medial side.

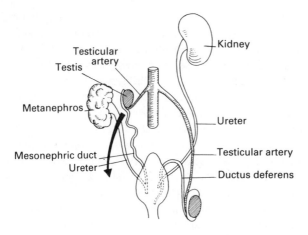

**Fig. 12.10** A diagram to explain the adult relation of the testicular artery and ductus deferens to the ureter. The arrow shows the path of testicular descent.

Secondly, there are a number of longitudinally running structures on the anterior surface of the muscle that must be recognized by the surgeon exploring the retroperitoneal tissues. These are: (i) the *genitofemoral nerve*, which lies deep to the psoas fascia; (ii) the *ureter*, which lies on the psoas fascia but comes away with the peritoneum and shows peristalsis; (iii) the *gonadal vessels*, which cross in front of the ureter because of the way the testis and ovary descend (Fig. 12.10); (iv) on the left side only, and rather medially, the *inferior mesenteric vessels*, especially the vein; and (v) lying at the anterior border of psoas, close to the aorta, the *sympathetic trunk*.

## THE AUTONOMIC NERVOUS SYSTEM IN THE ABDOMEN

A general account of this is given in Chapter 15 so that only the easily recognized anatomical features will be mentioned here. The sympathetic trunk enters the abdomen by passing with psoas major under the medial lumbocostal arch and then runs down along the anterior border of psoas. It usually has four ganglia, from which grey rami pass to the lumbar nerves but white rami only pass to the ganglia from the first two nerves (p. 292). The rami are very long, since they have to pass with the lumbar vessels under the fibrous arches from which psoas major arises. Branches from the ganglia pass: (i) to the various abdominal sympathetic plexuses; and (ii) to the iliac arteries directly and to other arteries via the plexuses.

The *coeliac ganglia* and the *coeliac plexus* surround the origins of the coeliac and superior mesenteric arteries, the main ganglia lying between the suprarenals. Offshoots from the main plexus travel with all the major arteries. The plexuses are joined by the three *splanchnic nerves* (p. 191), by branches of the vagus nerves (parasympathetic) and by branches from the sympathetic trunk. They thus carry, via the arteries, both sympathetic and parasympathetic fibres to the viscera, although the stomach gets its parasympathetic supply directly from the gastric nerves (vagus).

The plexus around the aorta is continued downwards across the aortic bifurcation and in front of the promontory of the sacrum to form the *superior hypogastric plexus*. This may form a fairly wide ribbon of nerves, which is often known as the *presacral nerve*. It divides into the right and left *inferior hypogastric plexuses*, which are joined by *pelvic splanchnic nerves* (parasympathetic) from the 2nd, 3rd and 4th sacral nerves. These plexuses lie at the sides of the main pelvic viscera and supply them with both sympathetic and parasympathetic fibres. The parasympathetic fibres from the pelvic splanchnic nerves supply the colon up as far as the junction between the midgut and hindgut, i.e. about two-thirds of the way along the transverse colon.

# Chapter 13
# The Pelvis and Perineum, Lymphatics

Although the pelvic cavity is part of the general abdominal cavity, it does form a rather special region — for example, gynaecologists' activities are almost entirely confined to it — and it is therefore the subject of a separate chapter. The pelvic cavity is a rather confined space so that large tumours may occupy much of the cavity and may affect other pelvic structures. Fluids such as pus or blood in the abdomen may gravitate to the pelvis and form abscesses there, while, to the obstetrician, the diameters of the pelvic cavity are of profound importance.

## THE BONY PELVIS

This is bounded by the two hip bones and the sacrum. The hip bones have already been described in Chapter 6 and the sacrum will be dealt with in Chapter 14. It is therefore only necessary to mention some of the features of the pelvis as a whole. The right and left pubic bones are united anteriorly by means of a secondary cartilaginous joint (Chapter 1), i.e. each bone is covered by a layer of cartilage and there is a thick disc of fibrocartilage between them. The joint between the hip bone and the sacrum is a plane synovial joint, although it does not look like one and only a very small degree of movement is possible.

### The sacro-iliac joint

This joint is on the main line of weight transmission from the spine to the hip bone and thence to the femur. It is thus one of the strongest joints in the body and its ligaments are arranged so as to allow a very small amount of movement but to resist any major displacement of the sacrum. The joint surfaces are covered by interlocking lumps and depressions and, as in other synovial joints, are covered with articular cartilage. The most important ligaments are the *interosseous sacro-iliac ligaments*, which pass between the back of the sacrum and the strong and craggy posterior parts of the ilia (Fig. 13.1). There is a much weaker *anterior* or *ventral sacro-iliac ligament*, which is a thickening in the fibrous capsule of the joint. Accessory ligaments are the *sacrospinous* and *sacrotuberous ligaments*, which pass respectively from the spine and the tuberosity of the ischium to the side of the sacrum. They convert the greater and lesser sciatic notches to *greater* and *lesser sciatic foramina*. The *iliolumbar* ligament also helps to bind the hip bone to the spine; it passes from the transverse process of the 5th lumbar vertebra to the posterior part of the iliac crest. Figure 13.1 shows that the line of weight transmission passes through the joint to the ilium and thence to the femur so that the ventral part of the pelvis formed by the pubis and ischium is not involved. This part acts as a tie-beam to stop the ilia being forced apart by the weight of the body so that there is no need for any very powerful ligaments around the symphysis pubis. The figure also shows that the sacrum is, in effect, suspended from the iliac

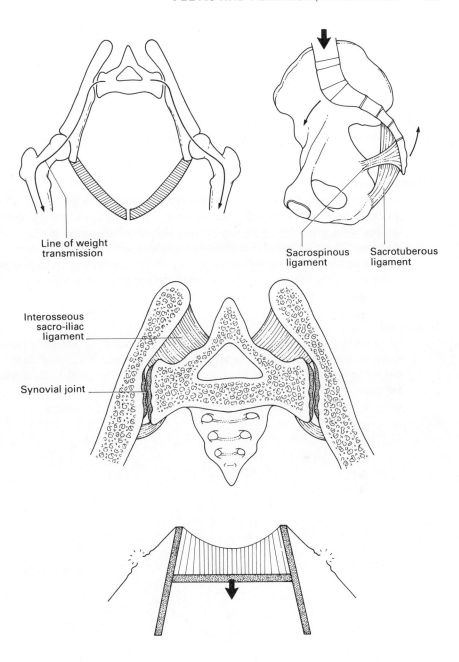

**Fig. 13.1** *Top.* The pubic bones act as a tie-beam between the two sides of the hip bone. The sacrotuberous and sacrospinous ligaments prevent rotation of the sacrum. *Centre.* The body weight, acting on the sacrum, tends to pull the hip bones together and put pressure on the sacro-iliac joint, thus stabilizing it. *Bottom.* A suspension bridge is stabilized in a similar way.

bones by its ligaments in much the same way as a suspension bridge is supported from its pillars. Thus the body weight pulls the two ilia together and the interlocking surfaces of the sacro-iliac joint are pressed together firmly. In the side view it is obvious that there is also a rotatory effect on the sacrum. This is resisted by the sacrotuberous and sacrospinous ligaments.

## MUSCLES OF THE PELVIS

The bony pelvis is lined by a layer of muscles, which considerably diminishes its capacity.

### Obturator internus

This muscle covers the bony wall of the pelvis, leaving a gap in the upper part of the obturator foramen (Fig. 13.2). Its tendon leaves via the lesser sciatic foramen, bending through almost a right angle to pass laterally to the greater trochanter of the femur. There is a large bursa between the tendon and the bone. The muscle is covered by a layer of fascia (*obturator fascia*), which is particularly dense over the upper part of the muscle. It is a lateral rotator of the thigh and is supplied by the nerve to obturator internus, which will be described later.

### Levator ani

Quadrupeds move their tails from side to side by means of a muscle that passes from the side walls of the pelvis to the caudal vertebrae, and up and down by erector spinae and a muscle that runs backwards from the pubis to the caudal vertebrae. Man does not (normally) have a tail to wag but, because of his upright posture, needs a muscular floor to the pelvis to support the viscera. Nature's

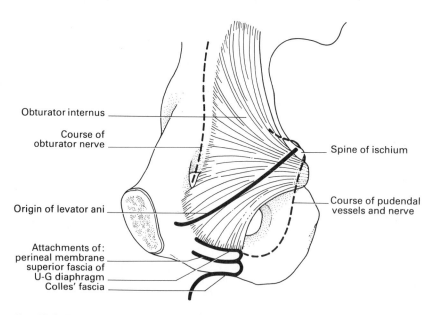

Fig. 13.2 The side wall of the pelvis.

response is obvious and the human levator ani (along with coccygeus) is the result. In animals, the corresponding muscle arises from the brim of the pelvis but in man, in order to form the pelvis floor, it has migrated downwards, leaving a dense trail of almost aponeurotic fascia behind it. This is the dense fascia over the upper part of obturator internus referred to above.

Levator ani arises from the back of the pubis lateral to the symphysis, from the spine of the ischium and, in between those two origins, from a fibrous thickening in the obturator fascia (Fig. 13.2). From this origin the muscle sweeps downwards, medially and backwards to form the 'basin' after which the pelvis is named. The most anterior fibres from the back of the pubis, sometimes called the *levator prostatae*, pass backwards and downwards just below the prostate, which sits on the muscle like a billiard ball in a funnel (see Fig. 13.18), to be inserted into a fibrous nodule called the *perineal body* (Fig. 13.3). In the female the muscle surrounds the vagina to form a *sphincter vaginae*. The perineal body is very important in the female (see below) and, although it is difficult to distinguish from the surrounding connective tissue by dissection, it can be felt between a finger in the vagina and a thumb on the perineum. Behind this part of the muscle, other fibres pass around the recto-anal junction and blend in with the various parts of the anal sphincter, as will be described later. This part of the muscle is called *puborectalis* or the 'recto-anal sling'. Further back again, the muscle meets its opposite number to form the *anococcygeal raphe*, and the most posterior fibres are inserted into the coccyx. Levator ani, therefore, together with coccygeus, described below, forms a complete floor for the pelvis, which can be tightened up voluntarily by contraction of the muscle in what has been aptly described as a 'perineal shrug'. The insertion into the perineal body is essential for this and the effect of tearing the perineal body, which may happen during a complicated birth, can easily be imagined (Fig. 13.4).

Levator ani and coccygeus together are sometimes known as the *pelvic diaphragm* since they separate the pelvis from the ischiorectal fossa, which will be

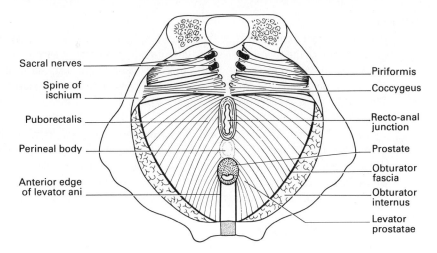

**Fig. 13.3** The floor of the pelvis, from above.

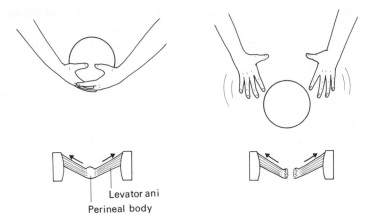

Levator ani
Perineal body

**Fig. 13.4** The importance of the perineal body to the integrity of the pelvic diaphragm.

described later. Anything which has to pass from the thorax to the abdomen (oesophagus, inferior vena cava, aorta, etc.) has to pass through the (thoracoabdominal) diaphragm and anything which needs to pass from the pelvis into the perineum must pass through the (pelvic) diaphragm. This includes the urethra, the vagina and the rectum. The pudendal nerves and vessels, however, make a special detour to avoid the diaphragm (p. 265). The nerve supply of levator ani is from S4 and the inferior rectal nerve.

### Coccygeus

This small muscle is immediately deep to the sacrospinous ligament (see Fig. 13.1) and has similar attachments to the spine of the ischium and the coccyx. The nerve supply is from S4.

### Piriformis

This takes origin from the middle three pieces of the sacrum, closely related to the anterior rami of the sacral nerves that emerge from the anterior sacral foramina. It passes through the greater sciatic foramen, along with such structures as the sciatic nerve and the superior and inferior gluteal vessels and nerves (see Fig. 6.23), to be inserted into the greater trochanter of the femur. The nerve supply is from adjacent sacral nerves.

These three muscles, levator ani, piriformis and coccygeus, form an almost complete muscular lining to the pelvis, since they lie more or less edge to edge. They are covered by a layer of condensed fascia, above which lies the rather looser pelvic fascia. This surrounds the pelvic viscera and is condensed here and there to form 'ligaments' such as the uterosacral and puboprostatic ligaments (see below).

## THE SACRAL PLEXUS

This will be a convenient place to describe the sacral plexus, which is derived from L4, 5, S1, 2 and 3, although only half of L4 is involved (see Chapter 12). The lower half of L4 joins L5 to form the *lumbosacral trunk*, a large nerve that crosses the brim of the pelvis to join S1. (Remember that *below* the 7th cervical vertebra each nerve lies below the corresponding vertebra so that the 5th lumbar nerve is between the 5th lumbar and lst sacral vertebra.) Fortunately, the sacral plexus does not have the complicated arrangement of the brachial plexus, and its nerves simply pick up branches from the appropriate spinal nerves: for example, the sciatic nerve from L4, 5, S1, 2 and 3 and the superior gluteal nerve from L4, 5 and S1. As was described in Chapter 6, the sciatic nerve is, in fact, a combination of two nerves which may arise separately. The *tibial nerve*, which will supply the back of the leg and the sole of the foot, arises from ventral divisions of the spinal nerves and the *common peroneal*, destined for the lateral and anterior (extensor) muscles, comes from dorsal divisions (compare with the femoral and obturator nerves). A full list of the nerves derived from the sacral plexus will now be given and you should remember the root values of at least the two parts of the sciatic nerve. The further course of all these nerves has been described in Chapters 6 and 7.

| | |
|---|---|
| *Tibial part of sciatic* | L4, 5, S1, 2, 3 (ventral divisions) |
| *Common peroneal part of sciatic* | L4, 5, S1, 2 (dorsal divisions) |
| *Superior gluteal* (above piriformis) | L4, 5, S1 |
| *Inferior gluteal* (below piriformis) | L5, S1, 2 |
| *Posterior cutaneous of thigh* | S1, 2, 3 (part dorsal and part ventral divisions) |
| *Pudendal* | S2, 3, 4 |
| *Nerves to levator ani, coccygeus and anal sphincter* | S4 |

There are other small nerves to quadratus femoris, obturator internus, the gemelli and piriformis.

## THE PELVIC VISCERA

These comprise the bladder, uterus, etc. and rectum in the female and the bladder, prostate, etc. and rectum in the male. They are partly covered by peritoneum on their upper aspects but below they are extraperitoneal. The peritoneal relations are surgically important and are shown in Fig. 13.5. In the male pelvis, the empty bladder is covered above by peritoneum, which is reflected off the back of the anterior abdominal wall. When it fills, it rises above the pubic symphysis and strips the peritoneum from the anterior abdominal wall, so that a needle can be passed into the bladder from the suprapubic region without opening the peritoneum. Laterally, the peritoneum passes across the side wall of the pelvis and posteriorly it dips downwards over the back of the bladder and the upper parts of the seminal vesicles before being reflected back on to the rectum to form the *rectovesical pouch*. In the female, the relation of peritoneum to the bladder is similar but it dips

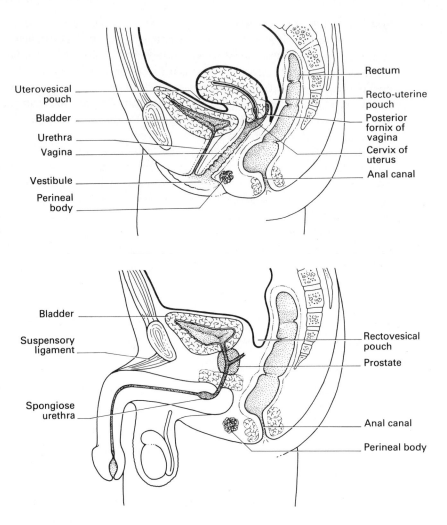

**Fig. 13.5** Midline sections of the male and female pelves.

down to the level of the internal os (the junction between the uterine body and cervix) to form the *uterovesical pouch*. This is separated from the vagina by pelvic fascia, which, in the region of the uterus, is called the *parametrium*. Behind the uterus, the peritoneum is reflected on to the rectum quite low down so that not only is the whole posterior surface of the uterus covered but also the uppermost part of the vagina. This is the *recto-uterine pouch,* or *pouch of Douglas,* and it is a sobering thought that the peritoneal cavity is so close to the outside world as represented by the lumen of the vagina. Laterally, the peritoneum is draped over the uterine tubes and ovary to form the *broad ligament* of the uterus and the *suspensory ligament* of the ovary, which will be described in more detail later.

## The bladder

The empty bladder is shaped rather like a truncated boat with important structures attached at the four corners (Fig. 13.6). It fits in neatly behind the symphysis pubis anteriorly and between the two levator ani muscles at the side. When full, it rises above the symphysis and in the young child it is normally an abdominal rather than a pelvic organ. It has a wall of smooth muscle (the *detrusor muscle*) arranged in three layers although they are not very clear-cut. The mucous membrane, lined by transitional epithelium, is mostly thrown into folds but over the triangular base of the bladder it is smooth. This area is known as the *trigone* and has the two ureters and the urethra opening at its three corners. This part of the bladder is derived from the lower end of the mesonephric duct (mesoderm) rather than the endo-derm of the rest of the bladder. The ureters pass very obliquely through the bladder wall and through the mucosa, while the longitudinal muscle of the ureter spreads out to form part of the superficial muscle layer of the trigone. When the detrusor muscle contracts, the intramural part of the ureter is compressed and its orifice is drawn medially. This prevents vesico-ureteric reflux.

The urethral orifice is surrounded by a prolongation downwards of the circular layer of bladder musculature, which, together with some elastic tissue, forms the *sphincter vesicae* (not to be confused with the *sphincter urethrae*, which is striated muscle). It is doubtful whether it acts as a true sphincter as far as the urine is concerned. It is probable that its main function is to prevent reflux of semen during ejaculation.

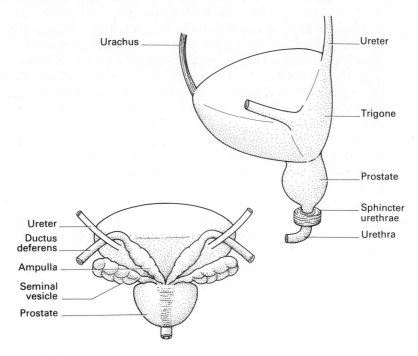

**Fig. 13.6** Side and posterior views of the bladder.

The bladder is surrounded below the peritoneum by pelvic fascia, which is thickened anteriorly to form *puboprostatic ligaments* (male) or *pubovesical ligaments* (female) and laterally, along the sides of the bladder, is connected to the obturator fascia, forming the *lateral true ligaments*. The bladder receives its nerve supply from S2, 3 and 4 via the nervi erigentes and pudendal nerves and also from the sympathetic system.

## The prostate

In the male, the internal urethral orifice leads directly into the *prostatic urethra*, the gland itself being immediately below the bladder. The prostate is an exocrine gland with 15–20 ducts and is characterized by the large amount of smooth muscle in its stroma, which makes it easily recognizable in a histological section. It is surrounded by a fibrous capsule and a prostatic venous plexus and the prostatic urethra runs through its substance, rather anteriorly. In fact, the part of the prostate in front of the urethra (the *isthmus*) is purely fibromuscular and devoid of glandular tissue. There are two ways of subdividing the gland into lobes. Firstly, in a rather unscientific way, the pyramid-shaped part of the gland above and between the two ejaculatory ducts (described later) is often known as the *'middle lobe'* and when this becomes enlarged it projects upwards to form a swelling inside the bladder behind the urethral orifice. This enlarged 'middle lobe' may be big enough and mobile enough to be pushed down during micturition and to obstruct the urethral orifice. The remainder of the gland may be subdivided into two *lateral lobes* by a shallow vertical groove that can be felt *per rectum*. A more practical way of subdividing the gland is by the distribution of its ducts. The *peripheral zone* of the gland, whose ducts open lateral to the verumontanum, comprises about 70% of the glandular tissue and is the usual site of carcinoma. Another 20% or so of tissue forms a *central zone*, whose ducts open on to the edge of the verumontanum; this is probably of different embryological origin from the peripheral zone although it is difficult to determine this in the embryo since, unlike the pictures in embryology textbooks, embryonic ectoderm, mesoderm and endoderm do not come in different colours. Finally, and most importantly, around the upper part of the prostatic urethra there is a zone of smaller glands and this is the site of benign prostatic hypertrophy ('enlarged prostate'), which affects a large proportion of males over 50 and is a common cause of urine outflow obstruction.

### The prostatic urethra

Surprisingly, this is the widest and most easily dilatable part of the male urethra. Its posterior wall has a central streamlined elevation called the *colliculus seminalis* or, more commonly, the *verumontanum*, on to which open the ejaculatory ducts and the prostatic utricle, together with, peripherally, the prostatic ducts draining the central zone. At the sides of the verumontanum open the ducts of the peripheral zone.

In the female, a number of *peri-urethral glands* are said to represent the prostate.

## The seminal vesicles

The ductus (vas) deferens, whose course will be described later (see also Chapter

9), leaves the side wall of the pelvis to hook over the ureter (see Fig. 12.10). It then approaches the opposite ductus behind the bladder. Between the two lie the *seminal vesicles* (Fig. 13.6). These are two sacculated organs about 5 cm long and each consists of a single branched tube. The uppermost parts of the seminal vesicles are covered by the peritoneum of the rectovesical pouch. They secrete a fluid that contributes to the semen (but are *not* a storehouse for spermatozoa, as has been stated). The duct of the seminal vesicle joins the ductus deferens to form the *ejaculatory duct* and the ducts of each side converge to open on to the verumontanum on either side of the *prostatic utricle*.

The prostatic utricle is a minute, midline, blind-ended diverticulum that passes into the prostatic substance. It represents (almost) the only surviving remnant of the paramesonephric (Müllerian) ducts in the male and its only importance is that, in cases of pseudohermaphroditism, it may be enlarged to form a mini-uterus and vagina.

## The uterus and vagina

The uterus has a body, a fundus, a cervix and a pair of uterine or Fallopian tubes (Fallopius was a friend of Eustachio, p. 398). It is composed of smooth muscle that is extremely powerful, especially when it hypertrophies during pregnancy. The uterine cervix is larger than the body in the young child (in whom the uterus, like the bladder, is an abdominal organ) but only about half the length of the body in the adult. The lumen of the cervix is rather smaller than that of the uterus and there is a constriction, the *internal os*, between the two. The opening of the cervix into the vagina is called the *external os*; this is circular at first but after the birth of the first child it is a transverse slit with anterior and posterior lips. The cervix is subdivided into vaginal and supravaginal portions. The former projects into the vagina and is surrounded by the anterior, posterior and lateral *vaginal fornices*, while the latter is above the level of the vagina and is surrounded by the pelvic connective tissue, which in this region is called the *parametrium*, just as the uterine muscle is called the *myometrium* and the mucous membrane the *endo-metrium* (Greek, *metra* = womb).

The cavity of the uterus is not unlike the trigone of the bladder, being triangular in coronal section with the uterine tubes opening at the two upper corners and the internal os at the lower angle. The cavity of the cervix is fusiform, being wider in the middle. The endometrium of the upper third of the cervix undergoes some menstrual changes similar to those in the body of the uterus. During the first stage of labour, this part of the cervix gets taken up into the uterus to form what obstetricians call the *lower uterine segment*.

The normal uterus as a whole is said to be *anteflexed* and *anteverted*. The former term means that the uterus (and cervix) is curved along its long axis with the concavity directed downwards and forwards. Anteversion indicates that the uterus is tilted forwards on the vagina so that the cervix points backwards and downwards. The position is not fixed, however, since the uterus becomes less and less anteverted as the bladder fills, because its normal position is tilted forwards over the bladder (see Fig. 13.5).

The uterine tubes pass laterally towards the side wall of the pelvis, lying in the upper margin of the *broad ligament*. The function of the tubes is to convey the ovum from the ovary to the uterus, fertilization normally occurring in the *ampulla*. This is the longest portion of the tube and it receives the ovum from the *infundibulum*, the funnel-shaped, fimbriated end, which is very closely related to the surface of the ovary. The ampulla leads into the *isthmus*, a narrow part that leads via an *intramural section* into the uterine cavity. The ovarian end of the tube passes over the upper pole of the ovary and down its posterior border; its fimbriated end covers the medial surface of the ovary (Fig. 13.7).

### The broad ligament

This is a double fold of peritoneum thrown over the uterus and the uterine tubes and extending to the side wall of the pelvis. Owing to the anteverted position of the uterus, the broad ligament has superior and inferior surfaces rather than anterior and posterior (Fig. 13.7). The ovary is attached to the upper surface by the *mesovarium*, while the portion of the broad ligament between this and the tube is the *mesosalpinx*. The name of the fold of peritoneum that connects the ovary and the end of the tube to the side wall of the pelvis has been called the *infundibulo-pelvic ligament* but is officially known by the totally misleading title of the *suspensory ligament of the ovary*. I shall use the former term.

There are a number of structures enclosed between the layers of the broad ligament, in addition to the uterine tubes. The *round ligament of the uterus* and the *ligament of the ovary* both traverse the broad ligament; they represent the female gubernaculum (p. 205). There are also some remnants of the mesonephric tubules and ducts, called the *epoöphoron* and *paroöphoron*, which can give rise to

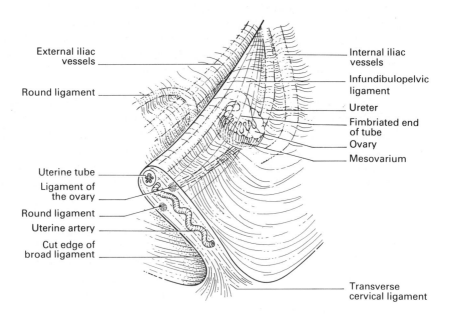

**Fig. 13.7** The right broad ligament, divided at its attachment to the uterus.

broad ligament cysts. The *uterine artery* runs near the base of the broad ligament and ascends between its layers along the side of the uterus, while also in its base are the *transverse cervical ligaments*.

## Supports of the uterus

Prolapse (or abnormal descent) of the uterus is a common condition but the way in which the uterus is kept in its normal position is imperfectly understood, at least as far as the relative importance of the various contributory factors are concerned. Certainly the integrity of the pelvic floor, particularly the perineal body, is important, as was described at the beginning of this chapter. The normal ante-verted position of the uterus may also be important, since it is partly supported by its position above the bladder. The round ligaments help to keep it anteverted but attempts to bring a retroverted uterus back to its normal position by shortening the round ligaments have failed, because the ligaments simply stretch. The cervix is securely anchored in its central position in the pelvis by a number of 'ligaments', which are actually condensations of the pelvic fascia along with some smooth muscle fibres. The *uterosacral ligaments* extend back from the cervix to the sacrum, passing on either side of the rectum. They are each enclosed in a fold of peritoneum, called the *recto-uterine fold*; the two folds form the lateral boundaries of the recto-uterine pouch and can be palpated *per rectum*. Another condensation of fascia extends from the cervix to the side wall of the pelvis. This is the *transverse cervical ligament* (more commonly known as the *cardinal* or *Mackenrodt's ligament*). Anteriorly, the *pubovesical ligaments* have already been mentioned and parts of these extend backwards through the fascia to blend with the fascia around the cervix to form the *pubocervical ligaments*.

A combination of all these factors keeps the uterus in place, but it is easy to understand how a difficult birth may stretch and distort these ligaments, alter the normal position of the uterus and interfere with the integrity of the pelvic floor, so that the uterus, bladder or rectum may descend and bulge into the vagina from without.

## The ovary

This, before the first pregnancy, lies on the side wall of the pelvis with its long axis vertical in the angle between the external iliac vessels above and the internal iliac arteries and the ureter behind (arteries shown in Fig. 13.9). It is attached to the upper surface of the broad ligament by the mesovarium. Students often ask whether the ovary is covered by peritoneum or not. The answer is no, since the peritoneum of the mesovarium gives way to a low cubical epithelium, the *germinal epithelium*, which is not really peritoneum although it obviously forms part of the lining of the peritoneal cavity. The important thing to understand is that when the Graafian follicle enlarges and finally ruptures (gradually, it does *not* burst like a balloon) the ovum is expelled momentarily into the peritoneal cavity, before it enters the fimbriated end of the uterine tube. Beneath the germinal epithelium there is a rather thin *tunica albuginea*. The surface of the mature ovary is irregular because of the presence of maturing follicles, corpora lutea and scars of old corpora lutea.

## The vagina

The vagina passes upwards and backwards from the vestibule to the cervix of the uterus, which, owing to the anteverted position of the uterus, projects through the anterior wall of the upper part of the vagina. It is lined by stratified squamous epithelium, which undergoes some minor variations during the menstrual cycle. It is surrounded by smooth muscle in an outer longitudinal and an inner circular layer. The smooth muscle is reinforced by encircling striated muscle fibres (*bulbospongiosus*) to help to form the *sphincter vaginae*. There are no glands in the vagina, lubrication being derived from the cervical mucus.

## Blood supply

The female internal genitalia are supplied by the uterine, vaginal and ovarian arteries. The main blood supply to the uterus is from the *uterine artery* itself. This is a branch of the internal iliac artery and has a very important relationship to the ureter and lateral fornix of the vagina. It passes medially to the side of the supravaginal cervix and *above* the ureter ('water under the bridge'), which is itself passing forwards on either side of the cervix before inclining medially to open into the outer angle of the trigone (Fig. 13.8). The ureter is only a very short distance above the lateral fornix of the vagina. The *vaginal artery* usually replaces the inferior vesical artery of the male and passes below the ureter to supply the vagina. The *ovarian artery* rises from the aorta in the upper abdomen, passes from the side wall of the pelvis in the infundibulopelvic fold and anastomoses with the uterine artery in the broad ligament. It supplies the ovary and part of the uterine tube but does not do much for the uterus. During pregnancy, the uterine artery hypertrophies but the ovarian does not.

## THE VESSELS AND NERVES OF THE PELVIS

The common iliac arteries divide at the brim of the pelvis into internal and external

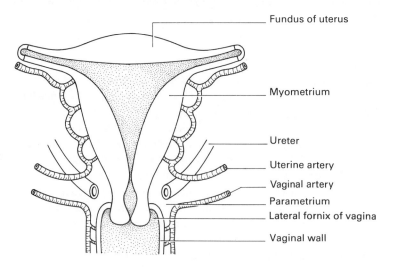

Fundus of uterus

Myometrium

Ureter

Uterine artery

Vaginal artery

Parametrium

Lateral fornix of vagina

Vaginal wall

**Fig. 13.8** The relation of two arteries and the ureter to the lateral fornix of the vagina.

iliacs (see Fig. 13.10). The latter have already been described. The internal iliac enters the pelvis by travelling downwards and backwards towards the lower part of the greater sciatic foramen. Its anterior branch continues this course but the posterior branch diverges towards the upper part of the foramen (Fig. 13.9). In the fetus, the internal iliac artery and its anterior branch are relatively huge because they continue up out of the pelvis as the umbilical artery (p. 201). When this closes down after birth, its first part remains patent as the stem of the first branch of the adult anterior division, namely the *superior vesical artery*. The *obliterated umbilical artery* therefore appears to be a 'branch' of the superior vesical artery. The next branch is usually the *obturator artery* (it must be stressed that the arrangement of the pelvic arteries is subject to a great deal of variation). This joins the obturator nerve, which has descended from the medial border of psoas along the edge of obturator internus (see Fig. 13.2) and they pass through the upper part of the obturator foramen along with the veins. The artery commonly arises from the inferior epigastric and runs down behind the pubis to reach the obturator foramen. It may lie behind the lacunar ligament and be in danger in operations for femoral hernia. The *inferior vesical* and *middle rectal* arteries arise from a common trunk and are small, and the former is usually replaced by the vaginal artery in the female. Remember that the middle rectal artery supplies only the outer coats of the rectum (p. 236). In the female, the *uterine artery* follows: its course has already been described. Finally, two terminal branches leave the pelvis through the greater sciatic notch below piriformis. These are the *internal pudendal* and the *inferior gluteal*. The former vessel only goes this way to avoid piercing the origin of levator ani (see Fig. 13.2). Having crossed the spine of the ischium in the gluteal region it immediately re-enters below the levator ani origin, i.e. it passes

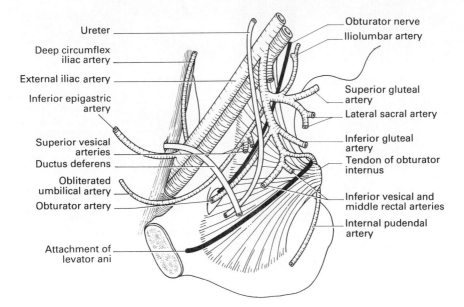

**Fig. 13.9** Structures on the side wall of the male pelvis.

into the ischiorectal fossa, where its further course will be described quite soon. The inferior gluteal leaves the pelvis between the 1st and 2nd sacral nerves and enters the gluteal region.

The posterior branch of the internal iliac artery breaks up into three non-visceral branches, which leave the pelvis respectively by climbing up over its brim, by entering the sacral foramina and by going through the greater sciatic foramen. The *iliolumbar artery* ascends in front of the sacrum (Fig. 13.10) and anastomoses with the lumbar arteries. The *lateral sacral* may be duplicated. It enters the sacral foramina to supply the contents of the sacral canal. The *superior gluteal* leaves the pelvis between the lumbosacral trunk and S1.

The *ductus deferens* (or the round ligament of the uterus in the female) and the ureter may now be superimposed on the vessels since they lie immediately subjacent to the covering peritoneum. The ureter crosses the external iliac vessels and the forward-running branches of the internal iliac artery, lying close to the internal iliac itself at first (see Fig. 13.9). It then inclines medially, passing just above the lateral fornix of the vagina in the female, and enters the bladder. The ductus deferens enters the abdomen at the deep inguinal ring, crosses the external

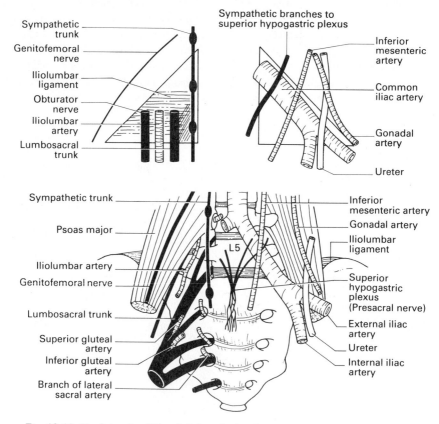

**Fig. 13.10** The 'triangle of Marcille', described in the text.

iliac artery by passing lateral to the inferior epigastric artery (see Fig. 9.7) and then crosses the obliterated umbilical artery, the obturator artery and finally the ureter (see Fig. 12.7).

The ovary, in the nulliparous woman (i.e. before the first pregnancy), lies in the ovarian fossa, which is bounded by the external iliac vessels and the obliterated umbilical artery above and in front and the ureter, backed up by the internal iliac vessels, behind. The obturator nerve and vessels are lateral to it.

### The triangle of Marcille

You are now in a position to appreciate this useful triangle (Fig. 13.10), which gives a helpful insight into the relations around the 5th lumbar vertebra. It is bounded by the side of this vertebra medially, the medial border of psoas major laterally and the upper border of the sacrum below. In the floor of the triangle are the *iliolumbar* and *lumbosacral ligaments*. In front of this there is an artery between two nerves, namely the *iliolumbar artery* as it climbs out of the pelvis with the *obturator nerve* laterally and the *lumbosacral trunk* medially. The triangle is framed by two nerves, the *sympathetic trunk* on the vertebral body medially and the *genitofemoral nerve* on the anterior surface of psoas laterally. Also lateral to the triangle are the *gonadal vessels*. In front of the triangle on the left lie the *common iliac artery and vein*; they bifurcate at the lower outer angle of the triangle into the internal and external iliac vessels. There is a slight difference between the two sides here; the right common iliac artery crosses the body of the vertebra and both common iliac veins overlie the triangle. Also, on the left side only, the common iliac vessels are crossed by the *inferior mesenteric/superior rectal* vessels. The *ureter* crosses the external iliac artery on both sides.

## THE PERINEUM

Levator ani has been described as forming the floor of the pelvis but since it is attached quite high up on the obturator internus (see Fig. 13.2) there is obviously a space below it, bounded laterally by the lower part of obturator internus. This space is known as the *ischiorectal fossa*, although, as will be described later, this name is often reserved for the posterior part of the space on either side of the rectum (hence the name of the fossa). Although students often find difficulty in understanding this region, it is quite easy as long as you understand that the fossa is not only between the ischium and rectum but extends forwards right up to the back of the body of the pubis.

### Bones and ligaments

The skeletal basis of this region is the *outlet of the pelvis*. It is bounded anteriorly by the symphysis, the inferior ramus of the pubis and the ramus of the ischium. Posteriorly are the lower part of the sacrum and the coccyx, which are connected to the tuberosity of the ischium by the *sacrotuberous ligament* (p. 252). The outlet is thus diamond-shaped (Fig. 13.11) and can be subdivided into two triangles, the *anal triangle* and the *urogenital triangle*. The superficial part of the whole region is called the *perineum* and includes the external genitalia (in the urogenital triangle) and the anus (in the anal triangle). It is also called the *pudendal region*

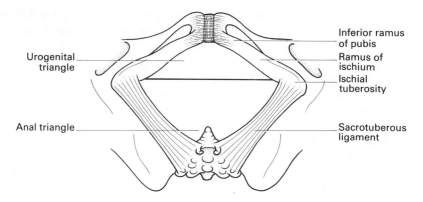

**Fig. 13.11** The outlet of the pelvis.

and is principally supplied by the pudendal nerves and vessels. ('Pudere', to be ashamed. Hence impudent, meaning unashamed, and the old name for the middle finger, *pudicus*, because it is the natural finger with which to scratch the perineum.)

## Muscles of the region

Obturator internus and levator ani have already been described but we are now looking at them from a different viewpoint. Since the two levator ani muscles are converging towards the midline, to blend in with the perineal body and the anal sphincters, there is a wedge-shaped space, edge uppermost, between levator ani and obturator internus or, strictly speaking, between the obturator fascia and the fascia covering the under-surface of levator ani and the outer surface of the sphincter ani externus (Fig. 13.12). This space is the ischiorectal fossa and its

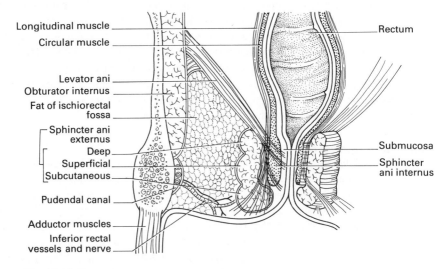

**Fig. 13.12** The ischiorectal fossa in coronal section.

principal content is fat. It is important surgically because an abscess here (ischiorectal abscess) may rupture internally into the anal canal and externally on to the skin so that a *fistula* (abnormal communication between two cavities or between a cavity and the outside world) is formed. Infection makes these difficult to treat. *If you look at Fig. 13.2 you will see that both of these muscles extend forwards to the back of the body of the pubis, so the ischiorectal fossa must have a similar disposition.*

### Sphincter ani externus and internus

The former is the striated muscle that surrounds the *anal canal* (contrary to traditional medical student usage, these two words do *not* rhyme) and, together with the smooth muscle of the *sphincter ani internus* (a thickening of the circular muscle of the intestine), are responsible for the maintenance of continence. John Hilton (of Hilton's law, p. 11) referred to it, justly, as 'that gallant and indomitable little sphincter'. It is traditionally described as consisting of three parts, *subcutaneous, superficial* and *deep*, and as such it will be described here, even though sometimes it is not easy to identify clearly these subdivisions which may merely be the result of anatomists' passion for neatly demarcated areas or zones (such as the anal and urogenital triangles mentioned above).

The internal sphincter of circular smooth muscle extends downwards to enclose the upper three-quarters of the anal canal and is itself enclosed by the greater part of the sphincter ani externus (Fig. 13.12). The *subcutaneous part* of the external sphincter surrounds the lower end of the anal canal just below the lower edge of the internal sphincter. The *superficial part* is more fusiform and is attached to the coccyx behind and the perineal body in front. The *deep part* again is circular in form and is continuous with the puborectalis part of levator ani. These two masses of striated muscle, along with the smooth muscle of the internal sphincter form the *anorectal ring*. This is easily palpable with a finger in the canal and it forms a firm ring immediately above which the finger enters the rather dilated ampulla of the rectum. The anorectal ring must not be divided or the result will be incontinence. The other palpable landmark in the anal sphincter is the *intersphincteric groove*, a depression that can be felt in between the lower end of the internal sphincter and the subcutaneous portion of the external sphincter.

Between the sphincter ani internus and its enclosing collar of the external sphincter is a longitudinal coat. This is a continuation downwards of the longitudinal muscle coat of the rectum together with some fibres from puborectalis, but it becomes more fibrous, with elastic tissue in it, as it descends. Some of its strands pass inwards through the internal sphincter to form a well-marked layer in the submucosa. It finally ends by splitting into laminae that spread through the subcutaneous part of the external sphincter, forming a number of compartments, which is why infections and other peri-anal conditions are so intensely painful.

### Other muscles

The remaining muscles of the perineum are small and are concerned mainly with the external genitalia and the urethra. They will therefore be described with the urogenital triangle.

## The ischiorectal fossa

The ischiorectal fossa is usually described as the space between levator ani, together with the sphincter ani internus, medially and obturator internus and the ischium and pubis laterally. However, as has been mentioned above, it has a prolongation forwards that is deep to the *urogenital diaphragm*, a double-decker sandwich which will be described later. The main part of the fossa is limited posteriorly by the sacrotuberous ligament and the origin of the gluteus maximus from this ligament. Its main content is fat and it is crossed by the inferior rectal vessels and nerve, which leave the side wall to be distributed to the anal canal (not the rectum). These are branches of the internal pudendal vessels and the pudendal nerve, which lie enclosed in a tunnel of fascia known as the *pudendal (Alcock's) canal*. The fascia is continuous with that covering obturator internus.

## The urogenital triangle

The most important structure to understand in this region is the *perineal membrane*, which you will find surgeons call the *triangular ligament* and up-to-date anatomists call the *inferior fascia of the urogenital diaphragm*. Call it what you will, it is a very strong triangular membrane that stretches across the urogenital triangle from the rami of the pubis and ischium at the sides (see Fig. 13.2). It is thus rather similar in disposition to the mylohyoid muscle that forms the floor of the mouth (Fig. 13.13 and p. 337). There is a small gap at the apex of the triangle behind the pubic symphysis, through which pass the *deep dorsal vein* and the *dorsal nerve of the penis*. The perineal membrane is also pierced by the urethra, the vagina in the female, the ducts of the bulbo-urethral glands and the deep and dorsal arteries of the penis (or clitoris).

You will remember (p. 197) that the membranous layer of superficial fascia of the abdomen surrounds the penis (as far as the corona) to form the tube-like *fascia penis*. It also extends downwards into the thigh to fuse with the fascia lata a short distance below the inguinal ligament and then backwards to surround the scrotum,

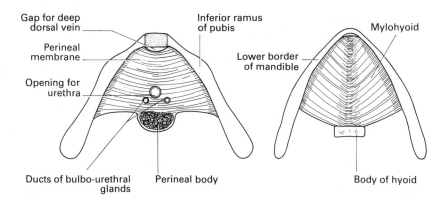

**Fig. 13.13** The perineal membrane (inferior fascia of the urogenital diaphragm) fills in the space between the pubic bones in much the same way as the mylohyoid fills in the space between the two sides of the mandible.

and finally, in the perineum, it fuses with the posterior edge of the perineal membrane (see Fig. 13.2). The perineal part of this layer is called *Colles' fascia*. There is thus formed a *superficial perineal pouch*, bounded above by the perineal membrane and below by *Colles' fascia* (see Fig. 13.16). It is closed off posteriorly by the continuity of these two layers. It contains the roots of the penis and some muscles but before describing these, a brief description of the penis itself becomes necessary.

### The penis (Fig. 13.14)

This consists essentially of three components, the *corpus spongiosum* and the two *corpora cavernosa*. The latter lie close together, being enclosed in a common fibrous sheath, but they are separated by a septum. The corpus spongiosum is enlarged at the end to form the *glans*, and the *spongiose* or *penile urethra* runs through it to open at the external urethral orifice. The whole is enclosed in a condensation of superficial fascia, the *fascia penis*, and the weight of the penis is partly supported by a *suspensory ligament*, which blends with the fascia penis and is attached to the pubic symphysis. The *superficial dorsal vein* is in the midline and, since it drains the skin, is outside the fascia penis. It drains into a tributary of the great saphenous vein on one or other side. The *deep dorsal vein*, also in the midline, drains the deep structures and so is within the fascia penis. It passes through the gap in the front of the perineal membrane and joins the prostatic venous plexus. The single vein is flanked by the paired deep dorsal arteries and nerves in the usual order (NAV, p. 116). The skin is thin and loose and a fold passes over the glans to form the *prepuce*.

### The superficial perineal pouch

The largest structures here are the roots of the penis, namely the continuation backwards of the corpora cavernosa, which are known as the *crura*, and the continuation of the corpus spongiosum, which is the *bulb of the penis* (Fig. 13.15). The crura are attached firmly to the inferior ramus of the pubis and the ramus of the ischium, while the bulb is attached to the under-surface of the perineal

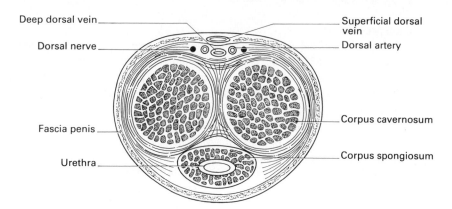

**Fig. 13.14** Cross-section of penis.

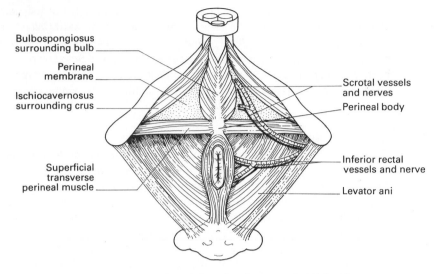

Fig. 13.15 The perineum seen from below after the removal of Colles' fascia.

membrane. The urethra has to pierce the membrane in order to enter the bulb. The three components of the penis consist of erectile tissue, i.e. tissue containing large cavernous vascular spaces into which arteries open. The vascular arrangements are not fully understood and, although erection of the penis is often said to be a purely vascular phenomenon, it is difficult to believe this when the pressure in the cavernous spaces during erection is higher than the systolic blood pressure.

Each crus is surrounded by an *ischiocavernosus* muscle and the bulb by *bulbospongiosus*. The *transversus perinei superficialis* muscle extends from the ischium to the perineal body. Bulbospongiosus helps in expelling the last drops of urine from the urethra during micturition and also contributes to ejaculation but the other functions of these muscles are uncertain.

The superficial perineal pouch also contains the *scrotal* (or *labial*) *vessels* and *nerves*, which will be described later.

## The deep perineal pouch

This is bounded below by the perineal membrane and above by a rather indefinite layer of fascia called the *superior fascia of the urogenital diaphragm* (Fig. 13.16). This, like Colles' fascia, is continuous with the posterior edge of the perineal membrane and is attached to the side wall of the pelvis (see Fig. 13.2). The deep pouch contains the *membranous urethra,* a short, narrow and not very dilatable portion of the urethra. This is surrounded by the *sphincter urethrae*, which is composed of striated muscle. It is not, however, a very well-marked muscle, particularly in the female, where it may sometimes be difficult to find at all, and the main skeletal muscle in this region (and in the prostatic urethra) is incorporated into the wall of the urethra itself. Also in the deep perineal pouch are the *bulbo-urethral (Cowper's) glands*. Their ducts pierce the perineal membrane to open into the urethra in the bulb of the penis. *Transversus perinei profundus* lies in

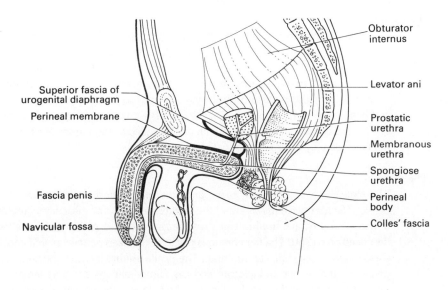

**Fig. 13.16** Midline sagittal section of the male pelvis.

the deep pouch in a position corresponding to that of superficialis in the superficial pouch. Various nerves and vessels in this region will be described later.

### The female perineum

There are many points of similarity between the male and female perineum, which may be appreciated from their development. In the early embryo, the region has a pair of *genital swellings*, within which are a pair of *urethra folds* that lie on either side of the opening into the *urogenital sinus*, which will later form the bladder and urethra. In the male, the genital swellings form the scrotum, after the descent of the testes, while the urethral folds are 'zipped up' from behind to form a floor for the penile urethra. The external urethral orifice is thus transferred from the perineum to the base of the *genital tubercle*, a swelling lying anteriorly. The latter enlarges greatly to form the penis and the urethral orifice is finally transferred to its tip. In the female, the genital swellings form the labia majora and the urethral folds remain 'unzipped' to form the labia minora. The genital swelling forms the clitoris, which has similar components to the penis but is much smaller and does not contain the urethra.

Thus the opening between the labia minora leads into the *vestibule*, a shallow cavity into which open the urethra and the vagina. The opening of the vagina is guarded, rather half-heartedly, by the *hymen*, a thin fold of mucous membrane. The *bulb of the vestibule* is equivalent to the bulb of the penis but is in two parts that lie on either side of the vestibule. The equivalent of the bulbo-urethral glands are the *greater vestibular glands* (Bartholin's glands). These lie close to the posterior ends of the bulbs of the vestibule and open into the vestibule just below the hymen.

## Hypospadias

Developmental errors in the 'zipping up' process are not uncommon. The simplest form is a failure of the final stage of transfer of the orifice to the tip of the glans so that the urethra opens on the under-surface of the glans and the meatus is often narrowed. In more severe forms there may be one or more urethral orifices further back in the perineum ('watering-can urethra') while if the 'zipper' fails completely the opening of the urogenital sinus remains wide open. If, as often occurs, this condition is associated with undescended testes, the sex of a baby may not be evident by external examination.

## The urethra

The male urethra is in three parts, *prostatic, membranous* and *spongiose*, or penile. The narrowest part is either the membranous part or the external urethral orifice. The prostatic urethra traverses the prostate and its features have been described on p. 260. The membranous part is in the deep perineal pouch and it is surrounded by the sphincter urethrae. The penile urethra is within the bulb of the penis and the corpus spongiosum and has two dilatations along its length, the *intrabulbar fossa* at the proximal end of the bulb and the *navicular fossa* in the glans. In the anterior part of the urethra there are a number of shallow diverticula, called *lacunae*, into which open urethral glands. The male urethra is obviously exposed to injury in the perineum and it is interesting to follow the spread of extravasated urine if the urethra is ruptured, for example by falling astride a narrow object. It will pass into the superficial pouch but cannot extend backwards because of the attachment of Colles' fascia to the perineal membrane (Fig. 13.16). It passes forwards, deep to Colles' fascia and will therefore infiltrate the scrotum, the penis as far as the neck, and the anterior abdominal wall deep to the membranous layer of superficial fascia. It will pass down into the thigh for a short distance before being stopped by fusion of the membranous layer with the fascia lata.

The female urethra is relatively short (4 cm) and simple. It runs behind the pubic symphysis, embedded in the anterior wall of the vagina, and opens about 2 cm behind the clitoris. It contains in its wall a number of para-urethral glands (p. 260).

## The urogenital diaphragm

Having described the structures in the urogenital triangle in some detail, it remains to fit the urogenital system into the bony and muscular pelvis. This is easily done if the complex urogenital diaphragm is regarded as a triangular, double-decker sandwich (Fig. 13.17). The bottom layer is Colles' fascia and the middle layer the perineal membrane. Between the two are the roots of the penis (or clitoris) and their associated muscles and vessels. The topmost layer is the superior fascia of the urogenital diaphragm and between it and the perineal membrane are the membranous urethra, the bulbo-urethral glands and the sphincter urethrae, such as it is. This whole sandwich may now be inserted into the urogenital triangle with its apex behind the symphysis pubis and its sides attached to the pubis and ischium.

Figure 13.12 shows a coronal section through the anal triangle. To convert this to a section through the urogenital triangle you have only to erase the rectum

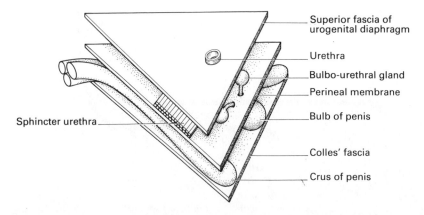

**Fig. 13.17** The three layers of the urogenital diaphragm, with the superficial and deep perineal pouches between them.

and anal canal and insert the urogenital sandwich (Fig. 13.18). A slight alteration also needs to be made to the bony wall of the pelvis. You will now understand how the ischiorectal fossa is continued forwards above the urogenital diaphragm and on either side of the prostate.

## The vessels and nerves of the perineum

We now have to return to the gluteal region. Emerging from the lower end of the greater sciatic notch are, from medial to lateral, the *pudendal nerve*, the *internal pudendal artery* and the *nerve to obturator internus*. All three cross the spine

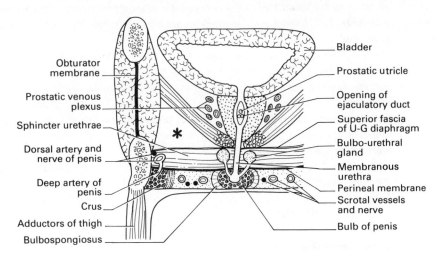

**Fig. 13.18** Figure 13.12 is a coronal section through the anal triangle of the perineum. This is a similar section further forward, through the urogenital triangle. Compare the two. The asterisk marks the forward prolongation of the ischiorectal fossa.

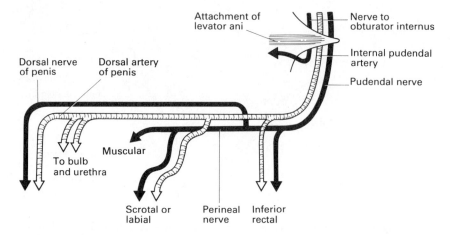

**Fig. 13.19** A comparison of the branches of the pudendal nerve and the internal pudendal artery. They enter the region by crossing the spine of the ischium, along with the nerve to obturator internus. *Note:* There are three places where an artery lies between two nerves in relation to the hip bone. Two of these are shown in this diagram and the third in Fig. 13.10.

of the ischium, thereby bypassing the origin of the levator ani, and enter the perineum through the lesser sciatic notch. The nerve to obturator internus immediately plunges into that muscle, while the other nerve and the vessels enter the pudendal canal.

Both nerve and artery give an *inferior rectal* branch that leaves the pudendal canal, passes medially through the ischiorectal fossa and supplies the lower part of the anal canal and its associated muscles. The nerve is mixed, supplying sensory branches to the lower half of the anal canal and motor branches to the sphincter ani externus and levator ani. The muscles are also innervated by a separate branch from S4.

The pudendal nerve then divides into the *dorsal nerve of the penis* (clitoris) and the *perineal nerve*, which lie on either side of the artery (Fig. 13.19). The perineal nerve gives *scrotal* (or *labial*) branches, and *muscular branches* to the muscles of the region. The dorsal nerve of the penis (clitoris) enters the deep perineal pouch, keeping to its lateral wall, and finally pierces the perineal membrane and passes on to the dorsum of the penis (clitoris). The internal pudendal artery gives *scrotal* (or *labial*) *branches* and then enters the deep perineal pouch. It gives branches to the bulb of the penis (clitoris) and the urethra and ends, still in the deep pouch, by dividing into the *dorsal* and *deep arteries* of the penis (clitoris). These pierce the perineal membrane and pass respectively on to the dorsum of the penis and into the crus.

### Vaginal and rectal examination

Never insult the vagina by putting a finger in the rectum first. Vaginal examination always precedes rectal examination because of the possibility of transferring infection, and a knowledge of the structures that can be felt is largely a matter of applied anatomy. The vagina is directed upwards and backwards and in the angle between it and the floor of the perineum the *perineal body* can be felt between the finger and a thumb outside. At the top of the vagina the intravaginal cervix is easily felt. The cervix is directed backwards if the uterus is normally anteverted so that the anterior fornix is much shallower than the posterior. Through the anterior fornix can be felt the supravaginal part of the cervix and the body of the uterus. They can be felt more clearly in a bimanual examination in which one hand is placed on the lower abdomen. It may then be possible to feel the ovary and the tubes in a thin person. If the bladder is full, it may be felt through the anterior wall of the vagina, while the sacral promontory may be palpable through the posterior fornix.

A finger in the anal canal can palpate the intersphincteric groove and the anorectal ring (p. 269). Above the latter the finger enters the rectum. Through its anterior wall can be felt the prostate in the male and the cervix of the uterus in the female. Above the prostate the finger will be in relation to the base of the bladder and the seminal vesicles, but these structures cannot be palpated *per se* unless they are diseased. Lower down is the bulb of the penis and a finger in the rectum may be helpful in turning the tip of a catheter upwards round the corner into the membranous urethra. Laterally, in the female, the uterosacral ligaments are palpable, and in both sexes (perhaps) the ischial spine and tuberosity. Posteriorly is the concavity of the sacrum. Many other structures may be felt *per rectum* or *per vaginam* when they are diseased, for example enlarged lymph nodes or seminal vesicles, and, in particular, tenderness may be elicited in the presence of inflammation. A pelvic abscess in the recto-uterine or rectovesical pouch may be felt by either type of examination. Thus these examinations are often essential in making a diagnosis of pelvic conditions — 'If you don't put your finger in, you may put your foot in it.'

## LYMPHATICS OF THE ABDOMEN
Now that the anatomy of the abdomen has been completed, the lymphatics can be described.

### Cisterna chyli
This marks the beginning of the thoracic duct. It is a thin-walled sac lying in front of the bodies of the 1st and 2nd lumbar vertebrae. It receives lymph from the abdomen and lower limbs and, from it, the thoracic duct ascends into the thorax through the aortic opening of the diaphragm.

### The lymph nodes
These are closely related to the main arteries so you will find it helpful to revise these first. There are two main groups of aortic nodes, the *pre-aortic*, which are related to the three ventral branches of the aorta, and the *lateral aortic* (or para-aortic), which are related to the lateral branches. There are also some *retro-aortic* nodes.

### Pre-aortic nodes

The *coeliac nodes* around the stem of the coeliac artery drain lymph from outlying groups related to the main arteries of the region. They are therefore called the *left gastric*, (pancreatico-)*splenic* and *hepatic*, with further groups around the gastroduodenal branch of the hepatic artery, namely *pyloric* and *right gastro-epiploic* (Fig. 13.20). The lymph drainage of the stomach is best understood if the stomach is divided into four sectors that drain to the nearest appropriate group (Fig. 13.20). In general, the direction of lymph flow is from right to left along the lesser curvature and left to right along the greater curvature. The major part of the liver drains to the hepatic group but some of those of the upper surface drain to some nodes around the termination of the inferior vena cava and to the left gastric group (from the left lobe). The pancreas drains mostly into the splenic nodes but some enter the superior mesenteric nodes.

The *superior* and *inferior mesenteric* nodes also have outlying groups along their main branches, and the lymph drainage of the various parts of the gut follow the appropriate arteries of supply. The lymph from the lower part of the rectum and the upper part of the anal canal passes laterally with the middle rectal artery to the *internal iliac nodes*, but the lower part of the anal canal drains, of course, to the superficial inguinal nodes. From the pre-aortic nodes, lymphatics converge to form a number of *intestinal trunks* that open into the cisterna chyli.

### The lateral aortic nodes

These lie on either side of the aorta and some are therefore closely related to the inferior vena cava. Lower down, there are *iliac nodes* (common, internal and external), along with some outlying groups along their branches, including *sacral nodes* in the concavity of the sacrum.

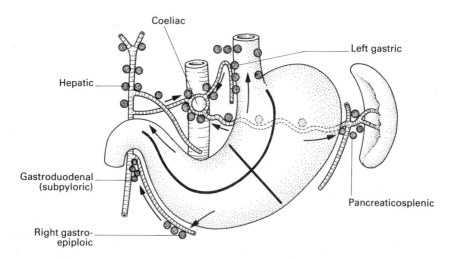

**Fig. 13.20** For the purpose of studying its lympathic drainage the stomach is divided into four quadrants by the thick black lines.

The kidneys and suprarenals drain into the local lateral aortic nodes, as do the gonads, via lymphatics that follow the course of the gonadal vessels. The lymph from the bladder and the lower end of the ureter, the ductus deferens and the seminal vesicles goes mainly to the external iliac nodes but also the internal iliacs. The prostate drains to the internal iliac and sacral nodes.

The cervix of the uterus drains to the external and internal iliacs and the sacral nodes. The body of the uterus drains into the external iliac nodes and upwards directly to the lateral aortic and pre-aortic nodes. Some lymphatics accompany the round ligament to the superficial inguinal nodes.

The deeper parts of the abdominal wall have lymphatics that follow the lumbar vessels to the lateral aortic and retro-aortic nodes; others run upwards to the parasternal and downwards to the external iliac nodes.

# Chapter 14
# The Spine

The spine is made up of a number of vertebrae (seven cervical, twelve thoracic, five lumbar, five sacral and three to five coccygeal), which are held together by ligaments, intervertebral discs, and synovial joints between the articular processes. Although the movement between any two vertebrae is slight, movement in the spine as a whole is over quite a wide range. Just as typical spinal nerves are found in the thoracic region, so the thoracic vertebrae form a good basis for the description of vertebrae of other regions.

## THORACIC VERTEBRAE

Figure 8.1 shows a thoracic vertebra from the middle of the series. The body is often described as 'heart-shaped', meaning a playing card heart rather than a human heart. Do not use the term 'centrum' as being synonymous with 'body' (see p. 284). The body is the weight-bearing part of the vertebra, the neural arch and its processes being merely for the protection of the spinal cord and its nerves, for the attachment of muscles, for articulation with the ribs and for regulating the types of movement that can occur in any region. One might add also that the spines of the vertebrae make useful landmarks (p. 288). The internal structure of the body reflects its weight-bearing function, since the trabeculae of cancellous bone are appropriately arranged to give the maximum strength. The vertebral bodies are also one of the sites of red bone marrow so they are drained by large veins, the *basivertebral veins*, which occupy the large foramina in the back of the bodies.

The other components of the neural arch of the vertebrae are shown in Fig. 8.1. The transverse processes in the thoracic region are long, for articulation with the ribs, and the spines are also long and downturned so that each overlaps that of the vertebra below. The articular facets form a synovial joint with the facets of the vertebrae above and below. The most positive way of identifying a vertebra as thoracic, however, is the presence of facets for the ribs. As has been described in Chapter 8, a typical rib articulates by means of its tubercle with the transverse process of its own vertebra and with the two demifacets on the bodies of its own vertebra and the one above. The inferior articular facets of the 12th thoracic vertebra are lumbar in type, which makes this vertebra easy to identify.

Note particularly that the superior articular facets face backwards and slightly outwards and their plane lies on a circle centred on the vertebral body (see Fig. 14.10). The significance of this will be discussed later.

## CERVICAL VERTEBRAE

The body is smaller than that of thoracic vertebrae (not surprisingly, since it has less weight to carry) and it overlaps the vertebra below anteriorly. The spine is smaller and bifurcate. The most obvious feature, however, is the transverse process, which has *anterior* and *posterior tubercles* and a *foramen transversarium*. This is easy to understand if you remember that all vertebrae, not only the thoracic, have 'ribs' attached to them. In the cervical, lumbar and sacral vertebrae, however, the ribs, or *costal elements*, are incorporated into the vertebra. In the lumbar and sacral regions this costal element is merged with the transverse process and cannot be easily recognized, but in the cervical region you can recognize the articulations between the head of the 'rib' and the vertebral body and between the tubercle and the true transverse process (Fig. 14.1). It is important to understand this because occasionally the costal element of the 7th cervical vertebra is enlarged to form a *cervical rib*, which may cause vascular and possibly nervous symptoms and signs (see Chapter 18).

The 7th cervical vertebra is intermediate in type and does not have anterior and posterior tubercules on its transverse process, although it does have a foramen transversarium. The 1st and 2nd cervical vertebra are completely atypical since they are modified to support the skull and to control its movements. The atlas is not much more than a ring of bone, its centrum having fused during development with the centrum of the axis to form the *dens*. This therefore projects upwards through the ring of the atlas and is held in place by a *transverse ligament* (Fig. 14.2). Do not worry about the significance of the anterior arch of the atlas,

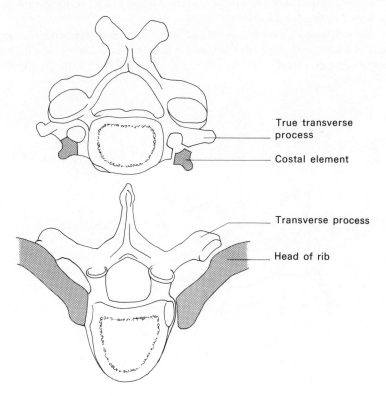

True transverse
process

Costal element

Transverse process

Head of rib

**Fig. 14.1** Showing how the foramen transversarium corresponds to the space between the neck of a rib and the vertebra.

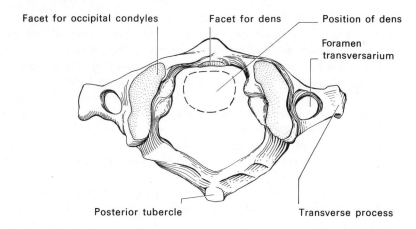

Facet for occipital condyles    Facet for dens    Position of dens

Foramen
transversarium

Posterior tubercle    Transverse process

**Fig. 14.2** The upper surface of the atlas.

it is beyond the scope of this book. It is a little difficult to tell which is the upper surface of the atlas. Look at the shape of the articular facets. On the upper surface they are oval or kidney-shaped for articulation with the occipital condyles but on the under-surface they are more or less circular. Look also for the groove on the upper surface for the vertebral artery.

The old name for the axis was the *os chelonii* because of its resemblance to the head of a tortoise. Look at the dens and you will see how appropriate the name was, much better than axis. The axis has a massive spine for the attachment of

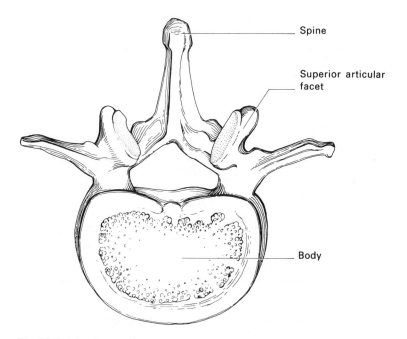

Spine

Superior articular
facet

Body

**Fig. 14.3** A lumbar vertebra.

muscles but the atlas has merely a posterior tubercle. A spine would interfere with extension of the skull such as takes place on looking up at the sky. On the other hand, the atlas has particularly long transverse processes. They act as levers for the muscles that turn the head.

## THE LUMBAR VERTEBRAE

These are the most massive of all the vertebrae (Fig. 14.3) and their bodies increase in size as they are followed downwards. The spines are large and project backwards; since they do not overlap like the thoracic spines, it is easy to insert a needle between them in the operation of *lumbar puncture*, described below. The transverse processes are also strong and in the 5th lumbar vertebra the transverse process merges with the body rather than the pedicle, so it is easy to identify. The superior articular facets face medially and slightly backwards.

## THE SACRUM

There are normally five sacral vertebrae fused together to form one composite bone. It is completed below by the coccyx. With five sacral vertebrae, there will be

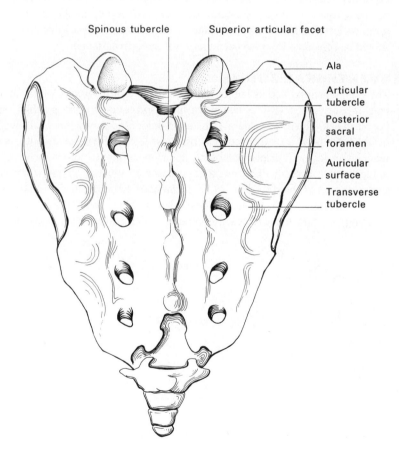

**Fig. 14.4** The sacrum and coccyx, posterior view.

four intervertebral foramina between them but, since the vertebrae are fused, these take the form of anterior and posterior *sacral foramina*, which transmit the anterior and posterior rami of the sacral nerves (Figs 13.10 & 14.4). The spines, articular facets and transverse processes of the sacral vertebrae are represented by tubercles on the posterior surface. The lower end of the continuous vertebral canal (*sacral canal*) is open and is called the sacral hiatus. This provides a route for the introduction of anaesthetic into the vertebral canal. At the sides, the sacrum has an auricular surface for articulation with the ilium (p. 103).

## DEVELOPMENT OF THE VERTEBRAE

The vertebrae ossify in cartilage from three primary centres, one for the *centrum* and one for each *neural arch*. At birth, each vertebra consists of a bony centrum with two neural arches, held together by cartilage (Fig. 14.5). Note the expanded lower ends of the neural arches, which form the lateral parts of the future body when the cartilage later disappears. The various forms of spina bifida are due to failure of the mesoderm that precedes cartilage to grow over the neural tube to form a neural arch so that in extreme cases an open neural tube may remain exposed on the surface. In the simplest form, however, there is just a gap between the two halves of the bony neural arch and an absence of the spine. This is symptomless and hence is called *spina bifida occulta*. Textbooks of embryology should be consulted for a full account of vertebral development.

## INTERVERTEBRAL JOINTS

The weight-bearing bodies are joined by secondary cartilaginous joints, i.e. each body is covered by a layer of hyaline cartilage and between them is a thick disc of fibrocartilage. The intervertebral disc has an outer fibrous ring (*annulus fibrosus*), in which the fibres are arranged in a criss-cross fashion for extra strength, and an inner softer region (*nucleus pulposus*), rich in glycosaminoglycans, which ensure a high water content. The nucleus pulposus is situated rather towards the posterior part of the disc (Fig. 14.6). In the upright position, the weight of the body compresses the disc (you are actually less tall at the end of the day than in the morning) so that the nucleus pulposus is under considerable pressure. With

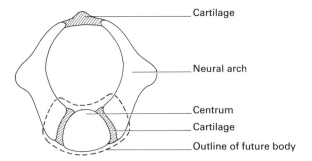

**Fig. 14.5** A vertebra at birth. The future body (dotted outline) is made up of the centrum plus the bases of the neural arches.

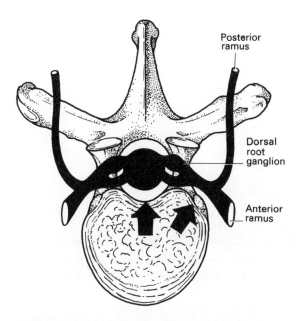

Posterior
ramus

Dorsal
root
ganglion

Anterior
ramus

**Fig. 14.6** The arrows indicate how a disc herniation posteriorly may affect the spinal cord while a posterolateral herniation may affect a spinal nerve.

increasing age, the water content of the discs diminishes and degenerative changes occur so that the nucleus pulposus may herniate through a defect in the annulus fibrosus. Owing to the position of the nucleus, herniation (or prolapse) of the nucleus usually occurs backwards, and Fig. 14.6 shows how this may press on a spinal nerve or on the spinal cord. Herniation may also occur in a vertical direction, eroding the body of an adjacent vertebra to produce a *Schmorl's node*. A 'slipped disc' does *not* mean that a whole disc has slipped out of place.

As well as the discs, the vertebrae are held together by ligaments. The easiest way to remember these is to realize that there are two ligaments per component of the vertebra, i.e. paired components are each joined by one ligament while unpaired components are joined by two. Thus the spines are joined by two ligaments, *supraspinous* and *interspinous* (Fig. 14.7). The bodies are joined by two ligaments, *anterior* and *posterior longitudinal*, the latter being within the vertebral canal. The paired articular processes are each united to their partner by a *capsular ligament* (since these are synovial joints), the transverse processes are united by *intertransverse ligaments* and the laminae are united by the *ligamenta flava* ('flavum' means yellow, since this ligament contains much yellow elastic tissue). There are, of course, no ligaments joining the pedicles, since if there were the spinal nerves would not be able to emerge.

The atlanto-axial joint is quite different since the atlas does not have a body. The dens articulates with the anterior arch of the atlas by a plane synovial joint and is held in place by a *transverse ligament*. There is also a pair of articular facets. The atlanto-axial joint is a pivot joint, the atlas rotating around the dens and carrying the head with it. The atlanto-occipital joint is between the kidney-shaped facets on

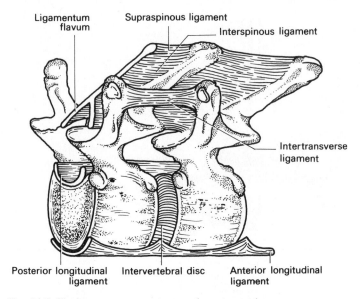

Ligamentum flavum

Supraspinous ligament

Interspinous ligament

Intertransverse ligament

Posterior longitudinal ligament

Intervertebral disc

Anterior longitudinal ligament

**Fig. 14.7** The ligaments connecting two thoracic vertebrae.

the upper surface of the atlas and the occipital condyles. Nodding movements (i.e. flexion and extension) take place here. The anterior longitudinal ligament is attached to both axis and atlas and is continued upwards to be attached to the base of the skull (occipital bone). The posterior longitudinal ligament is similarly continued upwards to cover the posterior surface of the dens and to be attached to the occipital bone inside the skull as the *membrana tectoria*.

## THE SPINE AS A WHOLE

The spine has a series of curves that help to maintain the centre of gravity in a position above the feet. These are best understood by studying their development (Fig. 14.8). Although there are faint signs of the spinal curves before birth, the

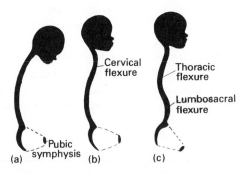

Cervical flexure

Thoracic flexure

Lumbosacral flexure

Pubic symphysis

(a)  (b)  (c)

**Fig. 14.8** Showing the development of the curvatures of the spine. (a) Fetal condition. (b) Development of the cervical flexure. (c) Development of the lumbosacral flexure and forward tilting of the pelvis.

newborn baby's spine is curved throughout, with a forward concavity. When lifting a baby, it is important to support the head, otherwise it lolls about alarmingly; a baby is quite unable to balance its head on top of the spine. A few months later the baby begins to hold its head up and look around and it is at this stage that the forward convexity of the cervical spine appears. Some time before its first birthday, the baby sits up and begins to stand, so that the development of another secondary curve in the lumbar region is necessary for balance. The primary curves, in the thoracic and sacral regions, are mainly due to the shape of the vertebrae and the secondary curves to the intervertebral discs being slightly thicker in front than behind. The discs are thicker in the regions of secondary curves so that it is in the cervical and lumbar regions that the greatest movement is permitted. Movement in the thoracic region is limited by the presence of the ribs.

Movements of the spine are shown in Fig. 14.9. They are *flexion* and *extension, side* or *lateral flexion* and *rotation*. A short-sighted person peering at a notice-board uses a combination of flexion of the cervical spine and extension at the atlanto-occipital joint. If you try to flex your own neck to one side, you will understand how the disposition of the articular facets in the cervical region cause this movement to be associated with rotation to the opposite side. In the thoracic region, rotation is free but in the lumbar spine it is almost non-existent because of the orientation of the articular facets (Fig. 14.10).

The curves and the movements of the spine play an important role in posture, particularly in keeping the centre of gravity in the right place and keeping the head upright. Thus, if the centre of gravity is displaced forwards, as occurs in pregnancy, an increased lumbar curve can compensate for this, just as a general flexion of the spine as a whole can help to keep the balance of the body when carrying a heavy rucksack. Similarly, side flexion of the spine accompanies the carrying of a heavy suitcase and it can also compensate for a sideways tilt of the pelvis caused, perhaps, by a discrepancy in the length of the lower limbs (see Fig. 6.4).

Flexion          Rotation          Lateral flexion          Extension

**Fig. 14.9** The movements of the spine.

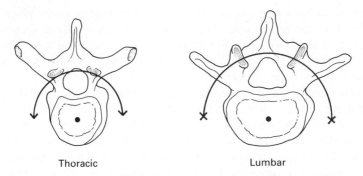

Thoracic                    Lumbar

**Fig. 14.10** The possibility of rotation depends upon the direction of the articular facets.

## Landmarks

If the trunk is flexed, the spines of most of the vertebrae can be seen or palpated, although the first six cervical spines are not accessible owing to the forward convexity of the cervical spine. The first easily palpable spine is therefore that of the 7th cervical vertebra, although the spine of T1 may be equally prominent. The *scapula* normally lies between the levels of the 2nd and 7th thoracic spines. The *oblique fissure* of the lungs begins about T4. The *supracristal plane* (a line joining the highest part of the iliac crest on each side) passes through L4, and a line joining the two posterior superior iliac spines is at the level of S2. The *spinal cord* ends (in the adult) at the lower border of L1 and the *subarachnoid space* at S2. *Lumbar puncture* is usually carried out between the 3rd and 4th lumbar spines, the needle passing through or alongside the supra- and interspinous ligaments and the dura and arachnoid, into the subarachnoid space.

## Muscles of the back

Do not, on any account, try to learn the subdivisions or attachments of the various parts of the main composite muscle of the back, *erector spinae*, except for a few that end up by being attached to the skull and will be described in Chapter 17. They are shown, for your entertainment, in Fig. 14.11 but the points you should notice are these. Erector spinae fills up the groove between the spines of the vertebrae, their transverse processes and the angles of the ribs, forming the principal edible part of a chop. The muscle starts in the sacral region and enlarges and becomes very powerful in the lumbar region. Its relationship to the thoracolumbar fascia is shown in Fig. 12.2. Higher still, it splits into columns of separate muscle bundles that are attached to spines, transverse processes and ribs. Obviously the longest bundles must be the most superficial, while the shortest, which may pass only from one vertebra to the next, are deep down in contact with bone. The large uppermost part of the muscle is *semispinalis capitis*, which, covered by trapezius, forms the thick ridge on either side of the midline at the back of the neck. The nerve supply of erector spinae is derived from the posterior rami of all the nerves.

The most important action, as its name implies, is extension of the spine, as in rising from a stooping position. Its antagonists (the flexors) are the neck flexors such as sternocleidomastoid and the scalene muscles, psoas major and the anterior abdominal muscles, especially rectus abdominis. Remember, however,

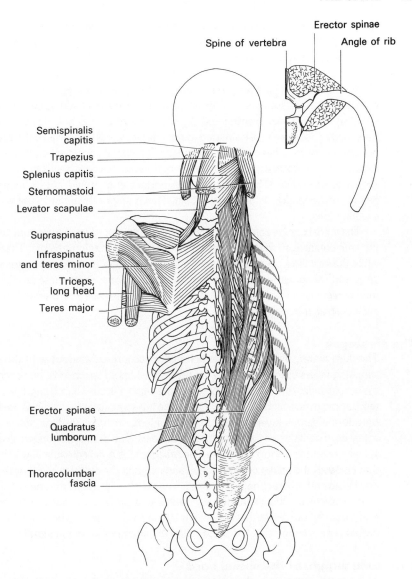

**Fig. 14.11** The muscles of the back. The components of erector spinae have been separated. The small diagram shows the anatomy of a chop.

that in bending forwards gravity is the prime mover and erector spinae supports the trunk by its eccentric action (Chapter 1). Contraction of erector spinae on one side will produce side flexion, while some of its oblique, deep components will produce rotation, acting along with the oblique muscles of the abdomen (Chapter 9).

## Spinal cord

In the fetus, the spinal cord and the vertebral column are the same length, but the latter continues to grow in length for a longer period than the former. Thus, since the upper end of the cord is anchored by the brain within the skull, the lower end retreats up the vertebral canal. At birth, therefore, the spinal cord ends at the lower border of L3 and in the adult at the lower border of L1. This also explains why the upper cervical nerves pass horizontally from the cord to the intervertebral foramina but, as one proceeds downwards, the nerve roots become more and more oblique until below L1 they pass vertically down the vertebral canal to form the *cauda equina*. Be sure you understand that the cauda equina is composed of the anterior and posterior (ventral and dorsal) *nerve roots* and not of spinal nerves. The dorsal root ganglia are situated in the intervertebral foramina so the roots do not unite until they have passed this point. The spinal cord itself ends in a slight expansion, the *conus medullaris*, and then continues as a thin fibrous strand, the *filum terminale*. This is totally unimportant and should not be mentioned first in a description of the cauda equina. This is one of the commonest errors in examinations.

The spinal cord is enlarged in the regions where the spinal nerves contribute to the limb plexuses so that there are *cervical* and *lumbar enlargements*. Therefore, while the vertebral canal is relatively small and circular in the thoracic spine, it is large and triangular in the cervical and lumbar regions. Since these are the most mobile regions of the spine, the large vertebral canals have a further advantage in allowing free movements.

### The meninges

The dura mater, pia mater and arachnoid mater are continuous with the corresponding layers inside the skull, as is the subarachnoid space. It is, however, only the serous layer of dura that is present, the outer fibrous layer being the periosteum lining the cranium (p. 376). Just as the intracranial venous sinuses are found outside the serous layer of dura in the cranium, so the extradural space in the spinal cord contains an extensive plexus of veins. The dura is prolonged along the spinal nerves for a short distance. The arachnoid is a delicate tube-like structure that encloses the spinal cord and the cauda equina down as far as S2, where the arachnoid and subarachnoid space end. The pia closely invests the spinal cord and forms sheaths for the spinal nerves. Laterally, it projects outwards in a series of sharp pointed processes that anchor the spinal cord to the dura between the spinal nerves. This toothed sheet of pia is called the *ligamentum denticulatum*.

### The blood supply of the spinal cord

The arteries that supply the spinal cord and cauda equina are derived from the spinal branches of the vertebral, intercostal, lumbar and lateral sacral arteries but these are not all the same size. Each divides into dorsal and ventral branches that follow the corresponding nerve roots and contribute to the longitudinal *anterior spinal artery* in the anterior median fissure of the cord and to the paired *posterior spinal arteries* running on either side of the origins of the dorsal nerve roots. Some of the anterior root arteries are larger than others (see p. 193) but some do not reach the anterior spinal artery at all so that this vessel is not the same calibre along its length. The nutrition of the spinal cord may therefore depend to a considerable extent on the presence of the larger root arteries, especially in the lower thoracic and upper lumbar regions.

# Chapter 15
# The Autonomic Nervous System

Autonomous means self-governing. In other words, the autonomic nervous system is not, in general, under voluntary control but carries on with its functions of controlling the circulation, digestion, etc. even while you are asleep. Some parts are, however, capable of conscious control. For example, the focusing of the eye by the ciliary muscles and the stimulation of the lacrimal gland to produce tears, which can be done voluntarily by skilled actors.

The autonomic system is subdivided into *sympathetic* and *parasympathetic* systems and, although textbooks of physiology should be consulted for functional details, it helps with the anatomy to know that the sympathetic system prepares the body for emergency action(the 'flight or fight' reaction) while the parasympathetic system takes over under more relaxing conditions. You can get a general idea of the distribution of the sympathetic nerves if you carefully study your reactions before your next oral examination. The face is pale, because of cutaneous vasoconstriction, the pupils are dilated, the heart rate and the blood pressure are increased, the digestive glands cease to secrete (noticeable because of the dry mouth) and the skin (particularly of the palms of the hands) is sweaty. There may also be a prickly sensation because of contraction of the arrectores pili muscles (the 'hair standing on end'). The bronchial smooth muscle is relaxed, as is the smooth muscle of the gut (except for the sphincters), although these effects are not so noticeable.

Anatomically, there are important differences between the sympathetic and parasympathetic systems (Fig. 15.1). Both systems have peripheral ganglia but in the case of the sympathetic system most of these are not far from the spinal cord so the preganglionic fibres that leave the central nervous system to pass to the sympathetic system are relatively short while the postganglionic fibres have to travel a long way to reach their target viscera. *The sympathetic ganglia are seen in the dissecting room.* The preganglionic fibres of the parasympathetic system are long because the ganglia, which are microscopic (except for four in the head), are embedded in the viscera themselves, so the postganglionic fibres are short. *The parasympathetic ganglia are seen in the histology laboratory.*

The regions where the preganglionic fibres leave the central nervous system (the *outflow*) are also different. The *sympathetic outflow* is in the middle and the *parasympathetic outflow* at each end. To put this into more precise form, sympathetic preganglionic fibres leave the spinal cord via the anterior nerve roots of spinal nerves T1 to L2 (or 3). They travel as passengers in the anterior rami, which they leave as *white rami communicantes* to reach the sympathetic trunk. The parasympathetic preganglionic fibres are carried in a *cranial* and a *sacral outflow*. The fibres of the cranial outflow are carried as passengers in cranial nerves 3, 7, 9, 10 and 11; those of the sacral outflow are in sacral nerves 2, 3 and 4.

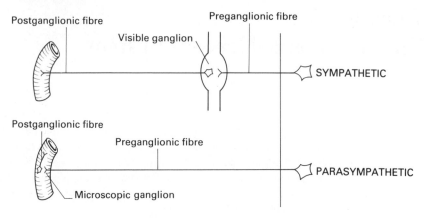

**Fig. 15.1** Comparison of the basic layout of the sympathetic and parasympathetic nervous systems.

## THE SYMPATHETIC NERVOUS SYSTEM

The most obvious anatomical component of the sympathetic system is the ganglionated *sympathetic trunk*, which extends the whole length of the trunk on each side of the midline, from the base of the skull to the tip of the coccyx. In the cervical region it lies on the bodies of the vertebrae behind the carotid sheath (Chapter 18), in the thoracic region along the heads of the ribs just outside the pleura, in the lumbar region along the anterior border of the psoas major and on the sacrum just medial to the sacral foramina. Developmentally there are the same number of ganglia as there are spinal nerves but many of them fuse to form compound ganglia. Thus, in the neck there are only three ganglia, a *superior cervical ganglion* (connected to the first four cervical nerves), a *middle ganglion* (connected to C5 and 6) and an *inferior ganglion* (C7 and 8). In the thoracic region there are usually 10–12 but the first is often fused with the inferior cervical ganglion to form the *cervicothoracic* or *stellate ganglion*. The lumbar ganglia are usually four in number, as are the sacral. The two sympathetic trunks end near the coccyx in the midline *ganglion impar*.

The sympathetic outflow leaves the anterior rami of T1–L2 as *white rami communicantes* (white because these fibres are myelinated). When these preganglionic fibres reach a ganglion, they may follow **three** different routes (Fig. 15.2):

**1**   They may synapse with a nerve cell in the ganglion itself.

**2**   They may pass up or down the sympathetic trunk to a different level before synapsing with a ganglion cell.

**3**   They may pass straight through the ganglion, maintaining their preganglionic status, until they reach an outlying ganglion such as the coeliac ganglion.

From the sympathetic trunk, the postganglionic fibres are distributed in **three** ways (Fig. 15.2):

**A**   By means of *grey* (unmyelinated) *rami communicantes*. Each spinal nerve receives one or more grey rami from the corresponding ganglion. For example,

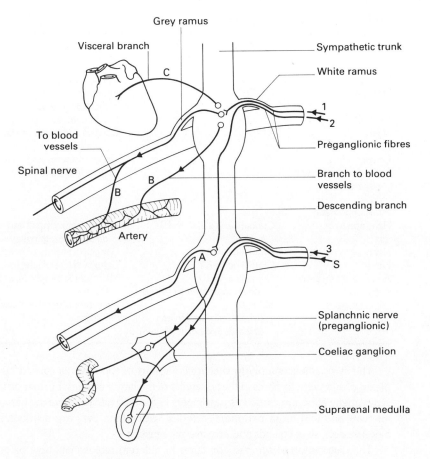

**Fig. 15.2** The possible courses of postganglionic sympathetic fibres. See text for description. The fibre marked 'S' is mentioned on p. 295.

the first four cervical nerves each receive at least one ramus from the superior cervical ganglion.

**B**  By fibres passing to adjacent arteries. These form a plexus around the artery and are distributed with its branches.

**C**  By fibres passing directly to various viscera in distinct, and sometimes named, branches, for example the cardiac branches of the cervical ganglia.

The important thing to understand is that there are no white rami (outflow fibres) above T1 or below L2 so that the cervical ganglia and the lower lumbar and sacral ganglia receive their sympathetic supply by means of preganglionic fibres that have 'turned the corner' to ascend or descend from the thoracic and upper lumbar regions respectively. Thus, if the sympathetic trunk is divided above the cervico-thoracic ganglion, the head and neck are completely without a sympathetic supply, while the lower limbs would be deprived of a sympathetic supply if the trunk were divided below L2. The origins of the preganglionic (or connector) fibres for various regions of the body are shown in Table 15.1.

**Table 15.1.** The autonomic system

| Region | Origin of connector fibres | Site of synapse |
|---|---|---|
| *Sympathetic* | | |
| Head and neck | T1–T5 | Cervical ganglia |
| Upper limb | T2–T6 | Inferior cervical and 1st thoracic ganglia |
| Lower limb | T10–L2 | Lumbar and sacral ganglia |
| Heart | T1–5 | Cervical and upper thoracic ganglia |
| Lungs | T2–4 | Upper thoracic ganglia |
| Abdominal and pelvic viscera | T6–L2 | Coeliac and subsidiary ganglia |
| | | |
| *Parasympathetic* | | |
| Head and neck | Cranial nerves 3, 7, 9, 10 | Various parasympathetic macroscopic ganglia |
| Heart | Cranial nerve 10 | Ganglia in vicinity of heart |
| Lungs | Cranial nerve 10 | Ganglia in hila of lungs |
| Abdominal and pelvic viscera | Cranial nerve 10 (down to transverse colon) | Microscopic ganglia in walls of viscera |
| | S2, 3, 4 (from transverse colon onwards and pelvic viscera) | Microscopic ganglia in walls of viscera |

The sympathetic supply to the proximal part of the limbs is carried by the plexus of sympathetic fibres in the adventitia of the larger arteries. Further distally, sympathetic fibres are carried as passengers in the limb nerves, leave the nerves at intervals and join more peripheral arteries, with which they are distributed to blood vessels, sweat glands and arrector pili muscles.

The distribution of sympathetic fibres in different regions will now be given briefly, to supplement information in the regional chapters.

## Cranial sympathetic

The sympathetic supply to the head is mainly carried in the *internal carotid nerve*, a prolongation upwards of the sympathetic trunk above the superior cervical ganglion. This accompanies the internal carotid artery, forming a carotid plexus, which continues as plexuses around the anterior and middle cerebral arteries and the ophthalmic arteries. This supply is supplemented by a plexus around the vertebral artery, which is derived from the cervicothoracic ganglion. Externally, branches of the external carotid artery convey fibres from the superior cervical ganglion. As well as blood vessels and sweat glands, the cranial sympathetic system supplies dilator pupillae and some smooth muscle in levator palpebrae superioris (the muscle that lifts the upper eyelid and so opens the eye). Thus, if the cervicothoracic ganglion is removed (ganglionectomy), there will be absence of sweating (*anhidrosis*), drooping of the upper eyelid (*ptosis*) and constriction of the pupil (*myosis*). In addition, the eyeball will be less prominent than usual (*enophthalmos*) because of paralysis of some smooth muscle within the orbit. This combination is known as *Horner's syndrome*. If it is necessary to remove the sympathetic supply (*sympathectomy*) to the upper limb, the cervicothoracic ganglion is left intact to avoid Horner's syndrome and the sympathetic trunk is divided below the 3rd thoracic ganglion. All grey rami from at least the 2nd and 3rd ganglia are removed as well as the *nerve of Kuntz*, if present. This is a grey ramus from the 2nd ganglion to the lst thoracic nerve.

### Cervical sympathetic

The three cervical ganglia give grey rami to all the cervical nerves as well as arterial branches to the arteries in the vicinity. They also give cardiac branches, which descend into the thorax to join the superficial and deep cardiac plexuses, and branches to the larynx and pharynx, mainly for the supply of the vessels. There is also a sympathetic supply to the salivary glands via the arteries, but the significance of this is not clear since the main secretomotor fibres seem to be derived from the parasympathetic system.

### Thoracic sympathetic

Each intercostal nerve is connected to the sympathetic trunk by a white ramus and at least two grey rami. From the ganglia, postganglionic fibres also pass to a plexus around the aorta and its branches (particularly the posterior intercostal arteries), to the *posterior pulmonary plexus* (from the 2nd to the 4th ganglia) and to the *cardiac plexuses* (from the upper five ganglia). The largest branches, however, are the preganglionic *splanchnic nerves, greater, lesser* and *lowest*, which arise respectively from ganglia 5–9, 10–11 and 12. They pass on to the sides of the vertebral bodies, enter the abdomen through the crura of the diaphragm and synapse in the *coeliac ganglia*. Some fibres even pass through these without synapse and remain as preganglionic fibres until they reach the *suprarenal medulla* (S in Fig. 15.2).

### Lumbar sympathetic

White rami pass from L1 and 2 to the sympathetic ganglia while all ganglia send grey rami to each of the lumbar nerves. *Lumbar splanchnic nerves* connect the ganglia to the coeliac plexus. Further branches join the aortic and iliac plexuses and others pass in front of the common iliac vessels to join the *superior hypogastric plexus*.

The coeliac ganglia and their associated plexus surround the coeliac artery and send offshoots along the branches of the aorta and the ventral arteries (coeliac, superior and inferior mesenteric). Postganglionic branches supply the intestine and associated structures.

### Pelvic sympathetic

The sympathetic ganglia send grey rami to the sacral nerves and to the local arteries. The superior hypogastric plexus lies in front of the promontary of the sacrum between the two common iliac arteries and is sometimes called the *presacral nerve*. It divides into right and left *inferior hypogastric plexuses*. These are joined by the pelvic splanchnic nerves and give vascular branches and branches to the pelvic viscera.

## THE PARASYMPATHETIC SYSTEM

The parasympathetic fibres of the cranial outflow travel as passengers in the *oculomotor* (3), the *facial* (7), the *glossopharyngeal* (9) and the *vagus and cranial accessory* (10, 11) nerves. As has already been mentioned, the cranial outflow fibres synapse in four relatively large ganglia, although only two of them (*submandibular* and *pterygopalatine* or *sphenopalatine*) are large enough to be seen easily in the dissecting room, the other two (*otic* and *ciliary*) being rather small and indistinct. The cranial parasympathetic system in the 3rd, 7th and 9th nerves supply secretomotor fibres to the salivary and lacrimal glands and the ciliary muscle and constrictor pupillae in the eye. The vagus and its accessory nerve (this is why the 11th nerve is called the accessory) combine to form the largest part of the cranial outflow since the vagus supplies most of the abdominal viscera, including the colon as far as about two-thirds of the way along the transverse colon, i.e. about the junction of the midgut with the hindgut (p. 211). This, in fact, is why the nerve is called vagus, being a similar word to vagrant and even vague.

The course of the parasympathetic fibres in the head region will be described later, as will the vagus nerve in the neck. Its intrathoracic course is described in Chapter 8 and its intra-abdominal course in Chapter 10. Here it is only necessary to mention its distribution, at least as far as its parasympathetic components are concerned, for it has a number of other components as well. In the neck it gives upper and lower *cardiac branches* and *oesophageal branches* to the smooth muscle and glands (at the lower end) of the oesophagus. In the thoracic region additional branches pass to the *cardiac plexuses*. The fibres of the plexuses synapse in microscopic ganglia and then supply the atria, the atrioventricular bundle and the coronary arteries. The vagus splits up to form the *pulmonary plexuses*, again with synapses in minute ganglia, and the postganglionic fibres supply the bronchial smooth muscle, the glands and the blood vessels. In the abdomen, the *anterior* and *posterior gastric nerves* (p. 222) supply the stomach and, via branches that join the coeliac plexus, the intestine and its related structures. The parasympathetic system is secretomotor to the glands of the stomach and intestine, motor to the smooth muscle, but inhibitory to the sphincters. In the gut, the ganglia are situated in *Auerbach's* and *Meissner's* plexuses.

*The sacral outflow*, in S2 and 3 (and possibly 4), joins the inferior hypogastric plexus by means of the *pelvic splanchnic nerves* and these supply the smooth muscle of the rectum and bladder with motor fibres (but with inhibitory fibres to the sphincters), the uterus with vasodilator and possibly inhibitory fibres and the erectile tissue of the penis and clitoris with vasodilator fibres (ejaculation, however, is under the control of the sympathetic system). Some parasympathetic fibres climb back out of the pelvis around the inferior mesenteric artery and, via its branches, supply the distal part of the transverse colon and the descending and sigmoid colon.

## AFFERENT AUTONOMIC PATHWAYS

It has to be admitted that the afferent pathways have not been mapped out with the precision that is a feature of the efferent pathways. In general, it may be said that as far as conscious sensation is concerned, pain from the viscera is poorly localized, often referred, and caused by stretching or violent muscle contraction rather than cutting or other forms of trauma. Other afferent fibres concerned with sensations, such as consciousness of fullness of bladder, rectum or stomach, probably travel in the vagus or pelvic splanchnics, although some pain fibres from the viscera travel with the sympathetic system. Thus extirpation of the superior hypogastric plexus may relieve pelvic pain from incurable malignant disease and pain from the passage of a stone down the ureter (ureteric colic) is also carried in the sympathetic system. Autonomic afferents are also part of a number of reflex activities, such as the Hering–Breuer reflex and reflexes involving the carotid sinus and carotid body.

# Chapter 16
# The Skull

The skull consists essentially of two parts, the cranial skeleton and the facial skeleton. The former is a strong and rigid container for the brain while the latter is a rather fragile and lightweight basis for the face, including the eyes, nasal cavities and mouth. You will not have much of a problem with joints when studying the skull since most of the joints are immovable, being of the fibrous type; there are only three synovial joints, two of which are minute, the joints between the ear ossicles (p. 400), and one of which is large, important and not very stable, the temporomandibular joint. You will, however, have a little difficulty in understanding some of the bones, particularly those of the face.

A little comparative anatomy may make the distinction between the cranial and facial skeletons easier to understand. In the skull of a quadruped such as the dog, the face is in front of the cranium so that the cribriform plate, which transmits the olfactory nerves from the nose, is more or less vertical. The spinal cord emerges from the skull horizontally so that the occipital bone, which includes the foramen magnum, is almost vertical (Fig. 16.1). Since the head has to be held in a horizontal plane, large muscles need to be attached to the back of the skull in order to support its weight. Now look at an ape's skull, if you have one to hand. The facial skeleton is now below and in front of the cranium so the cribriform plate is oblique. The foramen magnum is also oblique since the head is at the upper rather than the anterior end of the spine. However, gravity still tends to flex the heavy head so powerful extensor muscles are attached to the back of the skull. In the human skull, the face is below the anterior part of the cranium so the cribriform plate is horizontal. The skull descends in two steps to the foramen magnum, which is in a more or less horizontal plane. The cervical curve of the spine (p. 286) enables the skull to be balanced on top of the spine with little effort, so the extensor muscles need not be particularly powerful. Even so, if they relax, the head does nod forwards, as can easily be seen during a boring lecture.

## THE DEVELOPMENT OF THE SKULL

You do not need a detailed knowledge of this but a little study helps. In very general terms, one can say that the base of the skull ossifies in cartilage and the flat bones of the vault ossify in membrane. The bones of the face are a mixture. Some bones, such as the temporal, form part of the base and of the vault and so ossify in both ways. Textbooks of embryology should be consulted for a detailed account.

The brain and the cranial nerves start to develop before the skull, so when the cartilaginous skull base appears (the *chondrocranium*) the components form around the nerves, leaving foramina through which the nerves pass. The chondrocranium then becomes ossified from a number of centres. The last piece of cartilage to ossify is that between the body of the sphenoid and the occipital bone,

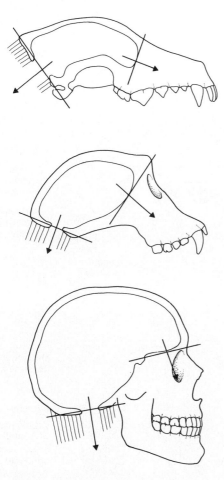

**Fig. 16.1** Skulls of a dog, an ape and a human to show the different planes of the foramen magnum and the cribriform plate. These depend on how the skull is balanced on the spine and whether the face is in front of or below the cranium.

a short distance in front of the foramen magnum. This is the *spheno-occipital synchondrosis*; it acts as a sort of epiphysial plate for growth in length of the skull base and it finally ossifies at the age of 25 years.

The bones of the vault of the skull begin to ossify from separate centres and the process spreads until the bones meet to form sutures. This process is not completed until the age of about 3 years. The features of the newborn skull will be described at the end of this chapter.

## THE VAULT OF THE SKULL (CALVARIA)

The bones of the cranium and the underlying lobes of the brain have similar names. They are shown in Figs 16.2–16.5. The frontal bone begins to ossify from two centres and there are left and right frontal bones at birth, but normally they

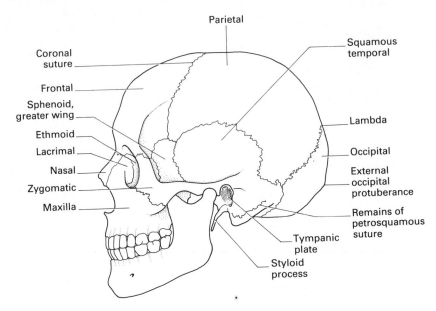

**Fig. 16.2** Side view of skull.

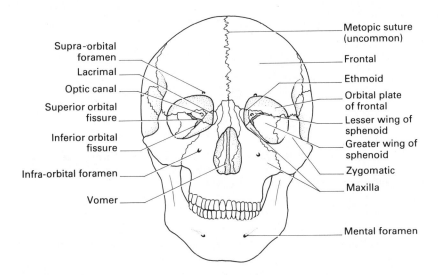

**Fig. 16.3** Front view of a skull with a metopic suture.

fuse to form a single large frontal bone. Occasionally they remain separate and the sagittal suture will therefore continue forwards to the root of the nose (the *metopic suture*). The parietal bones on each side meet in the midline to form the *sagittal suture* and the junction between the frontal and the two parietals is the *coronal suture*. At birth, these bones are quite widely separated so that there is a diamond-shaped area, bounded by the two frontal and two parietal bones, which

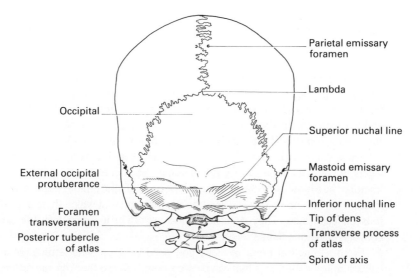

**Fig. 16.4** Rear view of skull, atlas and axis.

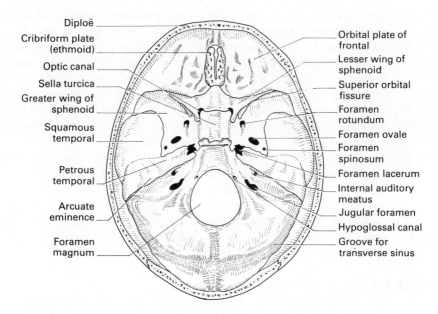

**Fig. 16.5** The inside of the base of the skull.

remains unossified. It can be felt as a soft area in the baby's skull, which pulsates and which bulges when the baby cries. This is the *anterior fontanelle* and it closes by the age of 18 months to 2 years. Further back, the two parietal bones and the occipital meet at an inverted Y-shaped suture which is therefore called the *lambda*. In the newborn skull they do not quite meet and there is a triangular *posterior fontanelle*. This is much smaller than the anterior fontanelle. At the sides the flat (squamous) part of the temporal bone contributes to the side wall of the calvaria in the region popularly known as the temple. The name of the bone, from *tempus* = time, is supposed to be derived from the fact that this is the region where time first greys the hair. A good story but perhaps the German *Schlafen-bein*, 'sleeping bone', is more descriptive. Finally, the side wall is completed by part of the greater wing of the sphenoid which forms an H-shaped suture with the frontal, parietal and temporal bones (Fig. 16.2). The region is called the *pterion* and is situated 4 cm above the midpoint of the zygomatic arch. It is the site for surgical exploration of the middle meningeal artery (see below).

These flat bones consist of a sandwich of cancellous bone (*diploë*) between two layers of compact bone so they are extremely strong, although the skull can be fractured by a blow or by compression of the head. As the diploë contains red bone marrow it is drained by large veins (the *diploic veins*), of which there are usually four on each side. They descend almost vertically to open into the nearest convenient venous sinus.

From a practical point of view, you need to know what structures in the calvarium leave markings that can be seen in an X-ray so that you can distinguish them from fractures. They are: (i) the sutures (do not forget a possible metopic suture); (ii) the diploic veins; and (iii) the middle meningeal artery (to be described later). The pineal gland, near the centre of the brain, may contain small calcareous granules (*corpora arenaceae*), which are opaque to X-rays, and, when seen, its position may be helpful in recognizing displacements of the brain. This is a convenient place to mention that the auricle of the ear can be seen as a semicircu-lar shadow on a lateral skull X-ray and should on no account be identified as a semicircular canal (it happens quite often!). Another feature of the flat bones of the skull is that in places they are perforated by emissary foramina that transmit *emissary veins* (p. 374). One is usually found in the parietal bone and one in the temporal bone behind the external auditory meatus.

From behind (Fig. 16.4) most of the occipital bone can be seen. The *external occipital protuberance* can be felt by running a finger up the midline groove at the back of the neck. Above it, usually, is the posterior pole of the skull, which is the part that will hit the ground first when falling backwards, and above this again is the lambda. Stretching laterally from the protuberance are the *superior nuchal lines* and, below them, the *inferior nuchal lines*. The area below the superior nuchal lines is roughened for attachment of the extensor muscles.

## THE BASE OF THE SKULL

The internal appearance of the skull base is shown in Fig. 16.5. Anteriorly, the floor of the cranium is rather flat and is called the *anterior cranial fossa*. It is separated from a deeper *middle cranial fossa* by a sharp ridge – the lesser wing of the sphenoid bone. The *posterior cranial fossa* is deeper still and the boundary between it and the middle fossa is the sharp upper border of the petrous temporal bone.

### The anterior cranial fossa

The frontal bone, which has been seen to form the basis of the forehead region (see Fig. 16.3), turns sharply back to form a large part of the roof of the orbit. This part of the bone is therefore called *orbital plate of the frontal*. Behind this is the small *lesser wing of the sphenoid*, while intercalated in a gap between the anterior parts of the orbital plates, is the *cribriform plate of the ethmoid*, with its *crista galli* (cockscomb) projecting upwards in the midline. The holes in the cribriform plate transmit olfactory nerves from the upper part of the nasal cavity to the brain. If you hold the skull up to the light, you can see how close the nasal cavity is to the inside of the cranium, so that the cribriform plate is a possible route for infection to be transmitted to the nervous system, while, in a fracture of this region, cerebrospinal fluid may drip from the nose.

At the root of the lesser wing on each side is the *optic canal*, which transmits the optic nerve and ophthalmic artery to the orbit. Just lateral to this there is an *anterior clinoid process*, which projects backwards towards the *posterior clinoid process* of the sphenoid. Occasionally they meet. The frontal lobe of the brain occupies the anterior cranial fossa.

### The middle cranial fossa

This is formed by the greater wing of the sphenoid and the temporal bone and is occupied by the temporal lobe of the brain. The two greater wings and the central body form a butterfly shape, the most posterior tip of the wing containing the *foramen spinosum*, so called because on the under-surface at this point is the spine of the sphenoid. The foramen spinosum is the most posterior member of a roughly semicircular array of foramina that transmit a number of important structures (Fig. 16.6). Just in front of the foramen spinosum is the *foramen ovale*, the reason for whose name is obvious. Further in front is an equally obviously name, *foramen rotundum*, which leads downwards and forwards into a remote region of the skull called the pterygopalatine fossa, which will be described in Chapter 23. Finally, under the overhanging shelf of the lesser wing, is the *superior orbital fissure*. This is not really a foramen but a gap between the lesser and greater wings of the sphenoid. It leads into the orbit just lateral to and below the optic nerve.

In the centre of the middle cranial fossa is the body of the sphenoid bone, which is hollowed out to form the *hypophysial fossa*, in which the pituitary gland is situated in life. It is supposed to look like a Turkish saddle, which is not much help, but the name, in Latin, is firmly established: the *sella turcica*. It is also

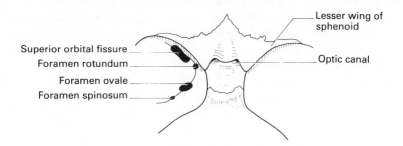

**Fig. 16.6** The four foramina in the greater wing of the sphenoid.

supposed to look like a four-poster bed, hence the names of the clinoid processes (as in 'clinical', the patients being in bed). It can clearly be seen in a lateral X-ray of the skull and its measurements (maximum length 14 mm and depth 8 mm) are important because pituitary tumours cause 'ballooning' of the sella. Just lateral to and behind the sella is a jagged hole called the *foramen lacerum*. It is not a true foramen but a gap between the apex of the petrous temporal bone and the body of the sphenoid. The carotid canal opens into it posteriorly (see Fig. 22.5).

The course of the *middle meningeal vessels* is clearly marked out by grooves on the inner surface of the skull; they are described on p. 376.

### The posterior cranial fossa

Bounded in front by the crest on the petrous temporal bone, the fossa is formed by the temporal and occipital bones and houses the cerebellum. The most obvious feature is the *foramen magnum*, which transmits the lower end of the medulla, etc. (*not* the spinal cord). A story is told of an examination candidate who, having failed to answer any questions, was shown the foramen magnum. 'Ah ha,' he said, 'I know that one. Many's the pint of beer I've poured through that!' In front of the foramen magnum the slope of the *clivus* leads up to the body of the sphenoid, while behind it the bone is marked by faint grooves for the *transverse sinuses*, which are venous channels. If one of the grooves is followed forwards it reaches the back of the petrous temporal bone and then pursues an S-shaped curve downwards and forwards (housing the *sigmoid sinus*) to end at the *jugular foramen*. As will be described later, the large *superior sagittal sinus* is usually continuous with the right transverse sinus and the smaller straight sinus with the left. Therefore the right jugular foramen is usually larger than the left. This foramen is between the occipital and temporal bones.

In the anterior margin of the foramen magnum is a small, anteriorly directed *hypoglossal* (or *anterior condylar*) canal, while posteriorly there is a *posterior condylar canal*. Finally, on the posterior aspect of the petrous temporal bone is the *internal auditory meatus*, which leads in to the region of the middle and inner ear. You may have noted the bulge on the upper surface of the petrous temporal (the *arcuate eminence*), which marks the position of the superior semicircular canal.

### The exterior of the base

Figure 16.7 shows the skull as seen from below. Many of the foramina you will recognize, such as the foramina magnum, spinosum, lacerum and ovale and the jugular foramen. Near the foramen spinosum is the *spine of the sphenoid*; the jugular foramen seen from below takes the form of a deep hollow, which is occupied in life by the *jugular bulb*, an expansion of the internal jugular vein. You should, however, learn to identify these foramina both on the inside and the outside of the skull. In an oral examination, the candidate should try to impress the examiner by his or her confidence as well as knowledge, and nothing looks worse than uncertain attempts to push a knitting needle through a foramen inside the skull to see where it comes out on the outside. Some of them don't, which makes matters worse.

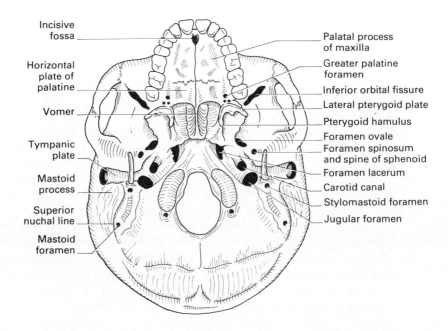

**Fig. 16.7** The outside of the base of the skull. See also Fig. 19.10.

From those five foramina already known, it is easy to identify the other main features. Just lateral to the jugular foramen is the *styloid process*, probably already knocked off in the skull that you are examining, and posterolateral to its stump is the *mastoid process*, which you can easily feel on yourself deep to the lobe of the ear. Between the two is the *stylomastoid foramen* for the facial nerve. In front of the jugular foramen, and therefore in the petrous temporal bone, is the *carotid canal*, so you may deduce that, at the base of the skull, the internal carotid lies just in front of the internal jugular vein. You will look in vain for the inner opening of this foramen: investigation with a piece of bent wire will show that the carotid canal runs forwards and medially in the temporal bone and opens into the posterior wall of the foramen lacerum (p. 381).

The spine of the sphenoid lies immediately medial to the mandibular fossa. This is part of the temporal bone, so the jaw joint is known officially as the *temporomandibular joint* (or TMJ). The foramen ovale lies at the base of a downwardly projecting plate of bone, the *lateral pterygoid plate*. The *medial pterygoid plate* is obvious. The pterygoid plates are part of the sphenoid. From the pterygoid region, the *inferior orbital fissure* leads forwards into the orbit. At the tip of the medial plate is a little hook, the *pterygoid hamulus*.

Further forwards, the base of the cranium is, of course, obscured by the facial skeleton, mainly the maxilla. The hard palate is formed partly by the horizontal *palatine process of the maxilla* and, posteriorly, by the *horizontal plate* of the palatine bone. Note the three foramina on the palate (Fig. 16.7).

Looking into the posterior opening of the nasal cavity you will see the single midline *vomer*, which is attached by a fibrous joint to the under-surface of the body of the sphenoid.

### The skeleton of the face

This is shown in Figs 16.2 & 16.3; the bones are labelled. In the orbit (see Fig. 22.8) you will recognize the superior orbital fissure and the optic canal and therefore you will recognize also the greater and lesser wings of the sphenoid. Most of the lateral wall is formed by the *zygomatic bone* and the floor by the upper surface of the *maxilla*. The main part of the maxilla is hollow, as you will appreciate if you place a small torch in your mouth in a dark room and close your lips over it, when the light will be seen coming through your cheeks unless your maxillary sinus is full of mucus. Most of the medial wall is made up of *ethmoid*, in front of which is the small and fragile *lacrimal bone*.

There are a number of foramina on the face but most of them are very small. Three of them are worthy of mention here. These are the *supra-orbital, infra-orbital* and *mental foramina* and they lie on a vertical line. These transmit the three major cutaneous nerves on the face; the nerves have the same names as the foramina and each is a branch of a separate division of the trigeminal nerve, respectively the ophthalmic, maxillary and mandibular divisions.

## DIFFICULT BONES OF THE SKULL

The above account is intended only as an outline and covers most of the questions that you are liable to be asked in an oral examination on the skull. Much more detailed descriptions will be given of particular regions of the skull in the chapters that follow, including descriptions of a number of bones not mentioned above (especially the mandible). Several of the skull bones are difficult to understand, as they have a number of parts, and there follow, by way of revision, descriptions of three of these.

### The sphenoid bone

This bone is difficult because you cannot see its anterior surface at all in an intact skull, but otherwise, from whichever aspect you look at the skull, you are bound to see part of this bone. Inside the skull, you have already seen the body, with the sella turcica, and the greater and lesser wings in the orbit, with the superior orbital fissure between them. From behind and from below you can see the body, the under-surface of the greater wings and the pterygoid plates. Although you cannot get a direct view of the anterior surface (to be described in more detail in Chapter 23), you can get an idea of its whereabouts by passing a piece of wire through the foramen rotundum. After a little manipulation you will see its end by peering through the slit between the lateral pterygoid plate and the back of the maxilla. This region is the *pterygopalatine fossa*. Now try to push a bent piece of wire through a small hole that opens into the anterior wall of the foramen lacerum (just as the carotid canal opens into the posterior wall). This is the *pterygoid canal* and its other end opens into the pterygopalatine fossa (see Fig. 22.5).

The body of the sphenoid is hollow, being occupied by the *sphenoidal air sinuses*. They open anteriorly into the back of the nose so it is possible to operate on the pituitary gland via the nose and the sinuses by removing the floor of the sella turcica.

## The ethmoid bone

This bone, again, puts in an appearance in various parts of the skull. To appreciate its complicated shape, you may make an enlarged model of the bone with an old desk (Figs 16.8 & 16.9). First place a vertical plate of wood in the midline, projecting above the desk surface for 15 cm. Bore a number of holes alongside it. Fit two shelves to the inner surfaces of the drawer units and finally fill the drawers

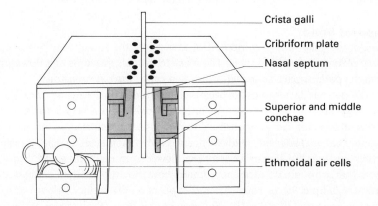

Crista galli

Cribriform plate

Nasal septum

Superior and middle conchae

Ethmoidal air cells

**Fig. 16.8** A 'mock-up' of the ethmoid bone. Compare with Fig. 16.9.

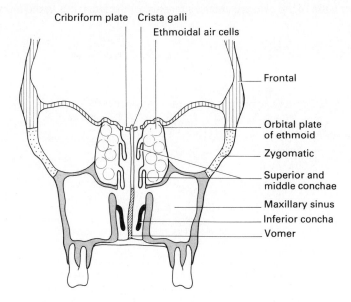

Cribriform plate    Crista galli

Ethmoidal air cells

Frontal

Orbital plate of ethmoid

Zygomatic

Superior and middle conchae

Maxillary sinus

Inferior concha

Vomer

**Fig. 16.9** The relations of the ethmoid bone and the maxillary sinus.

with blown-up balloons. The perpendicular plate represents the crista galli above the desk and the holes are the foramina in the cribriform plate. Below the desk, the plate is known simply as the *perpendicular plate* of the ethmoid and forms the upper part of the nasal septum (Fig. 16.9). The two shelves on each side are the *superior* and *middle conchae*, which can be seen indistinctly through the anterior or posterior nasal aperture.

*They are part of the ethmoid bone but the inferior concha, seen in Fig. 16.9, is a separate bone in its own right.* The outer surface of the drawer unit forms most of the medial wall of the orbit and the balloons represent the thin-walled *ethmoidal air cells*. Never pick up a skull by using the orbits as fingerholes: your fingers and thumb will crunch through the paper-thin air cells and ruin the skull.

## The temporal bone

Developmentally, this is composed of four parts, the *squamous, petrous, tympanic plate* and *styloid process*. The mastoid process develops as a downgrowth from the petrous part, so the whole is often called the *petromastoid*.

The petrous part is so called because it is particularly dense and 'rock-like' (cf. petrology, and the name Peter). It certainly shows up as a very dense shadow in X-rays. It develops in cartilage around the otocyst in the embryo, so that it encloses the inner ear, and later the middle ear cavity with its contained ossicles is included. On the surface, the squamous temporal develops in membrane and it grows to cover the petrous part externally. Later, however, the mastoid process grows downwards from the petrous part and appears on the surface below and behind the squamous part (Fig. 16.10). You may be able to see the remains of the suture between them on the surface of the mastoid process (see Fig. 16.2). The tympanic part of the temporal bone is a triangle of bone that forms the floor and anterior and posterior walls of the external auditory meatus. In the fetus, it is in the form of a

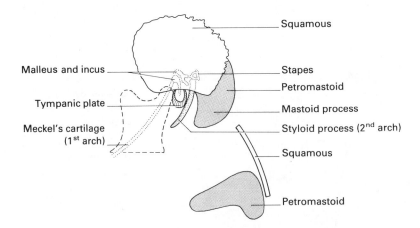

**Fig. 16.10** The four parts of the temporal bone: squamous, petromastoid, tympanic plate and styloid process. The small picture is an anteroposterior view and shows how the squamous part overlies the petromastoid, leaving only the mastoid process visible from the lateral side.

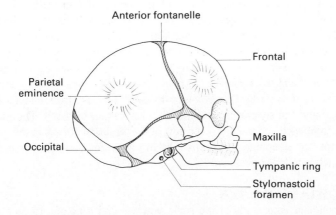

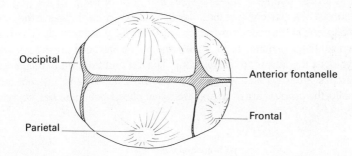

**Fig. 16.11** A fetal skull.

ring (Fig. 16.11) and, in the embryo, the cartilage of the 1st branchial arch is in front of it and that of the 2nd behind it. The mandible develops in membrane (mostly) around the 1st arch cartilage (*Meckel's cartilage*) and the 2nd arch cartilage becomes, *inter alia*, the styloid process. The upper ends of the cartilages become the malleus, incus and stapes.

The squamous temporal becomes part of the side wall of the cranium (often the thinnest part) and forms the socket for the head of the mandible. It also develops a forwardly projecting *zygomatic process*, which joins a similar process from the zygomatic to form the *zygomatic arch*.

## THE NEWBORN SKULL

The skull of a newborn baby (Fig. 16.11) is very different from that of the adult. The nervous system is always ahead of other parts of the body in development, the brain reaching 25% of its adult size at birth and 75% by the age of 4 years. The striking feature about the newborn skull is, therefore, the relatively huge size of the cranium, the face being a tiny collection of bones clustered under its anterior end. Other reasons for the relatively small size of the face are that the teeth have not erupted so that the alveolar parts of the maxilla and mandible are undeveloped, and the paranasal air sinuses do not develop properly until puberty so the maxilla is not yet inflated. The nasal cavity is thus largely between the orbits. As has been described already, the anterior fontanelle is closed only by membrane, the posterior fontanelle is still open, and there are various gaps between other bones of the cranium. All this helps in the moulding of the skull that takes place during its passage down the birth canal. The flat bones may, in fact, ride over one another. Many of the separate parts of the bones of the base of the skull are still united only by cartilage and the styloid process has only just begun to ossify. The mastoid process is unformed and the tympanic part of the temporal bone is in the form of a ring so that the tympanic membrane is relatively nearer the surface than in the adult and tends to face downwards.

## THE BRANCHIAL ARCHES

Before beginning your study of the neck, you will find it helpful to do a quick revision of the branchial region in the embryo. The *branchial arches* are actually *inverted* arches that lie in the side walls and floor of the pharynx (Fig. 16.12). They consist of a thickening of mesoderm, which produces bulges in the ectoderm externally, and in the endoderm, which lines the pharynx internally. There are five of them but they are numbered 1, 2, 3, 4 and 6 (you will need to ask a comparative embryologist if you want to know why). The 1st and 2nd are particularly large. The arches are separated by grooves on the external and internal surfaces. Internally, the grooves are deep and are called *pharyngeal pouches*, so that if one

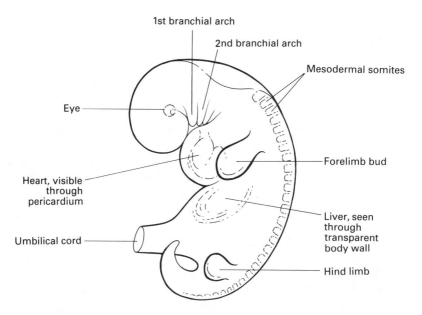

**Fig. 16.12** An embryo of about 33 days, before the invasion of the body wall by muscle developed from the mesodermal somites (p. 25). The branchial arches can be seen, covered with ectoderm and separated by the clefts.

could make a cast of the embryonic pharynx the pouches would appear as diverticula from the surface, shaped rather like wing nuts, with *dorsal* and *ventral wings* (see Fig. 18.4). Dorsal to the pharynx is the brain, so that it is *easy* for certain cranial nerves to grow down into the arches. The nerves are:

| | |
|---|---|
| Arch 1 | 5 |
| Arch 2 | 7 |
| Arch 3 | 9 |
| Arch 4 | 10/11 (superior laryngeal) |
| Arch 6 | 10/11 (recurrent laryngeal) |

The mesoderm of each arch differentiates into:

**1** A skeletal element that will form bones and ligaments.

**2** An artery, the *aortic arch* (p. 189).

**3** Striated muscle, supplied by the nerve of that arch. The muscle migrates but always drags its nerve supply with it, no matter how far it moves from its site of origin. Thus, the the mesoderm of the 2nd arch (7), for instance, supplies muscles as far apart as platysma and the occipital belly of occipitofrontalis.

# Chapter 17
# Posterior Triangle, Back of Neck

It is unfortunate that you had to dissect the neck (if you did dissect it) starting from the skin and working your way inwards. It is much easier to understand if you start with the deeper structures and work your way towards the surface, as you will see when you read about the pharynx in the next chapter. This is also true for the posterior triangle of the neck. Students are usually puzzled by the name of this triangle since all the important structures in it are on the front of the neck just above the clavicle. It is in fact a spiral triangle, whose shape you will understand if you cut out a triangle of paper and twist it through about 180° (Fig. 17.1). In this figure you can see that the boundaries are the anterior border of trapezius, the posterior border of sternocleidomastoid and the middle third of the clavicle. You have seen in Chapter 3 that the upper attachment of trapezius is to the medial part of the superior nuchal line and you will see later that sternomastoid is also attached along this line, extending as far as the origin of trapezius (see Fig. 17.7). The apex of the posterior triangle is therefore on the superior nuchal line, where these two muscles meet, but the base is on the clavicle. To appreciate relationships in this area you need to know the basic plan of the neck, which is very simple (Fig. 17.2). It consists essentially of a large posterior 'musculovertebral' block, consisting of the cervical vertebrae and the muscles that are grouped around them. This block is bound together by a dense layer of *prevertebral fascia*. This extends up to the base of the skull and down to where the anterior muscle, *longus colli*, ends, somewhere about the 3rd thoracic vertebra. Placed in front of this block and separated from it by loose connective tissue is another block, composed of the pharynx and larynx or, lower down, trachea and oesophagus. They are partially enclosed in a much thinner sheath of fascia, called the *pretracheal fascia*. The whole is enclosed in a musculofascial tube, composed of the sternocleidomastoid and trapezius muscles and the *investing layer* of deep fascia of the neck. This is the layer that envelops the neck immediately deep to the superficial fascia. It splits to enclose the two muscles. Trapezius is attached posteriorly to the spines of the vertebrae via the ligamentum nuchae. The relative positions of the two muscles will depend on the level at which the section shown in Fig. 17.2 is taken: the lower the section the further trapezius will extend forwards and the more anteriorly will sternocleidomastoid appear to be. Finally, filling in the angle between the vertebral block and the larynx–pharynx block is the *carotid sheath*, containing the internal jugular vein, the common or internal carotid arteries (depending on the level) and the vagus nerve between them posteriorly.

The arrows in Fig. 17.2 point to the posterior triangle between sternomastoid and trapezius. The floor of the triangle is thus formed by some of the muscles around the vertebrae, covered by the prevertebral fascia.

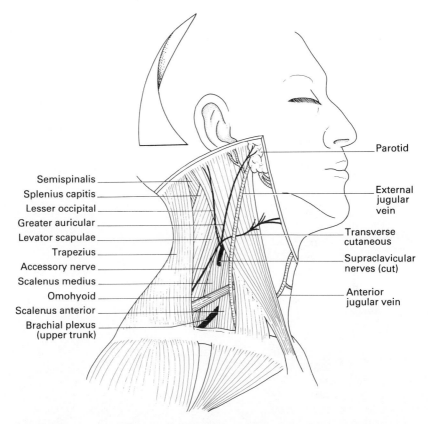

**Fig. 17.1** The major structures in the posterior triangle of the neck. Notice how five nerves seem to radiate out from a point on the posterior border of the sternocleidomastoid. The inset shows the twisted triangle mentioned in the text.

In order to make this basic structure more realistic, it is now not too difficult to fill in the details (Fig. 17.3). Firstly, the vertebra must be drawn in more accurately, remembering that each transverse process has an anterior and a posterior tubercle and a foramen transversarium. The muscles that surround the vertebrae can now be inserted. Immediately in front of the vertebral bodies is the *longus colli* muscle and that is all you need to know about it except that it goes up as far as the atlas. More laterally, the *scalenus anterior* arises from the anterior tubercles of all those vertebrae that have anterior tubercles, i.e. 3, 4, 5 and 6. *Scalenus medius* arises from posterior tubercles of the same vertebrae but also extends upwards and downwards for one vertebra to be attached to the transverse processes of C2–7. *Scalenus posterior* is small and unimportant. It bypasses the 1st rib and continues down to the 2nd (see Fig. 18.6). It is omitted in Fig. 17.3. From the posterior tubercles or transverse processes further back arises *levator scapulae* (C1, 2, 3 and 4) and, behind this again, *splenius capitis* (do not worry about which vertebra it is attached to). Finally, *semispinalis capitis*, with its vertical fibres, arises from various vertebrae and is inserted into the back of the skull between the two nuchal

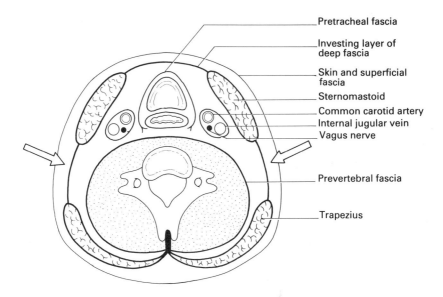

**Fig. 17.2** The basic plan of the neck, in cross-section. The arrows indicate the posterior triangle.

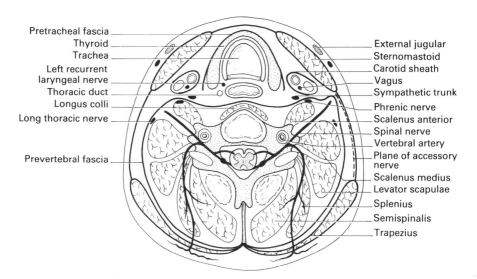

**Fig. 17.3** A more detailed section of the neck but basically similar to Fig. 17.2. There are still some structures omitted from the diagram for the sake of simplicity, such as the 'strap' muscles.

lines (see Fig. 17.7). These, then, are the muscles that form the floor of the posterior triangle, with their covering of prevertebral fascia. Their fibres, except for semispinalis, run downwards and backwards (Fig. 17.1). If you look at the tubercles on the transverse processes of a cervical vertebra, you will see that there is a deep groove between them. This is occupied by the spinal nerve (remember that, in the cervical region, each nerve is *above* the vertebra with the same number, except that the nerve C8 is *below* the 7th cervical vertebra because there are eight cervical nerves but only seven cervical vertebra). Running vertically through the foramina transversaria is the *vertebral artery* and its surrounding plexus of veins and sympathetic nerves. Does the nerve pass in front of, or behind, the vertebral artery? Figure 4.19 (supination) will remind you of the answer; it passes behind.

The spinal nerve anterior rami from C5 to T1 form the *roots* of the brachial plexus, C5 and 6 forming the *upper trunk*, C7 the *middle* and C8 and T1 the *lower trunk*. The roots are buried between scalenus anterior and medius and the trunks appear in the posterior triangle. Since these muscles arise respectively from anterior and posterior tubercles and since the spinal nerves are between the tubercles, the trunks surface between the two muscles. The lower trunk may be difficult to see because it lies right down on the 1st rib, sharing a shallow groove with the *subclavian artery* (Chapter 18). The *subclavian vein* is in front of scalenus anterior. These structures are deep to the prevertebral fascia. How then are they able to get down behind the clavicle into the axilla without piercing the fascia? They take a prolongation of the fascia down with them in the form of a tube that surrounds them during their passage from one region to the other (Fig. 17.4), and you have met this fascia already in Chapter 3, where it is referred to as the *axillary sheath*.

So much for the vertebral–muscle block of tissue. In front of it is the respiratory–digestive block, i.e. the trachea and oesophagus below the level of the

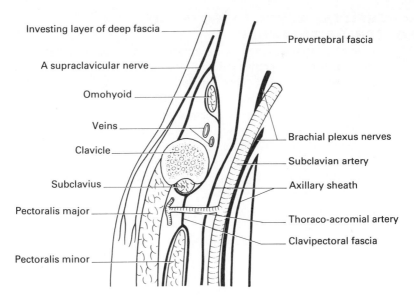

**Fig. 17.4** Cross-section of the clavicle to show the fascia in its vicinity. This is one of the 'boundary zones' and you need to study the posterior triangle and the axilla to understand it.

6th cervical vertebra (p. 324) and the larynx and pharynx above that level. The pharynx comprises the *nasopharynx, oropharynx* and *laryngopharynx*, which will be described in detail in the next chapter, and the laryngopharynx is continuous with the oesophagus behind the cricoid cartilage. The larynx becomes trachea at the same level. In the lower part of the neck, therefore, the trachea and oesophagus are together and are enclosed in a rather thin layer of *pretracheal fascia* whose only real importance is that, as well as the trachea, it encloses the *thyroid gland*. On the left side, the left *recurrent laryngeal nerve* occupies the groove between the trachea and oesophagus, while on the left side of the oesophagus is found the *thoracic duct* (pp. 174 and 193). You should always think of these four structures as a unit.

The final block is the carotid sheath and its contents, sometimes known irreverently as the 'carotid packet'. This contains, below the upper border of the thyroid cartilage, the *common carotid artery* and the *internal jugular vein*. Above this level the artery is the *internal carotid*. Lying within the carotid sheath between artery and vein and rather posteriorly is the *vagus nerve* and embedded in the anterior wall of the sheath is the *descendens hypoglossi*. This name has now been superseded by the 'superior ramus of the ansa cervicalis'. (Use the latter term if it appeals to you and if you want to be up to date). As you might expect, the tough fibrous tissue of the carotid sheath is very much thinner over the vein so as to allow for its expansion.

Outside these fascial compartments, the neck is made up of fat and loose connective tissue, in which run various branches of the main vessels and nerves and the sympathetic trunk.

## SUPERFICIAL STRUCTURES IN THE POSTERIOR TRIANGLE

As far as cutaneous nerves are concerned, you have only C2, 3 and 4 to contend with. You will see later than the anterior ramus of C1 joins the hypoglossal nerve and is distributed with it, while you have seen already that C5 joins the brachial plexus. From various combinations of C2, 3 and 4, therefore, arise the four cutaneous nerves of the neck. These all pierce the investing layer of deep fascia about halfway down the posterior border of sternomastoid. They are:

1  The *lesser occipital*, which supplies an area of skin behind the ear.

2  The *greater auricular*, which runs up close to the external jugular vein and supplies an area of skin over the parotid (the only cutaneous nerve on the face not derived from the trigeminal nerve).

3  The *transverse cutaneous nerve of the neck*, which runs forwards across sternomastoid.

4  The three *supraclavicular nerves*, medial, intermediate and lateral, which come from a rather thick, common trunk and cross the clavicle to supply skin down as far as the 2nd intercostal space. You should remember their root value (C3 and 4) to remind you of the origin of referred pain in the tip of the shoulder (p. 191).

Leaving the posterior border of sternomastoid in the same region is the *spinal accessory nerve*. This enters the neck through the jugular foramen (p. 304), enters sternomastoid about one-third of the way down its anterior border, supplies it and finally leaves the muscle as described above. It runs obliquely across the posterior triangle and enters trapezius two finger-breadths above the clavicle. It is a little difficult to dissect out because it is embedded in the investing layer of deep fascia. It supplies trapezius. Both sternomastoid and trapezius also receive proprioceptive fibres from the upper cervical nerves.

The final structure related superficially to the sternomastoid is the *external jugular vein*. This is formed near the upper end of the muscle by the union of the posterior branch of the *retromandibular vein* with a small (posterior auricular) vein from behind the ear (see Fig. 21.2). It crosses sternomastoid very obliquely to enter the posterior triangle and finally opens into the subclavian vein. It is joined at the base of the triangle by the *anterior jugular vein* and the *transverse cervical* and *suprascapular veins*, so that incisions above the clavicle meet a large number of vascular complications. The anterior jugular vein begins in the submental region (below the chin), descends to one side of the midline and then turns laterally, deep to sternomastoid, to meet the lower end of the external jugular.

The veins of the neck finally open directly or indirectly into the brachiocephalic veins and then into the superior vena cava and the right atrium. There are no valves at the caval orifice so that if a vein is opened with the patient in a head-up position it is possible for air to be sucked into the venous system and thence into the lungs (*air embolism*). With the body horizontal, however, the veins are distended. The external jugular can be used as a manometer: if the venous pressure in the right side of the heart is raised, the vein is visibly distended even when it is above the level of the heart.

Before leaving the superficial structures there is one more point to be considered. As has been seen, the investing layer of deep fascia covers both superficial and deep surfaces of sternomastoid and trapezius, forming a single layer between them. For about two finger-breadths above the clavicle, however, the two layers remain separate (see Fig. 17.4). The anterior layer, as you would expect, is attached to the upper border of the clavicle between the two muscle attachments, but the posterior layer passes behind the clavicle to become continuous with the clavipectoral fascia (p. 56). The posterior belly of the *omohyoid muscle* and its intermediate tendon are embedded in the posterior layer of fascia, which acts as a sling, tethering it down to the clavicle. This muscle belly crosses the lower part of the posterior triangle obliquely, on its way to the suprascapular notch (see Figs 17.1 & 17.4). It will be fully described later.

## THE STERNOCLEIDOMASTOID MUSCLE

This muscle, which is such a useful landmark in the neck, is attached below by a rounded tendon to the front of the manubrium sterni and by a flattened muscular sheet to the medial third of the clavicle. Above, it is attached to the mastoid process and to the lateral half of the superior nuchal line. Because of its attachment to the clavicle, the 'cleido' is part of its official name but it is more commonly referred to as the sternomastoid. Its nerve supply is the spinal accessory nerve (plus some cervical nerves), and its action, when both sides act together, is to flex the neck. It is probably the first muscle you use in the morning, after levator palpebrae superioris. When only one side is used, it will side-flex the head to the same side and rotate it to the other, i.e. the left sternomastoid approximates the left ear to the left shoulder. You can see this action well in a patient with the common condition of *spasmodic torticollis*, in which there is intermittent involuntary contraction of the muscle on one side.

## THE FLOOR OF THE POSTERIOR TRIANGLE

This is covered by the prevertebral fascia and, apart from the structures that have already been mentioned, there are only two structures that cross the triangle outside the fascia. These are the *superficial cervical* and the *suprascapular arteries*. The latter is large and may be tucked behind the clavicle and the former is small and higher up. They both arise from the *thyrocervical trunk*, which will be described in the next chapter. They cross scalenus anterior (outside the fascia); the superficial cervical artery breaks up to supply local structures and the suprascapular artery joins its nerve to pass through the suprascapular notch above the suprascapular ligament (see below). Frequently, there is a rather larger artery in place of the superficial cervical artery, which is then called the *transverse cervical artery*; this divides into the *superficial cervical artery* (superficial to levator scapulae) and a *dorsal scapular* (deep to levator scapulae), which runs down the medial border of the scapula and takes part in the scapular anastomosis (see Fig. 3.16). The dorsal scapular more commonly is a branch of the third part of the subclavian artery.

At this point, it is worth digressing for a moment to summarize the scapular anastomosis, since all the arteries concerned have now been described in this chapter or in Chapter 3. The *subscapular* is a branch of the axillary artery that follows the lateral border of the scapula, giving branches that enter the muscles clothing the bone and, in particular, the *circumflex scapular* artery, which goes through the triangular space to enter the infraspinous fossa. The *suprascapular artery*, from the thyrocervical trunk (itself a branch of the subclavian), goes through the suprascapular notch into the supraspinous fossa and then through the spinoglenoid notch to the infraspinous fossa. The *dorsal scapular* descends along the medial border of the scapula. All these arteries anastomose very freely (see Fig. 3.16), thus providing a connection between the subclavian and lower axillary arteries. Thus, if the main artery to the upper limb is tied or blocked anywhere between the origins of the thyrocervical trunk and the subscapular arteries, blood can still reach the upper limb by reversal of flow in the subscapular.

Deep to the prevertebral fascia are the muscles whose origins have already been mentioned. Their fibres run downwards and backwards except for semispinalis capitis, whose fibres are vertical. The muscles are, from anterior to posterior, scalenus anterior, scalenus medius, scalenus posterior (ignore it!), levator scapulae, splenius capitis and semispinalis capitis. Scalenus anterior is inserted by a tendon into the *scalene tubercle* on the medial border of the 1st rib and scalenus medius by muscle fibres into a large area on the upper surface of the rib (Figs 17.5 & 18.6). These will be mentioned again in Chapter 18 but here it will be enough to remember their origins and that, because of these, the trunks of the brachial plexus surface between scalenus anterior and scalenus medius. The upper trunk is easy to

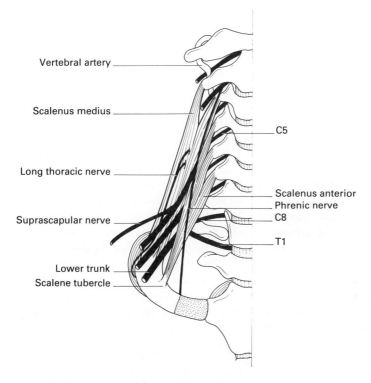

**Fig. 17.5** The supraclavicular part of the brachial plexus.

see, the middle trunk more difficult and the lower trunk almost impossible without some excavation behind the clavicle. This is because it lies in a groove on the 1st rib, which it shares with the subclavian artery.

The upper trunk (C5 and 6) forms the upper border of the brachial plexus and if you run your finger down this upper border it will be led backwards by the large *suprascapular nerve* (Fig. 17.5). This passes through the notch deep to the suprascapular ligament, supplies supraspinatus and then passes through the spinoglenoid notch to supply infraspinatus. The suprascapular artery follows it. There are other important nerves in this region. Branches from C3, **4** and 5 unite to form the *phrenic nerve* at the lateral border of scalenus anterior and the nerve itself is then a prominent feature on the anterior surface of this muscle (again, deep to the prevertebral fascia). Branches from C5, 6 and 7 ('Bells of Heaven', see Chapter 3) unite to form the *long thoracic nerve* (of Bell) on the surface of or in the substance of scalenus medius. This nerve runs down behind the brachial plexus and subclavian artery to reach serratus anterior, on which it lies in the medial wall of the axilla. You may find it difficult to link up the root of the neck with the upper limb but if you remember that serratus anterior arises from the upper eight ribs (therefore, obviously, partly from the 1st rib) you will understand that scalenus medius can lead the nerve down to its insertion, after which the nerve makes its own way on the surface of serratus anterior itself. The other supraclavicular branches of the brachial plexus are to the *scalene muscles*, the *dorsal scapular nerve* (to the rhomboids) and the *nerve to subclavius* (which you should now forget).

It may be helpful to remember that there are no important structures in the posterior triangle above the accessory nerve, except, perhaps, for the *occipital artery*. If, on dissecting the posterior triangle, you can see the vertical fibres of semispinalis capitis, you will see the occipital artery, which emerges from under cover of splenius and crosses semispinalis.

## LESIONS OF THE TRUNKS OF THE BRACHIAL PLEXUS

The upper trunk is sometimes damaged during a difficult delivery, due to excessive lateral flexion of the neck, or it may be affected by bullet or knife wounds. Since the 5th cervical nerve supplies the abductors of the shoulder, the flexors of the elbow and the supinators (Fig. 17.6), the arm will be hanging in the 'waiter's tip' position. This is known as *Erb's paralysis*.

The lower trunk may be affected in lesions affecting the thoracic outlet and possibly in cervical rib (p. 281). Since the first cervical nerve supplies the small muscles of the hand, there will be weakness and wasting of the thenar and hypothenar eminences and of the interossei, and the hand may be clawed (see p. 92). This is *Klumpke's paralysis*. There may also be sensory changes in the corresponding dermatome.

**Fig. 17.6** The beer-drinker's nerve — C5. This supplies the abductors of the shoulder, the flexors of the elbow and the supinators. When C5, or the upper trunk of the brachial plexus, is damaged, the upper limb will be adducted at the shoulder, the elbow extended and the forearm pronated — the 'waiter's tip' position. The three glasses are just to remind you of the arrangement of the three constrictors of the pharynx (p. 324).

# THE SUBOCCIPITAL REGION

It has been stated already that the posterior triangle spirals its way to the back of the neck in the occipital region, so this is probably the best place to describe the so-called *suboccipital triangle* in the region below the external occipital protuberance.

If you are dissecting this region, having removed skin and superficial fascia you will see the apex of the posterior triangle on the superior nuchal line, i.e. the meeting of the posterior border of sternomastoid and the anterior border of trapezius. Reflect trapezius and you will see semispinalis capitis, attached to the skull between the superior and inferior nuchal lines. You will also see the occipital artery emerging from under cover of splenius capitis and breaking up into branches to supply the back of the scalp. Two nerves are visible, both emerging through semispinalis and supplying the skin of the scalp as far as the vertex. They are the *greater occipital nerve* (the posterior ramus of C2) and the *third occipital nerve* (the posterior ramus of C3). Finally, when semispinalis is removed, the suboccipital triangle is revealed. You must first recognize the bony landmarks (Fig. 17.7). The atlas has no spine, only a posterior tubercle. It has, on the other hand, very long transverse processes, which are used as levers to rotate the head and atlas on the pivot of the dens. The axis does not have much in the way of transverse processes but it does have a very big spine. The landmarks to recognize are, therefore, the *transverse processes of the atlas*, the *spine of the axis*, the *posterior tubercle of the atlas* and the *superior and inferior nuchal lines* on the occipital bone.

## Structures forming and enclosed by the suboccipital triangle

*Rectus capitis posterior* (RCP) major arises from the spine of the axis and is inserted into the area below the inferior nuchal line lateral to *RCP minor*. The latter, being smaller, is attached to the posterior tubercle of the atlas and is inserted below the inferior nuchal line on each side of the midline. *Obliquus superior* is

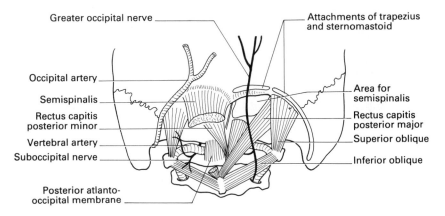

Greater occipital nerve — Attachments of trapezius and sternomastoid

Occipital artery —
Semispinalis —
Rectus capitis posterior minor —
Vertebral artery —
Suboccipital nerve —

Area for semispinalis
Rectus capitis posterior major
Superior oblique
Inferior oblique

Posterior atlanto-occipital membrane —

**Fig. 17.7** The suboccipital triangle. Trapezius, sternocleidomastoid and most of semispinalis have been removed.

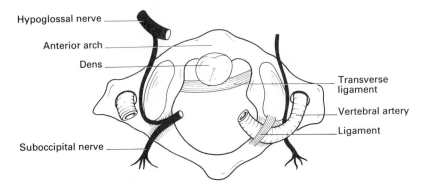

Hypoglossal nerve

Anterior arch

Dens

Transverse ligament

Vertebral artery

Ligament

Suboccipital nerve

**Fig. 17.8** Relations of the upper surface of the atlas.

from the transverse process of the atlas to the area between the two nuchal lines lateral to semispinalis. Finally, *obliquus inferior* is from the spine of the axis to the transverse process of the atlas. The actions of all these smaller muscles are obvious.

Of far more importance is the course of the vertebral artery. Having ascended through the foramina transversaria of the upper six cervical vertebrae (*not C7*), it emerges on to the upper surface of the atlas (Fig. 17.8). It then passes backwards, closely related to the articular facet, to reach a deep groove, which is sometimes converted to a foramen by a spicule of bone above the artery. It then passes upwards through the foramen magnum to meet its opposite number to form the *basilar artery*.

Emerging horizontally from the spinal cord and lying in the groove of the atlas between the vertebral artery and the bone is the 1st cervical nerve. Its posterior ramus (the *suboccipital nerve*) enters the suboccipital triangle and supplies the muscles. Its anterior ramus winds forwards round the articular facet and joins the hypoglossal nerve, which has just left the anterior condylar (hypoglossal) canal (Fig. 17.8). Between the atlas and the axis emerges the very thick posterior ramus of C2, which supplies muscles and then travels upwards, over the suboccipital triangle, to form the greater occipital nerve. The 3rd occipital, which is much smaller, follows a similar course between the axis and the 3rd cervical vertebra.

In view of the rather tortuous course of the vertebral artery in this region, it is not surprising that, if it is affected by atheroma, movement of the head and neck may affect the flow of blood through it and cause faintness or unconsciousness. This may also be caused by 'subclavian steal', which will be mentioned in Chapter 18.

# Chapter 18
# Lower Part of Neck

You have seen in Chapter 17 how there are two major blocks of tissue in the neck. The block containing the vertebrae and their surrounding muscles has been described in detail and it is now appropriate to describe the other block, which contains the trachea and oesophagus and the larynx and pharynx. Since the upper part of this block is complicated by the presence of the mandible, this chapter will deal only with the lower part.

## THE ANTERIOR TRIANGLE

First, as happened with the posterior triangle, the boundaries and subdivisions of the neck must be mentioned. The anterior triangle is the area enclosed by the anterior border of sternomastoid, the lower border of the mandible and the midline. It includes the submental region (below the chin) and the submandibular region, which will be dealt with in Chapter 20. The most obvious component, however, is the conspicuous prominence of the larynx, especially the thyroid cartilage, and especially in the male. The palpable components of the larynx are the *hyoid bone*, and the *thyroid* and *cricoid cartilages*, which lie respectively at vertebral levels C3, (4, 5), and 6. Of course the whole larynx and its associated structures move upwards on swallowing. The contents of the anterior triangle are obscured to some extent by the *platysma muscle*, a peculiar muscle which lies in the superficial fascia. You can see a similar cutaneous muscle at work in a horse standing under a tree on a hot day when it removes noxious insects by contraction of the muscle. In man, its fibres begin over the upper intercostal spaces and run up the neck to cross the lower border of the mandible. It then mingles in with the facial muscles around the mouth (see Fig. 21.1). In elderly thin people, the anterior borders of the two platysma muscles form a pair of sharp vertical ridges below the chin. It pulls the corners of the mouth and the skin of the face downwards and can often be seen contracting in athletes making a tremendous effort, since it has an antisphincteric action on the neck and so aids venous return. It is, in fact, the muscle used to ease a tight collar, but, more importantly, it is a valuable landmark to the surgeon.

## THE PHARYNX

### Exterior

It is important to understand that the pharynx, unlike the oesophagus, is not a complete tube. The pharyngeal wall surrounds the lateral and posterior aspects of the pharynx but anteriorly the cavity is open. At the top, it opens anteriorly into the nasal cavity and is called the *nasopharynx*. Lower down it opens into the mouth and is called the *oropharynx* whereas below it opens into the larynx, the *laryngopharynx*. You can thus view the posterior wall of the oropharynx through the open mouth and can inspect the naso- and laryngopharynx by using suitably angled mirrors.

The three constrictor muscles of the pharynx provide a more or less circular striated muscle coat. They overlap each other from below upwards: think of three glasses stacked on a bar (see Fig. 17.6). Three other muscles, *stylopharyngeus*, *palatopharyngeus* and *salpingopharyngeus*, spread out to form an inner longitudinal coat. They will be described later.

The pharynx is covered by a layer of fascia continuous with that over the buccinator muscle and therefore called the *buccopharyngeal fascia* (p. 335). Between the muscle and the mucous membrane (stratified squamous epithelium, of course) is another layer of fascia, called the *pharyngobasilar fascia*. These layers, together with the mucous membrane, close the gaps between individual muscles, which are not as large as the diagrams in this chapter suggest. The pharynx ends at the lower border of the cricoid cartilage (C6) by becoming oesophagus, just as the larynx, at this level, gives way to the trachea.

### The middle and inferior constrictors

The hyoid bone has a body in the midline, a pair of lesser horns (*cornua*) and a pair of greater horns. Although in a dissection of the neck with the chin pulled back it seems to be a long way down the neck, in fact it is tucked away under the mandible so that the mylohyoid muscle (see Fig. 19.3) is horizontal. It is a delicate bone and often fractured in cases of manual strangulation. Running from the styloid process (Chapter 16) to the lesser horn is the *stylohyoid ligament*, a remnant of the embryonic 2nd branchial arch. All this preamble is leading up to the origin of the middle constrictor, which arises from the lower end of the ligament, the lesser horn and the upper border of the greater horn (Fig. 18.1). The muscles travel round the pharynx on each side and meet in a midline raphe at the back, diverging upwards and downwards as shown in the diagram. The lower fibres are overlapped by the inferior constrictor and the upper fibres overlap the superior constrictor (the three glasses).

The inferior constrictor arises from the *oblique line* on the thyroid cartilage (Fig. 18.1), from the cricoid cartilage and, between these two attachments, from a small fibrous band that arches over the cricothyroid muscle. Its lower fibres are horizontal and are therefore often called *cricopharyngeus*. Its upper fibres ascend more and more steeply. Posteriorly, there is a potential weak area between the horizontal and the ascending fibres (Fig. 18.1). This is called *Killian's dehiscence* and it is possible for the mucous membrane to herniate out through it to form a pharyngeal diverticulum, which may press on the pharynx from the outside and cause difficulty in swallowing. The beginning of the oesophagus is one of the constricted regions mentioned in Chapter 8 (see Fig. 8.22).

### The infrahyoid muscles

These muscles, often called the *strap muscles*, are part of a longitudinal band of muscle on the front of the body that includes rectus abdominis and the occasional *sternalis* muscle that may be encountered on the front of the chest. They are

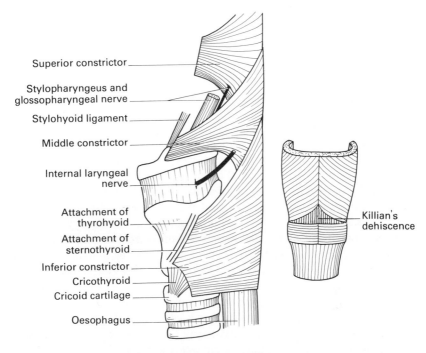

Superior constrictor

Stylopharyngeus and glossopharyngeal nerve

Stylohyoid ligament

Middle constrictor

Internal laryngeal nerve

Attachment of thyrohyoid

Attachment of sternothyroid

Inferior constrictor

Cricothyroid

Cricoid cartilage

Oesophagus

Killian's dehiscence

**Fig. 18.1** The constrictions of the pharynx (see also Fig. 19.1).

*sternothyroid, sternohyoid, omohyoid* and *thyrohyoid*. Two of these are, like the inferior constrictor, attached to the oblique line on the thyroid cartilage. Sternothyroid arises from the back of the manubrium and is attached above to the oblique line. Thyrohyoid is a continuation of this, from the oblique line to the hyoid bone. Its lateral border slopes laterally and extends on to the greater horn, so it is quadrilateral rather than strap-shaped. Sternohyoid, from the manubrium and extending laterally on to the back of the clavicle, must necessarily be superficial to sternothyroid and thyrohyoid on its way to the hyoid bone. The superior belly of omohyoid is lateral to sternohyoid and runs downwards and laterally, passing deep to sternomastoid. Its intermediate tendon, at the level of the cricoid, is hitched to the clavicle by fascia, which allows it to be angled as it gives way to the inferior belly, described in Chapter 17.

The infrahyoid muscles are supplied by the *ansa cervicalis*. From the hypoglossal nerve a small nerve usually called the *descendens hypoglossi* (but officially called the *superior ramus of the ansa cervicalis*) apparently departs from the main stem and descends to meet the *descendens cervicalis* from C2 and 3 to form a long loop, the *ansa cervicalis*. The hypoglossal branch is actually the anterior ramus of C1 (see Fig. 17.8), which travels as a passenger with the hypoglossal. Another similar 'branch' of the hypoglossal goes directly to thyrohyoid (and geniohyoid, to be described later). The strap muscles are therefore supplied by C1, 2 and 3. Their function is to steady the hyoid bone, acting as fixators so that the suprahyoid muscles can act, and they also probably lower the larynx after it has been raised during swallowing. The hyoid bone does not articulate with any other bone (which is why it is almost always missing from mounted skeletons) so should be regarded as a 'suspended' bone like the scapula, whose position is regulated only by the attached muscles.

The main interest of the strap muscles is in their relationship to the thyroid gland and its associated structures.

## THE THYROID GLAND

This is an endocrine gland producing three main hormones: tri-iodothyronine, thyroxine and calcitonin. Its isthmus crosses the midline on the 2nd, 3rd and 4th tracheal rings (some variation) and its lateral lobes lie at the sides of the trachea and oesophagus under cover of the strap muscles. If you push your finger upwards in this region, you will understand that the lateral lobe of the thyroid cannot normally rise above the level of the oblique line since it is caught in the 'pocket' of sternothyroid, between that muscle and the thyroid cartilage (Fig. 18.2). When enlarged, however, and it can get to be enormous, the strap muscles are stretched thinly over it and the gland displaces the carotid sheath laterally.

Occasionally there is a small upwards extension of the isthmus, called the *pyramidal lobe*, which may be continuous with a fibromuscular band that attaches it to the hyoid bone. Since the gland is enclosed in the pretracheal fascia, which attaches it to the trachea and larynx, it moves upwards on swallowing, an important diagnostic feature for lumps in the neck.

The thyroid is a very vascular structure, so much so that in some pathological conditions, such as thyrotoxicosis, the blood flow can be heard with a stethoscope as a bruit. The main arteries are the superior and inferior thyroid arteries (Fig. 18.3). The *superior thyroid artery* arises from the anterior surface of the external carotid immediately distal to the carotid bifurcation. It arches downwards,

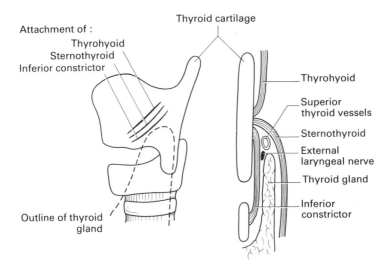

**Fig. 18.2** Relations of the lateral lobe of the thyroid gland.

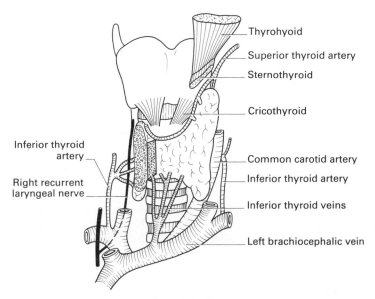

**Fig. 18.3** Blood supply of the thyroid gland.

giving a *superior laryngeal* branch that enters the larynx with the internal laryngeal nerve and then passes deep to the strap muscles to enter the sternothyroid
'pocket' (Fig. 18.2), before reaching the upper pole of the gland. Here it divides
into two main branches, one of which follows the upper border to the isthmus
while the other runs down the posterior border to anastomose with the *inferior
thyroid artery*. The latter is a branch of the thyrocervical trunk from the subclavian
artery. It ascends and turns medially at the level of the cricoid cartilage and then
enters the back of the thyroid gland some distance above the lower pole. It has
important relationships, which will be described in the last section of this chapter.
Both thyroid arteries are related to nerves, which must be avoided when tying the
arteries. A little distance behind the superior thyroid artery is the *external
laryngeal nerve*, while the *recurrent laryngeal* has a variable relationship to the
inferior thyroid artery (p. 333). These arteries anastomose freely with each other
and with the tracheal and oesophageal arteries. In the operation of partial or
subtotal thyroidectomy, all four arteries are usually tied and all but the posterior
parts of the lobes excised. The remains of the gland are two strips of tissue,
alongside the trachea, which contain the parathyroids, the whole being supplied
with blood by the anastomoses. The *thyroidea ima* is a small, occasional artery
from the brachiocephalic that may help to supply the thyroid.

The veins are three in number on each side. The *superior* and *middle thyroid
veins* pass laterally to enter the internal jugular but the *inferior veins* descend in
front of the trachea to reach the left brachiocephalic vein.

### Development of the thyroid gland

The gland begins as a diverticulum from the floor of the embryonic pharynx
(Fig. 18.4); this grows caudally, superficial to the branchial arches (and hence to

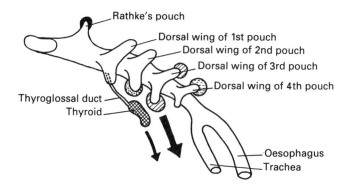

**Fig. 18.4** A schematic 'cast' of the embryonic pharynx to show its diverticula — the *pharyngeal pouches*. Proliferations of endothelial cells are associated with the dorsal and ventral wings. Rathke's pouch grows dorsally towards the brain.

the hyoid and larynx), before dividing into two lobes. The stem of the diverticulum, the *thyroglossal duct*, normally disappears, although traces may remain. As the lobes expand they come into contact with the ventral part of the 4th pharyngeal endodermal pouch, which contributes cells that become the *interfollicular ('C') cells*. After the tongue has developed, it can be seen that the point of outgrowth of the thyroglossal duct is the foramen caecum (p. 357). Aberrant thyroid tissue, or cysts derived from the duct (*thyroglossal cysts*) may appear anywhere between the foramen caecum (where the aberrant gland can be seen sitting on the tongue) and the normal position. They can be diagnosed because characteristically they move upwards when the patient puts his tongue out. Not surprisingly, the thyroid may sometimes descend too far and be found in the superior mediastinum, although a 'thoracic goitre' is usually caused by an enlargement of the lower part of the gland, which has grown downward. Infection of a thyroglossal cyst is common and may spread to a persistent thyroglossal duct, which must then be excised. Although the duct lies ventral to the body of the hyoid bone, it always passes up for a short distance behind the body (Fig. 18.5), which therefore has to be excised with the duct.

## PARATHYROID GLANDS

There are two parathyroid glands on each side. They are yellow-brown in colour, about the size of a small pea, and are important because of their role in calcium metabolism. The superior parathyroid is embedded in the posterior surface of the thyroid gland (but outside its capsule) a short distance above the entry of the inferior thyroid artery. The inferior is usually embedded behind the lower pole but is often found elsewhere. In about 5% of cases the inferior parathyroids are in the superior mediastinum. This is because the parathyroids develop from the endoderm of the 3rd (inferior gland) and 4th (superior gland) pharyngeal pouches (Figs 18.4 & 18.5). The thymus also develops from the 3rd pouch and may therefore carry the inferior parathyroid with it when it descends into the thorax.

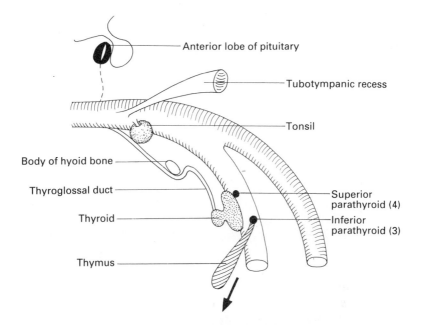

Anterior lobe of pituitary

Tubotympanic recess

Tonsil

Body of hyoid bone

Thyroglossal duct

Superior
parathyroid (4)

Thyroid

Inferior
parathyroid (3)

Thymus

**Fig. 18.5**  The fate of the endodermal proliferations. Rathke's pouch forms the
adenohypophysis. The 1st and part of the 2nd pouches form the *tubotympanic recess*,
which will become the auditory tube and the middle ear. The ventral wing of the 2nd pouch
forms the tonsil. The derivatives of the other pouches are described in the text.

Tumours of the parathyroids produce an excess of parathormone, which has the
effect of withdrawing calcium from bones (*osteitis fibrosa cystica*) with possibly
the formation of renal stones. There are, however, other causes of hyperparathy-
roidism.

## THE ROOT OF THE NECK AND THE THORACIC OUTLET

The area defined by the 1st thoracic vertebra, the 1st ribs and the manubrium
sterni is usually called the *thoracic outlet*, although sometimes the thoracic inlet,
since structures pass both into and out of the thorax through this space. Its
importance is that numerous structures, such as the carotid and subclavian
arteries, the trachea and oesophagus and large veins, are all packed together into
a relatively small and restricted space so that any decrease in the size of the
opening may have dire effects. This may be caused, for example, by tumours of the
apex of the lung (Pancoast tumours) or of the thyroid gland. A cervical rib
(p. 281) or the scalenus anterior muscle may affect the subclavian artery and the
brachial plexus as they enter the arm over the 1st rib. This region therefore merits
a detailed description.

### The scalene muscles

The origins of the scalene muscles have been described in the previous chapter
(p. 313). Their insertions are shown in Fig. 18.6. Scalenus anterior is attached by

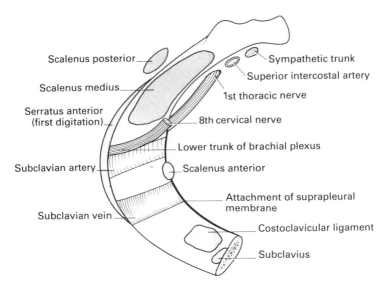

**Fig. 18.6** Relations of the upper surface of the 1st rib.

a narrow tendon to the scalene tubercle on the medial border of the 1st rib, while scalenus medius has a muscular, and therefore much larger, attachment to the upper surface. Scalenus posterior carries on down to the 2nd rib. The lower trunk of the brachial plexus and the subclavian artery are in contact with the 1st rib in a groove between scalenus anterior and medius, while the vein is in front of scalenus anterior. They are vulnerable here since they have to climb out of the thorax over the 1st rib before turning downwards. A *cervical rib* may be attached to the 7th cervical vertebra and to the manubrium but usually a short rib is connected to the manubrium by a fibrous band. The subclavian artery and the lower trunk have the same relationship to a cervical rib as they normally have to the 1st rib, so that they have to turn downwards even more sharply. Even without a cervical rib, the sharp edge of the scalenus anterior tendon may compress the artery and the nerve. This results in vascular disturbances (cold blue hands) and possibly wasting and weakness of the small muscles of the hands (T1 segment supplies these via the median and ulnar nerves), and pain or tingling along the medial border of the hand and forearm.

## The triangle of the vertebral artery

Another useful triangle! Its lateral margin is the medial border of scalenus anterior and its medial margin is the lateral border of longus colli (p. 313). The base of the triangle is the subclavian artery as it emerges from the thorax to reach the 1st rib. The artery is thus divided into three parts, medial to scalenus anterior, behind that muscle, and lateral to it, to end at the outer border of the 1st rib, where it becomes the axillary artery (Fig. 18.7).

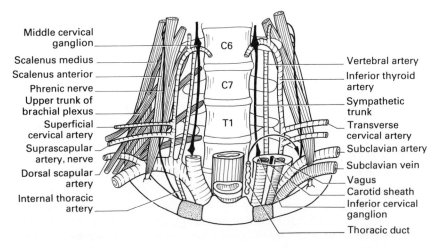

**Fig. 18.7**  The deep structures at the root of the neck. Longus colli has been omitted to show the vertebrae. The curved arrow on the right side indicates the plane of the terminal part of the thoracic duct.

### Subclavian artery

This arises, on the right side, from the brachiocephalic artery behind the sterno-clavicular joint, the other branch being the common carotid. On the left side it arises directly from the arch of the aorta just beyond the origin of the left common carotid. (In examination questions involving the course and relations of the common carotid and subclavian arteries, remember that, on the left side, *both have an intrathoracic course.*) On both sides, the plane of the common carotid is in front of that of the subclavian. The first part of the subclavian artery has three important branches, all shown in Fig. 18.7, and one less important branch.

The first branch is the *vertebral artery*, which climbs up the medial border of the triangle as far as its apex at the transverse process of C6. It enters the foramen transversarium here; the foramen transversarium of C7 only transmits some small veins. The vertebral vein itself is anterior to the artery.

The second branch is the *internal thoracic* (or internal mammary). This comes from the lower border and promptly disappears from view as it travels down into the thorax (p. 170).

The third branch is the *thyrocervical trunk*. This arises at the medial border of scalenus anterior and breaks into three branches, the *inferior thyroid, superficial cervical* (or transverse cervical, see p. 318) and *suprascapular* arteries. The latter two arteries cross scalenus anterior and the phrenic nerve to reach the posterior triangle, where their subsequent course has already been described. The inferior thyroid artery runs up the lateral border of the triangle and, at the level of C6, it arches medially across the vertebral artery to reach the thyroid gland.

The fourth and least important branch is the *costocervical trunk*. This arches back over the dome of the pleura and the suprapleural membrane (described later) and divides into a branch that goes backwards to supply muscles of the back, and a branch, the *superior intercostal artery*, that crosses the neck of the 1st rib (Fig. 18.6) to supply the upper two to three intercostal spaces.

Finally, there is usually one branch from the third part of the subclavian artery outside the triangle. This is the *dorsal scapular artery* (p. 318).

## Veins

The subclavian and internal jugular veins meet at the medial border of scalenus anterior to form the brachiocephalic vein. There is thus no part of the subclavian vein corresponding to the first part of the subclavian artery. The subclavian vein begins at the outer border of the 1st rib and passes in a shallow groove on the 1st rib in front of scalenus anterior. The internal jugular is the main vein draining the head and neck and is enclosed with the common and internal carotid arteries in the carotid sheath. Between the two vessels and situated posteriorly, is the vagus nerve (see Fig. 17.3). As can be seen in Fig. 18.7, both the common carotid artery and the internal jugular vein are on a plane anterior to the subclavian artery, so that the triangle of the vertebral artery is completely covered by the carotid sheath and its contents. These have necessarily been cut short in Fig. 18.7 in order to show the triangle.

## Nerves

The lower cervical and 1st thoracic nerves can be seen in Figs 18.6 & 18.7. They run in the floor of the triangle since they are behind the plane of scalenus anterior, which is arising from anterior tubercles. Behind scalenus anterior they form the *upper, middle* and *lower trunks* of the brachial plexus, which emerge between scalenus anterior and medius (see Fig. 17.5). The lower trunk is related directly to the 1st rib (Fig. 18.6).

Just lateral to the triangle, on the surface of scalenus anterior, the *phrenic nerve* descends to the thorax.

Running down the medial border of the triangle is the *sympathetic trunk* (Fig. 18.7). It may cross in front of or behind the inferior thyroid artery, or split to enclose it, and at this level it carries the small *middle cervical ganglion*. Lower down, the trunk crosses the neck of the 1st rib and about this level, and tucked behind the vertebral vessels, is the *inferior cervical ganglion* (or, if it is fused with the 1st thoracic ganglion, the *stellate ganglion*). These ganglia have been described in Chapter 15. It is helpful to remember that on the neck of the 1st rib (Fig. 18.6) there is an artery between two nerves, i.e. the superior thoracic artery between T1 and the sympathetic trunk (cf. the spine of the ischium, p. 275, the brim of the pelvis, p. 267 and the pudendal canal, p. 276).

## The thoracic duct

On the left side only, the thoracic duct has been seen to be ascending on the left edge of the oesophagus (Chapters 8 and 17). In the root of the neck it arches laterally, immediately behind the carotid sheath and its contents, and enters the junction between the internal jugular and subclavian veins. It therefore crosses the triangle of the vertebral artery transversely, crossing the vertebral vessels, the sympathetic trunk and probably the thyrocervical trunk (Fig. 18.7).

## The suprapleural membrane

This is a layer of fascia that is attached to the transverse process of the 7th cervical vertebra and spreads out like a tent, covering the dome of the pleura, to be attached around the medial border of the 1st rib. It therefore separates the structures in the root of the neck from the pleural cavity. The apex of the lung is thus closely related to the structures in the lower part of the neck, such as the subclavian vessels.

## The recurrent laryngeal nerve

Although not strictly contained within the triangle of the vertebral artery, this important nerve is nevertheless closely related to it. On the left side it 'recurs' around the ligamentum arteriosum (p. 190), but on the right side it leaves the vagus nerve just after that nerve has crossed the subclavian artery. It then runs upwards and medially so that on both sides the nerve ascends in the groove between the trachea and the oesophagus on its way to the larynx, although it often strays away from this groove. It passes close behind the thyroid gland and is therefore bound to encounter the inferior thyroid artery, which it crosses at right angles. Sometimes it crosses in front, sometimes behind, and sometimes the artery breaks up into branches before it reaches the gland and the nerve passes between the branches (Fig. 18.3). Finally the nerve enters the larynx just behind the tiny synovial joint between the inferior horn of the thyroid cartilage and the side of the cricoid cartilage.

## The trachea and oesophagus

The trachea is roughly 10 cm long and extends from the level of the 6th cervical vertebra down to its bifurcation at the lower border of T4. These levels are approximate since the larynx ascends during swallowing and the diaphragm descends in inspiration. The trachea, therefore, has to be extensible like a concertina, so its 'rings' of hyaline cartilage are separated by fibro-elastic tissue. Posteriorly, the rings are deficient, the gap being closed by similar tissue with the addition of smooth muscle fibres. If complete rings were present, they would interfere with the passage of food down the oesophagus, which lies directly behind.

It is important to remember that the trachea in children is extremely small — about 3 mm in diameter during the first year.

The trachea can be seen as a dark shadow in an X-ray (see Fig. 8.4) and can be palpated in the midline just above the upper border of the manubrium. This part of the trachea is crossed by the isthmus of the thyroid gland (which is usually divided during a tracheostomy), the inferior thyroid veins, the thyroidea ima artery, if present, and a transverse anastomosis between the anterior jugular veins.

The trachea is lined by pseudostratified ciliated columnar epithelium with seromucous glands. The cilia waft small particles upwards towards the laryngeal inlet; larger particles are removed by the cough reflex, mediated by the sensory nerve supply from the vagus and recurrent laryngeal nerves.

The trachea and lungs develop from a groove in the floor of the pharynx, which separates off to form the trachea and, later, the lung buds. Abnormal communications (*fistulae*) are therefore not uncommon and are often associated with narrowing or complete occlusion of the oesophagus (*atresia*). The commonest form of this anomaly (85%) is that in which the upper segment of the oesophagus ends blindly while the upper end of the lower segment communicates with the trachea.

The oesophagus in the neck begins just below the cricopharyngeus muscle, where the beginning of its outer longitudinal muscle can be seen. The muscle in the upper part is skeletal rather than smooth. The oesophagus lies on longus colli. The thoracic duct lies on its lateral border before arching over the triangle of the vertebral artery and the recurrent laryngeal nerves are in or near the groove between trachea and oesophagus. The lateral lobes of the thyroid gland and the common carotid arteries are also relations.

# Chapter 19
# Upper Part of Neck

Unfortunately, the upper part of the neck is more complicated than the lower part, mainly because it has more layers, the relationship between the layers being difficult to visualize. In particular, the mandible obtrudes on this region so that the relationships of the mandible provide a valuable insight into the structure of the region. Throughout this chapter, therefore, there will be repeated references to the final section and to Fig. 19.1, which describe and show the deep relations of the mandible. You are advised to refer constantly to this figure.

As in the previous chapter, it is easier to build up from the foundations rather than to work in from the outside, as you have to do during a dissection. This section will therefore begin with a description of the outside of the oro- and nasopharynx.

## SUPERIOR CONSTRICTOR

An ill-defined ligament or *raphe* (a fibrous band between muscles) runs from the pterygoid hamulus on the medial pterygoid plate to the mandible just behind the third molar tooth (Fig. 19.2). The raphe can be seen from the inside as a ridge when the mouth is opened wide. From the front of this *pterygomandibular ligament* (and the adjoining bone), the *buccinator* muscle takes origin and runs forwards in the cheek; running backwards from it is the *superior constrictor*. These two muscles are therefore continuous with each other, except for the intervention of the ligament, so that you will understand how the *buccopharyngeal fascia* can cover both muscles (p. 324). The lower fibres of the superior constrictor pass downwards, and the upper fibres pass from the pterygoid hamulus up to a small tubercle (the *pharyngeal tubercle*) on the occipital bone just in front of the foramen magnum. There is thus a gap between the upper border of the muscle and the base of the skull but this is filled in by the two fascial layers (p. 324) and, as will be described later in this chapter, the *tensor (veli palatini)* and *levator palati* muscles. The auditory (Eustachian or pharyngotympanic) tube passes through this opening.

On the outer surface of the superior constrictor is *stylopharyngeus*. This is one of three muscles that arise from the styloid process and run downwards and forwards (the other two are *stylohyoid* and *styloglossus*). It enters the pharynx between the superior and middle constrictors and then spreads out to form an inner longitudinal coat together with *palatopharyngeus* and *salpingopharyngeus*. Some of its fibres are attached to the posterior border of the thyroid cartilage so that one of its functions is to elevate the larynx during swallowing.

The mechanism of swallowing will be described in Chapter 21 but here it is necessary to mention the *palatopharyngeal sphincter*, a distinct ridge of circularly orientated muscle fibres that encircle the pharynx, as a thickening of the superior

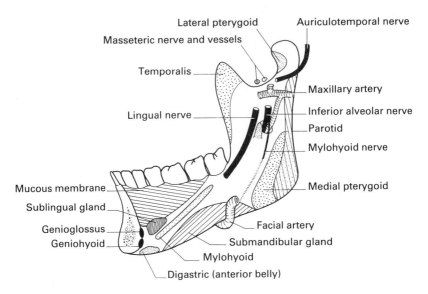

**Fig. 19.1** Structures related to the *medial* surface of the mandible.

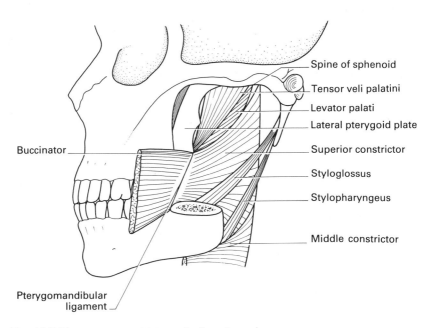

**Fig. 19.2** The superior constrictor and adjacent muscles.

constrictor, at the level of the palate. When these contract, they raise a ridge of mucous membrane known as the *ridge of Passavant*. This helps to close off the nasopharynx from the oropharynx, thus preventing reflux of food into the nasal cavity during swallowing.

## THE SUBMANDIBULAR REGION

Having described the superior constrictor, we must now return for a moment to the middle constrictor as a basis for the submandibular region. As was seen in the previous chapter, the main origin of the middle constrictor is from the greater horn of the hyoid bone. Just below this origin another muscle, *hyoglossus*, is attached to the hyoid and its fibres run upwards and forwards into the side of the tongue. Hyoglossus is thus superficial to the middle constrictor (Fig. 19.3).

The most anterior muscle in the submandibular region is *mylohyoid*. This region is usually depicted from the front with the chin raised, which gives one the impression that mylohyoid is a vertical muscle and that the hyoid bone is quite low down in the neck. In fact, the hyoid is tucked away under the mandible and mylohyoid is more or less horizontal, forming the floor of the mouth (see Fig. 13.13). It can thus be palpated between a finger in the mouth, between the tongue and the gums, and a thumb outside. It arises from the mylohyoid line on the mandible and its fibres run towards the midline to be inserted into the body of the hyoid bone, while above this the muscles of the two sides meet in the midline raphe.

The *digastric* muscle, as its name implies, has two bellies. The anterior belly arises from the inside of the front of the mandible (see Fig. 19.1) and the posterior belly from a groove medial to the mastoid process. The two bellies meet in an intermediate tendon, which is tethered to the upper surface of the hyoid bone by a loop of fascia. Associated with the posterior belly is a very small muscle, *stylohyoid*, which arises from the styloid process and is inserted into the hyoid, splitting to enclose the digastric tendon. It should be regarded as a separate part of the

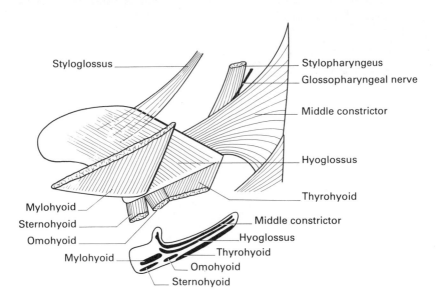

**Fig. 19.3** The three 'steps' in the submandibular region. The hyoid bone cannot be identified clearly because of the muscles, so it is shown again below, with the muscle attachments marked.

posterior belly, having the same relations, nerve supply and, therefore, development. The nerve supply of mylohyoid and the anterior belly is the *mylohyoid nerve*, a branch of the inferior alveolar nerve (see below), which itself is a branch of the mandibular division of the trigeminal. Both, therefore, are 1st branchial arch derivations. Stylohyoid and the posterior belly are supplied by the facial nerve since they are derived from the 2nd arch.

The submandibular region is best described in terms of three descending steps (Fig. 19.3). The first step is mylohyoid, the second is hyoglossus and the third is the middle constrictor. The major structures related to these steps will now be described, and all the structures on each of the steps will then be listed systematically.

## The submandibular gland

This large salivary gland is partially tucked up under the mandible but is prevented from going any higher by the attachment of mylohyoid to the mylohyoid line (see Fig. 19.1). The investing layer of deep fascia splits to enclose the gland and encloses also the submandibular lymph nodes, which are partly embedded in it. The gland lies mostly on the first two steps, i.e. mylohyoid and hypoglossus, but it also has a deep part that folds round the posterior border of mylohyoid (Fig. 19.4). From the anterior portion of the deep part, the *submandibular (Wharton's) duct* runs forwards on hyoglossus and finally opens at the top of a small *sublingual papilla*, which lies just to the side of the frenulum beneath the tongue. Inspect your own. If you have yawned while reading this so far tedious chapter, you may well find the page spattered with submandibular gland secretion expelled by the contraction of mylohyoid. The duct is double-crossed by the lingual nerve, described later.

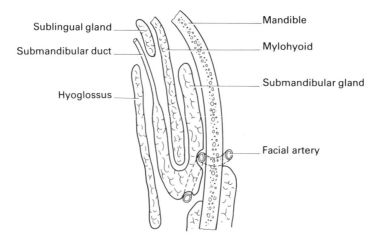

**Fig. 19.4** A horizontal section through the submandibular region to show the submandibular gland and its relations.

### The sublingual and other salivary glands

The sublingual gland is situated just to one side of the midline in the floor of the mouth and therefore deep to mylohyoid (see Fig. 19.1). It produces a ridge in the floor of the mouth and its numerous ducts open either on to the ridge or into the submandibular duct. Other small salivary glands are situated beneath the mucous membrane of the mouth. You can feel some of them with your tongue as small nodules on the inside of the lips.

The sublingual and submandibular glands receive a parasympathetic secreto-motor nerve supply, which reaches the region in the *chorda tympani*. This small nerve joins the lingual in the infratemporal region and its parasympathetic fibres (it also contains sensory fibres) synapse in the *submandibular ganglion* and pass directly to the submandibular gland and, via the lingual nerve, to the sublingual gland (see also p. 352).

### Branches of the external carotid artery in the region

Opposite the greater horn of the hyoid, the external carotid gives off the *lingual artery* and, a little higher, the *facial artery* (see Fig. 19.6). The two arteries often arise from a common trunk and are, of course, above the superior thyroid artery (p. 327).

The lingual artery loops upwards and down again, before disappearing from view *deep* to hyoglossus. It therefore lies entirely on the third step. A very useful landmark is the *hypoglossal nerve* crossing the loop of the lingual artery before coming to be *superficial* to hyoglossus. The artery gives branches to the tongue (*rami dorsales linguae*) and then turns up on the under-surface of the tongue to reach the tip. The corresponding veins are easily visible beneath the tongue.

The facial artery also loops upwards and is therefore deep to the mandible and embedded in the posterior surface of the submandibular gland (Figs 19.4 & 19.5). It then turns laterally and descends between the gland and the mandible before finally turning round the lower border of the mandible to reach the face (Chapter 21). It gives a *tonsillar branch* that pierces the superior constrictor just above styloglossus to reach the tonsillar bed.

From the posterior surface of the external carotid, two branches run posteriorly (Fig. 19.6). These are the small and unimportant *posterior auricular* and the larger *occipital artery*, which occupies a groove medial to the mastoid process and

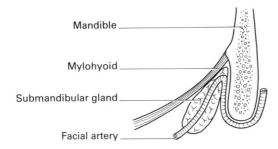

Mandible

Mylohyoid

Submandibular gland

Facial artery

**Fig. 19.5** A vertical section to show the route taken by the facial artery.

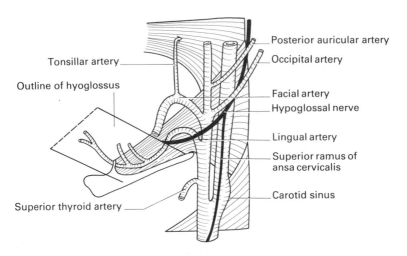

**Fig. 19.6** Some of the branches of the external carotid artery as it lies on the pharynx. The hypoglossal nerve has been cut off short to show the arteries. See Fig. 19.7 for its full course in this region. The superior ramus of the ansa cervicalis is the other name for the descendens hypoglossi (p. 325).

the posterior belly of digastric, finally to supply the back of the scalp. It is crossed by the hypoglossal nerve and is also related to the accessory nerve (see Fig. 19.9).

## Two major nerves

These lie on the second step (hyoglossus) and are the *lingual nerve* (above, and covered by the mandible) and the *hypoglossal nerve* (Fig. 19.7).

The lingual nerve enters the region by passing just behind the third molar tooth immediately deep to the mucous membrane of the gum (see Fig. 19.1). It crosses hyoglossus ('with a swerve' as the well-known students' poem puts it) and then turns upwards, deep to mylohyoid to reach the under-surface of the tongue. In the process, it crosses the submandibular duct superficially and then turns up deep to the duct (the double-cross referred to above). Attached to the lower side of the nerve is the *submandibular ganglion*, which lies deep to the submandibular gland and supplies it with secretomotor fibres that have synapsed in the ganglion. Other postganglionic fibres re-enter the lingual nerve, which transports them to the sublingual gland.

The lingual nerve itself is sensory to the anterior two-thirds of the tongue and the inner (lingual) surface of the gums. Its content of fibres belonging to the chorda tympani carry taste sensation from the same region. Remember that during part of its course the lingual *nerve* is separated from the lingual *artery* by hyoglossus.

The hypoglossal nerve crosses the external carotid artery and the upward loop of the lingual artery, runs across hyoglossus below the lingual nerve (Fig. 19.7), and turns up, deep to mylohyoid, to enter the tongue, where it supplies all the tongue muscles, intrinsic and extrinsic, except for palatoglossus, which is not really a muscle of the tongue anyway.

It is now time to review the three steps systematically.

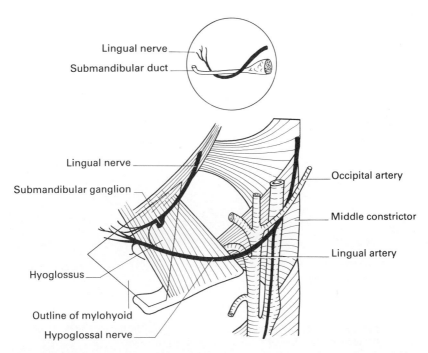

**Fig. 19.7** The two nerves on hyoglossus, the second 'step'. Inset, the lingual nerve double-crossing the submandibular (Wharton's) duct.

## Superficial structures

The submandibular region is crossed superficially by *platysma* (p. 323). The next to lowest branch of the facial nerve, the *marginal mandibular nerve*, crosses the lower border of the mandible near the angle and runs below it across the submandibular region before finally turning up to reach the region of the chin musculature. The *greater auricular nerve* crosses the posterior belly of digastric on its way to supply the skin of the face approximately over the area of the parotid gland.

## First step — mylohyoid

The major structures here are the *anterior belly of digastric* and most of the superficial part of the *submandibular gland* (Fig. 19.8). The *mylohyoid nerve* emerges from under cover of the mandible (see Figs 19.1 & 19.8) and supplies both muscles. Just below the chin, the *anterior jugular veins* begin by the joining up of a number of tributaries, and the *submental lymph nodes*, important in the lymphatic drainage of the tip of the tongue, also lie in this region. There is also a large *submental branch* of the facial artery. The *submandibular lymph nodes* lie in close relation to the submandibular gland.

## Second step — hyoglossus

The most important structures here are the two nerves, *lingual* above and *hypoglossal* below (see Fig. 19.7). There is usually a loop of communication

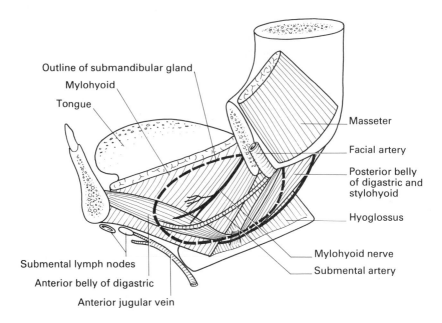

**Fig. 19.8** Structures on mylohyoid, the first 'step'. The broken line indicates the outline of the submandibular gland.

between them. The superficial part of the *submandibular gland* lies partly on the second step while its *deep part* and the beginning of its *duct* lie entirely on hyoglossus. Deep to the gland, and attached to the lingual nerve, is the *submandibular ganglion*. Some *large veins* accompany the hypoglossal nerve and they drain into the internal jugular vein. The intermediate tendon of *digastric* and the *stylohyoid muscle* are related to these structures superficially.

### Third step — middle constrictor

The external carotid gives off the *lingual* and *facial arteries* from its anterior aspect and the *occipital* and *posterior auricular arteries* from its posterior aspect. The *posterior belly of digastric* crosses the region. All these structures are separated from the medial surface of the mandible by the medial pterygoid muscle and the deep part of the parotid gland (see Fig. 19.1). Although not strictly on the third step, the *stylopharyngeus muscle* and the *glossopharyngeal nerve* both enter the pharynx just above the upper border of the middle constrictor.

### The submandibular muscles

This group of muscles, most of which have already been described, includes also *geniohyoid*. This is a narrow, but quite bulky muscle that lies alongside the midline deep to mylohyoid. It is attached to the lower genial tubercle on the mandible above, and to the hyoid bone below. It is really a deep member of the strap muscle group and is therefore supplied by the 1st cervical nerve via the hypoglossal. Geniohyoid, mylohyoid and the anterior belly of digastric all open the mouth when gravity is not sufficient. They therefore come into action during the 'down stroke' of chewing vigorously, in yawning and in singing. Since they act from the very mobile hyoid bone, they need the assistance of the infrahyoid muscles to act as fixators. If the mandible is fixed, they can help to elevate the larynx as in swallowing.

## INTERNAL CAROTID ARTERY

The bifurcation of the common carotid artery into internal and external carotids usually takes place at the level of the upper border of the thyroid cartilage. The internal carotid lies a little behind the external but, since it may have a very tortuous course, this relationship may be disturbed. Surgically, the external carotid is easily recognized because it has branches, whereas the internal carotid has no (normal) branches in the neck. At its origin, the internal carotid has a thin wall and is slightly dilated to form the *carotid sinus*. This is a baroreceptor, i.e. it responds to changes in blood pressure, and it should not be confused with the small *carotid body*, which is a chemoreceptor. Both, however, develop from the 3rd arch and are therefore supplied by the glossopharyngeal nerve.

The internal carotid is accompanied by the internal jugular vein, which lies behind it, and by the vagus nerve, which is between the artery and the vein. It finally enters the carotid canal in the petrous temporal bone, along with sympathetic nerve fibres from the superior cervical ganglion.

## VEINS

The veins of the region are large and may form a plexus. They are also subject to considerable variation. The *facial vein* begins on the face near the medial angle of the eye and it lies close to the corresponding artery at the lower border of the mandible. In the submandibular region it runs backwards, joins the anterior branch of the *retromandibular vein* to form the *common facial vein*, and this opens into the internal jugular vein. The *retromandibular* vein is formed in the parotid gland by the junction of the *superficial temporal* and *maxillary veins*. It receives various tributaries and ends behind the angle of the mandible by dividing into anterior and posterior branches. (The portal vein is not the only vein that has branches!) The anterior branch has already been described; the posterior joins the *posterior auricular vein* to form the external jugular vein (see Fig. 21.2).

## THE 'STYLO-MUSCLES'

Although these muscles are described elsewhere, it is helpful in remembering them to list them together. *Stylohyoid*, the most superficial, is associated with the posterior belly of digastric and is supplied by the facial nerve. *Styloglossus* passes into the side of the tongue, blending in with hyoglossus, and is supplied by the hypoglossal nerve. *Stylopharyngeus* is the deepest, entering the pharynx with the glossopharyngeal nerve, and is supplied by that nerve.

## THE LAST FOUR CRANIAL NERVES

These all leave the skull close together, the *glossopharyngeal, vagus* and *accessory* through the jugular foramen and the *hypoglossal* through the hypoglossal

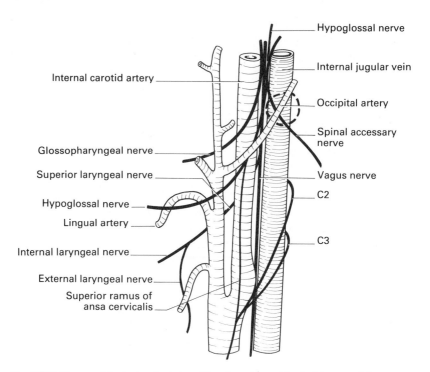

**Fig. 19.9** The carotid arteries, the internal jugular vein, and the last four cranial nerves. The broken line shows the position of the transverse process of the atlas, on the tip of which the occipital artery and accessory nerve cross on the internal jugular vein.

canal. At first they lie between the internal carotid artery and the internal jugular vein, where the cranial root of the accessory joins the vagus and is distributed with it. Thereafter, the glossopharyngeal passes forwards, across the internal carotid, the spinal accessory passes backwards across the internal jugular vein and the vagus passes straight down (Fig. 19.9). The hypoglossal is medial to the others, curves round behind the vagus and then passes forwards, superficial to almost everything.

### The glossopharyngeal nerve (9)

This nerve is mainly sensory, to the posterior third of the tongue and to the pharynx, as its name implies, as well as to the carotid body and sinus (p. 343). It also has an important tympanic branch (p. 401). Its sole motor branch supplies stylopharyngeus. It runs forwards and downwards, between the internal and external carotid arteries, along with stylopharyngeus. At first deep to that muscle, it then curves around its lower border, supplies it, and then enters the pharynx with it, to join the pharyngeal plexus.

### The vagus nerve (10)

This is the main part of the cranial parasympathetic outflow and it lies in the carotid sheath between the internal carotid and internal jugular. Just below the jugular foramen it has two (sensory) ganglia and is then joined by the cranial accessory nerve, whose axons take origin from many of the brainstem nuclei, which also give rise to vagal fibres, so that the nerve is truly 'accessory' to the vagus. Although it is possible to work out whether the nerve supply to a particular muscle, in, for example, the larynx, is derived from the vagus or from the cranial accessory, for the most part it is best to group the two together. In the neck, the vagus gives a *pharyngeal branch*, which passes between the internal and external carotid arteries and joins the pharyngeal plexus as its motor branch. The *superior laryngeal* nerve is deep to both carotid arteries and divides into *internal* and *external laryngeal* nerves. The interior laryngeal is sensory to the larynx, above the vocal cords, and to the epiglottis and the back of the tongue, while the external laryngeal supplies the cricothyroid muscle. The vagus also gives *superior* and *inferior cardiac branches*, which descend into the thorax to join the cardiac plexuses, and a peculiar *auricular branch* (p. 346).

### The accessory nerve (11)

The *spinal root* arises from the upper five segments of the spinal cord, ascends along the side of the cord, passes through the foramen magnum and then runs forwards to join the *cranial root*. The latter arises from the side of the medulla close to the vagus and is distributed with it. The spinal root leaves the vagus and runs downwards and backwards to cross the transverse process of the atlas, which is a prominent landmark. Also crossing the transverse process is the vertical internal jugular vein and the occipital artery, which runs upwards and backwards. The relations of the transverse process are therefore easy to remember (Fig. 19.9). The nerve passes through sternomastoid, which it supplies, and then crosses the posterior triangle (p. 317) to supply trapezius.

### The hypoglossal nerve (12)

After passing through the hypoglossal canal, the hypoglossal nerve is joined by the anterior ramus of the first cervical nerve (see Fig. 17.8). It curves round the back of the vagus, crosses it to its lateral side, and then runs downwards and forwards, superficial to both carotids. It crosses the loop of the lingual artery and its further course has already been described. Its apparent branch, the superior ramus of the ansa cervicalis (*descendens hypoglossi*), is actually composed of C1 fibres and this nerve is joined by the descendens cervicalis from C2 and 3 to form the ansa cervicalis. The strap muscles are therefore supplied by the first three cervical nerves. Two other branches, to *thyrohyoid* and *geniohyoid*, are also derived from C1.

## THE INFRATEMPORAL REGION

Before describing this region in detail, it is first necessary to look at the rather complicated arrangements at the base of the skull (Fig. 19.10). The two *pterygoid plates* are part of the sphenoid bone and the *pterygoid hamulus* is at the lower end of the medial plate. Near the base of the skull, the medial pterygoid plate has a small boat-shaped depression called the *scaphoid fossa*. A little further back you will easily recognize the *foramen ovale* and behind this again are the *foramen*

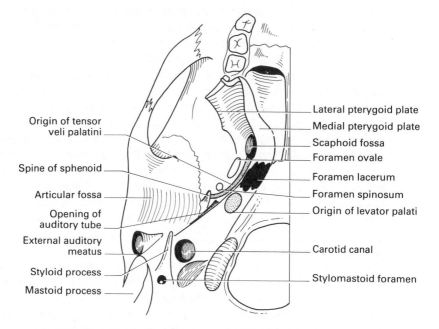

**Fig. 19.10** Structures related to the under-surface of the temporal, occipital and sphenoid bones. The foramen just behind the carotid canal is the jugular foramen.

*spinosum* and the *spine of the sphenoid*, the last two structures marking the most posterior part of the greater wing of the sphenoid. Between the scaphoid fossa and the spine there is a thin ridge of bone and medial to this is a groove in which lies the cartilaginous part of the auditory (Eustachian or pharyngotympanic) tube. The bony part of the tube opens at the posterior end of the groove. Medial to this again is the anterior part of the petrous temporal bone.

Immediately lateral to the spine of the sphenoid is the region of the temporomandibular joint, the fossa for the head of the mandible being part of the squamous temporal bone. Behind it is the external auditory meatus, whose floor is formed by the tympanic plate, and behind this is the mastoid process. Between the mastoid process and the styloid process is the *stylomastoid foramen* (for the facial nerve). By following this description you should be able, without any artistic ability, to construct a figure similar to Fig. 19.10, but do check all this yourself on a skull.

## THE TEMPOROMANDIBULAR JOINT

The temporomandibular joint is a favourite of junior medical students since in commercially prepared skulls the mandible is fitted with springs so that, when it is depressed and then released, the mouth shuts with a satisfying 'clunk'. If, however, the temporomandibular joint were a simple hinge joint like this, every time you yawn you would severely stretch the inferior alveolar nerve. The joint is therefore constructed so that the axis of rotation passes through the entry of this nerve into the mandibular foramen (see Fig. 19.1).

The joint is a synovial joint with an intra-articular disc but, unlike most synovial joints, the articular cartilage and the disc are composed of fibrocartilage or even fibrous tissue instead of hyaline cartilage. This may be because the bones ossify in membrane. In front of the glenoid fossa on the temporal bone there is an *articular eminence*, and the disc covers both the fossa and the eminence so that it is shaped rather like a jockey's cap (Fig. 19.11). The disc is attached medially and laterally to the sides of the mandibular condyle and anteriorly to the front of the condyle and the articular eminence. Posteriorly it is attached by two layers to the back of the condyle and to the glenoid fossa; the upper of the two layers is lax and composed of fibro-elastic tissue. The periphery of the disc is attached to the capsular ligament so that the joint is completely divided into two. The extensible attachment to the back of the glenoid fossa means that the disc, carrying the head of the mandible with it, can slide forwards on to the articular eminence when the mouth is opened. This forward movement is produced by the *lateral pterygoid* muscle, described later, which is attached not only to the neck of the mandible but also to the capsule and to the articular disc. Thus, when the mouth is opened, the head of the mandible rotates on the disc like the movement of a hinge joint but at the same time the disc, carrying the head of the mandible with it, moves forward on to the articular eminence. Thus, the axis of rotation passes horizontally through the region of the mandibular foramen and it may be that the *sphenomandibular ligament* forms a fulcrum for this movement. Alongside the mandibular foramen is the *lingula*, a small tongue of bone. The sphenomandibular ligament is a remnant of the skeletal element of the 1st branchial arch and is a wide band of dense fibrous tissue running from the spine of the sphenoid to the lingula.

When the mouth is open, the joint is very unstable and a blow on the mandible in this position may well cause a dislocation.

Closure of the mouth is associated with backward movement of the mandible and disc and this is mainly carried out by the horizontal posterior fibres of the temporalis muscle (see Fig. 19.13). The movements possible at the joint are as follows. Opening of the mouth is produced by gravity and by such submandibular muscles as mylohyoid and geniohyoid, with the lateral pterygoid producing the

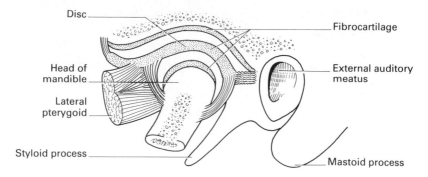

**Fig. 19.11** The left temporomandibular joint.

forward movement of the disc and head of the mandible. Closure of the mouth is produced chiefly by temporalis and masseter, together with the medial pterygoid, Closure is therefore a much more powerful action than opening, so, if attacked by a crocodile, try to hold its mouth closed! Other chewing movements are produced by the pterygoid muscles acting on either side alternately so that the head of the mandible moves forwards on first one side and then the other. Side to side movements are produced to some extent by the same muscles. Note the close proximity of the joint to the external auditory meatus. Clicking in the joint, which sometimes occurs, can be extremely disturbing to some patients.

## The muscles of the region

All these muscles are developed from the 1st branchial arch and are therefore supplied by the mandibular division of the trigeminal nerve. The most superficial muscle in the region is *masseter*, which is seen on the face but is better described here. It is attached above to the zygomatic arch, usually appearing to be composed of a deep and a superficial part (see Fig. 21.1). It is inserted into the outer surface of the mandible near its angle.

To see the full extent of the next muscle, *temporalis*, it is necessary to remove masseter and the zygomatic arch (Fig. 19.12). Temporalis arises from the lateral

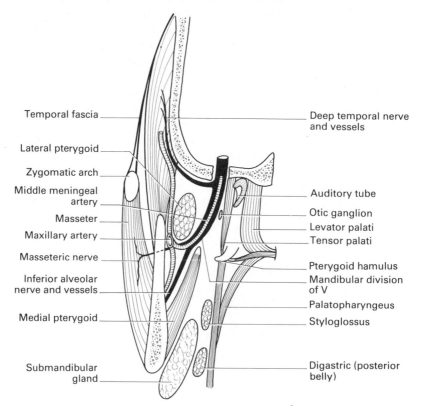

Temporal fascia — Deep temporal nerve and vessels

Lateral pterygoid —
Zygomatic arch —
Middle meningeal artery — Auditory tube
Masseter — Otic ganglion
Maxillary artery — Levator palati
Tensor palati
Masseteric nerve —
Inferior alveolar nerve and vessels — Pterygoid hamulus
Mandibular division of V
Palatopharyngeus
Medial pterygoid — Styloglossus

Submandibular gland — Digastric (posterior belly)

**Fig. 19.12** Coronal section through the infratemporal region.

side of the skull below the temporal line, extending forwards and backwards as far as it can go so that its most posterior fibres are horizontal (Fig. 19.13). In animals with a powerful bite, it extends even further upwards, as far as, and even beyond, the midline. Such animals contrive to do this by developing an upwardly project-ing keel, the sagittal crest, in the midline so that the temporalis origin can rise up its sides. The muscle is inserted not only into the coronoid process but also down the anterior border of the ramus of the mandible, almost as far as the third molar tooth (see Fig. 19.1).

If, now, temporalis and most of the ramus of the mandible are removed the remaining muscles, the medial and lateral pterygoids, are revealed (Fig. 19.14). The *lateral pterygoid* has two heads, one from the lateral surface of the lateral pterygoid plate and an upper head from the under-surface of the greater wing of the sphenoid. Its fibres run almost directly backwards to be inserted into the neck of the mandible and the intra-articular disc. The disc is said to be the degenerated tendon of this muscle. The *medial pterygoid* is attached to the medial surface of the lateral pterygoid plate, the plate thus separating the two pterygoid muscles. Its fibres run downwards and backwards, emerge from under cover of the lateral pterygoid, and are inserted into the deep surface of the mandible, more or less opposite the insertion of masseter (see Fig. 19.1). The angle of the mandible thus separates masseter and the medial pterygoid.

The actions of these muscles have already been described.

Deep to all these muscles is the superior constrictor muscle, styloglossus and stylopharyngeus on its surface, and all the other structures that have been described as lying on the surface of the superior constrictor. The gap between this muscle and the base of the skull (see Fig. 19.2) is mainly filled in by tensor veli palatini (tensor palati). The origin of this small muscle can be seen in Fig. 19.10. It arises from the scaphoid fossa at the base of the medial pterygoid plate, from the spine of the sphenoid, and, between the two, from the ridge of bone between these

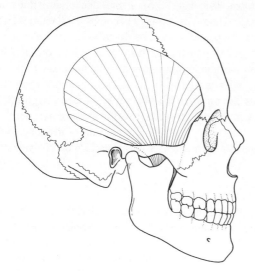

**Fig. 19.13** The temporalis muscle.

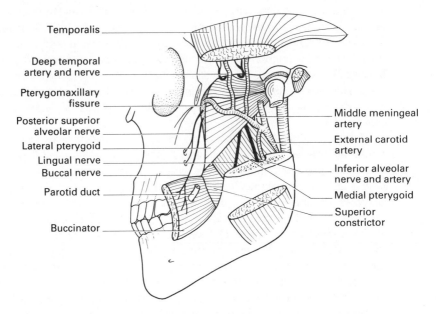

Temporalis

Deep temporal
artery and nerve

Pterygomaxillary
fissure

Posterior superior
alveolar nerve

Lateral pterygoid

Lingual nerve

Buccal nerve

Parotid duct

Buccinator

Middle meningeal
artery

External carotid
artery

Inferior alveolar
nerve and artery

Medial pterygoid

Superior
constrictor

**Fig. 19.14** The pterygoid muscles and the infratemporal fossa. Part of the mandible has been removed but the head remains in the glenoid fossa.

two points. Thus it can be seen from the diagram that the foramen ovale (and therefore the mandibular division of the trigeminal nerve) and the foramen spinosum (and therefore the middle meningeal artery) lie on the lateral surface of tensor veli palatini (Fig. 19.12). Some of the muscle fibres also arise from the lateral (membranous) side of the auditory tube. This part of the muscle is sometimes called *tensor tubae* (see Fig. 23.5). From this origin, the muscle converges towards the pterygoid hamulus, where it becomes tendinous and hooks around the hamulus to enter the palate. As will be described later, it elevates and tenses the palate and also pulls open the auditory tube. Thus, on climbing a mountain or going up in an aeroplane, your ears click at intervals, particularly when you swallow. The clicking is caused by the tympanic membrane resuming its normal shape as the air pressure on each side of it equalizes (see Chapter 23).

Levator palati arises from the under-surface of the petrous temporal, just medial to the auditory tube, and also from the medial side of the tube itself. It passes down to enter the palate from above and is therefore an elevator of the palate.

## The nerves and vessels

As you cannot fail to have gathered already from this chapter, this is the region of the 1st branchial arch, and the main nerve is the mandibular division of the trigeminal. This enters the region through the foramen ovale and immediately finds itself on the surface of tensor veli palatini. The lesser petrosal nerve also passes through this foramen (although it may have a small hole to itself) and it enters the *otic ganglion*, a tiny parasympathetic ganglion that lies between the mandibular nerve and tensor veli palatini. The nerve promptly splits into two divisions, anterior and posterior, of which the anterior gives rise to one sensory and three motor nerves while the posterior gives one motor and three sensory branches.

The anterior division gives the *pterygoid* and *masseteric* branches, the *deep temporal nerves* and the *buccal nerve*.

### Pterygoid branches

These pass straight from the anterior division to the medial and lateral pterygoid muscles. The nerve to the medial pterygoid gives one or two small branches that pass through the otic ganglion (without synapses) to supply the two tensors — *veli palatini* and *tympani*.

### Masseteric nerves

The mandibular nerve is deeply placed but masseter is on the face. The masseteric nerves solve the problem by passing through the mandibular notch between the neck and the coronoid process and entering the deep surface of the muscle.

### Deep temporal nerves

There are usually two of these and they cling to the bone. They pass directly laterally from the mandibular nerve as soon as it emerges from the foramen ovale. They thus lie on the under-surface of the greater wing of the sphenoid, passing through the upper origin of the lateral pterygoid muscle. They then turn the corner and pass upwards deep to temporalis, which they supply (see Fig. 19.12).

### Buccal nerve

This nerve is *sensory* and must not be confused with the buccal branch of the facial nerve, which is *motor*. It passes between the two heads of the lateral pterygoid muscle and runs forwards, deep to masseter. It emerges on the face from under cover of masseter, lying on buccinator, *which it does not supply*. It is sensory to the skin and mucous membrane of the cheek and to the outer surface of the gums posteriorly.

The posterior division of the mandibular nerve gives rise to the *auriculo-temporal*, *lingual* and *inferior alveolar nerves* and the latter provides the *mylo-hyoid nerve*.

### Auriculotemporal nerve

This passes backwards from the mandibular nerve at the base of the skull. Reference to Fig. 19.10 will show that it must immediately encounter the foramen spinosum and therefore the middle meningeal artery. Which side of the artery does it go? Both: the nerve splits to enclose the artery, passes medial to the neck of the mandible, and then turns upwards in front of the external auditory meatus to supply the scalp and parotid with sensory fibres (see Fig. 21.4). This nerve receives a ramus from the otic ganglion, containing postganglionic fibres that convey the parasympathetic secretomotor supply to the parotid gland. The preganglionic fibres travel in the lesser petrosal nerve (p. 401).

### Lingual nerve

This is another sensory branch, which emerges from under cover of the lateral pterygoid, lying on the medial pterygoid (see Fig. 19.14). It turns forwards and comes into direct contact with the mandible just below and behind the third molar tooth. It then lies on the hyoglossus (second step) and supplies ordinary sensory fibres to the anterior two-thirds of the tongue and the inner (lingual) surface of the anterior part of the gums. Associated with the nerve is the *chorda tympani*. In certain fishes the nerve of each branchial arch supplies a small *pretrematic branch* to the next arch in front. In the human embryo such a branch can be recognized in the 1st arch, i.e. the nerve of the 2nd (the facial) sends a small branch into the 1st arch. This pretrematic nerve is the chorda tympani. It leaves the facial nerve in the petrous temporal bone (Chapter 23), emerges from a minute fissure in the base of the skull medial to the spine of the sphenoid, runs downwards on the medial pterygoid, and finally joins the lingual nerve. *The chorda tympani is a mixed nerve.* It contains sensory fibres conveying taste sensation from the anterior two-thirds of the tongue and it also contains parasympathetic secretomotor fibres for the submandibular and sublingual glands. These have already been described in this chapter.

It may help in remembering the nerve supply of the salivary glands to note that the chorda tympani passes medial to the spine of the sphenoid, while the auriculotemporal nerve passes lateral to it (see Fig. 19.10). Theoretically, therefore, a lesion in the region of the spine could stop secretion of all three major salivary glands on that side. I do not think that this has *ever* happened, but anything that helps you to remember the complicated anatomy of this region is a bonus.

### Inferior alveolar nerve

This passes down behind the lingual nerve on the surface of the medial pterygoid and enters the mandibular foramen. It runs the length of the mandible in a canal inside the bone as far as *and beyond* the midline. It supplies the teeth, including the two incisors of the other side, which therefore have a double nerve supply. It gives off the mental nerve, which emerges from the mental foramen to supply the skin of the chin, the mucous membrane inside the lower lip and the outer surface of the gum. Just before entering the mandibular foramen, the inferior alveolar nerve gives its only motor branch, the *nerve to mylohyoid*. This grooves the mandible (see Fig. 19.1), emerges into the submandibular region on the first step and supplies mylohyoid and the anterior belly of digastric.

## Arteries

The artery of the region is the *maxillary artery*, one of the two terminal branches of the external carotid. The other is the *superficial temporal*, which will be described in the next chapter. The maxillary artery is, at first, embedded in the parotid gland. It runs forwards, deep to the neck of the mandible and on the surface of the medial pterygoid. It may be superficial or deep to the lateral pterygoid. In either case, it ends by entering the pterygopalatine fossa through the pterygomaxillary fissure. The artery is divided into three parts; part one is between its origin and the lower border of the lateral pterygoid, part two is superficial or deep to that muscle and part three is in the pterygopalatine fossa. Branches of part one all go through holes, and only the two largest will be mentioned here. The *inferior alveolar artery* follows the nerve of the same name and is distributed with it. The *middle meningeal artery* is also accompanied by a nerve but I dare not name it (see Preface to the First Edition). The artery passes deep to the lateral pterygoid and ascends to reach the foramen spinosum. Here it is enclosed by the two branches of the auriculotemporal nerve before it enters the cranial cavity. Its further course will be described in Chapter 22.

Branches of part two all go to muscles. The *deep temporal arteries* (usually two) lie on the surface of the lateral pterygoid, even if the maxillary artery is deep, and they ascend in the temporal fossa deep to the temporalis muscle. The masseteric, pterygoid and buccal arteries accompany the corresponding nerves.

Branches of part three all go through holes, but they will be described in Chapter 23.

## Some confusing ligaments

Although it means some repetition, there follows a brief summary of ligaments with similar names which often cause confusion.

### Stylohyoid ligament
From the tip of the styloid process to the lesser horn of the hyoid. Remnant of the 2nd branchial arch, skeletal element.

### Stylomandibular ligament
From the styloid process to the angle of the mandible. A thickening in the parotid sheath that separates the parotid and submandibular salivary glands.

### Sphenomandibular ligament
From the spine of the sphenoid to the lingula of the mandible. Remnant of the 1st branchial arch, skeletal element.

### Pterygomandibular ligament
From the pterygoid hamulus to the mandible just behind the third molar tooth. Origin of buccinator and the superior constrictor.

# Chapter 20
# The Pharynx, Palate, Mouth and Larynx

## PHARYNX

The outside of the pharynx and its related structures have already been described. Inside the roughly circular outer layer of striated muscle is an inner longitudinal layer. This is formed by the spreading out of three muscles, *stylopharyngeus*, *palatopharyngeus* and *salpingopharyngeus*. The first-named muscle arises from the styloid process and enters the pharynx between the superior and middle constrictors, along with the glossopharyngeal nerve (p. 342). Palatopharyngeus arises from the soft palate (below) and runs downwards beneath the mucous membrane, raising a longitudinal ridge called the *palatopharyngeal ridge* or, sometimes, the *posterior pillar of the fauces*. Salpingopharyngeus arises from the cartilage of the auditory tube and runs downwards. The fibres of these three muscles spread out and some are attached to the posterior border of the thyroid cartilage. They thus assist in raising the larynx during swallowing.

The mucous membrane of the oro- and laryngopharynx is covered by 'wear and tear' epithelium: stratified non-keratinized squamous. In the posterior wall of the nasopharynx is situated the *pharyngeal tonsil* or *adenoid*, part of a ring of lymphoid tissue around the back of the mouth that will be described later.

## THE PALATE

The hard palate is formed by the horizontal plates of the maxillae and the horizontal plates of the palatine bones. The bones are closely attached and may later fuse, as may the maxilla and palatine bones of the two sides (Fig. 20.1). Anteriorly there is an *incisive fossa*, into which open four small foramina that transmit the nasopalatine nerves and vessels. There is also an opening posteriorly, which is the *greater palatine canal* (p. 393). Smaller holes behind it are the *lesser palatine canals*.

The soft palate hangs down from the back of the hard palate and separates the nasopharynx from the oropharynx. Its basis is the *palatine aponeurosis*. You will remember tensor veli palatini, which arises from the scaphoid fossa at the base of the medial pterygoid plate, from the spine of the sphenoid, from the thin line of bone between them and from the cartilage of the auditory tube. Its fibres converge as they pass downwards and they form a thin tendon that hooks around the pterygoid hamulus (see Fig. 19.12). There is, of course, a tiny bursa between it and the bone. The tendon turns through about 90° round the hamulus and then spreads out to form the palatine aponeurosis. This is attached to the posterior border of the hard palate and forms a strong structure for the attachment of the palatine muscles.

Levator palati arises from the under-surface of the petrous temporal bone and the auditory tube and it runs downwards and medially to be inserted into the upper

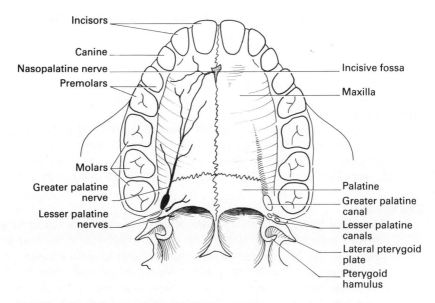

**Fig. 20.1** The bones and nerves of the hard palate.

surface of the palatine aponeurosis (see Fig. 19.12). Palatopharyngeus arises from the aponeurosis and runs down into the oropharynx, helping to form the inner longitudinal coat of muscle that has already been described. The palato-glossus muscle arises from the under-surface of the palatine aponeurosis and runs downwards into the side of the tongue. It, too, raises a ridge of mucous membrane, the *palatoglossal ridge* or *anterior pillar of the fauces*. Between this and the palatopharyngeal ridge is the *tonsillar fossa*.

All these structures are clearly visible when the mouth is opened widely. The final muscle of the soft palate is musculus uvuli, which is very unimportant.

The soft palate is covered by mucous membrane that contains tiny salivary glands and some taste buds. The mucous membrane is stratified squamous on the surface facing the mouth and pseudostratified ciliated columnar ('respiratory') epithelium on the nasal surface. The change-over is not at the edge of the palate. During swallowing the soft palate moves upwards and backwards so that its posterior surface meets the ridge of Passavant (Chapter 21); the stratified squamous epithelium therefore continues around the free border of the palate on to its posterior surface for a little way.

The muscles of the palate, as well as those of the pharynx, are all supplied by the pharyngeal plexus, except for tensor veli palatini, which is supplied by the mandibular division of the trigeminal nerve because it develops from the 1st branchial arch (this nerve supplies tensor tympani as well).

### The pharyngeal plexus

The pharynx receives nerves from three sources: the *pharyngeal branch of the vagus*, which also contains fibres from the cranial accessory nerve, the *glossopharyngeal nerve*, and *branches from the superior cervical ganglion of the sympathetic system*. The nerves form a plexus in the submucosa that supplies motor fibres to the pharyngeal and palatine muscles and the upper part of the oesophagus from the vagus/accessory nerves; the plexus is also sensory to the pharynx and part of the palate via the glossopharyngeal nerve.

## The nerves of the palate

The main (sensory) nerve supply to the hard palate is from the nasopalatine nerves anteriorly and the greater and lesser palatine nerves posteriorly. These nerves also supply the inner (lingual) surface of the gums and they are all derived from the maxillary nerve. Their course will be described in the next chapter.

## The tonsils

The official title for these is the *palatine tonsils* because there is also a *pharyngeal tonsil* (or adenoid) and a *lingual tonsil*, to be described later in this chapter. The palatine tonsils, like the rest of the lymphatic system, reach the peak of their development at, or just before, puberty. When enlarged at this time of life, they may almost meet in the midline, giving a characteristic nasal voice. Each tonsil lies in the tonsillar fossa between the palatoglossal and palatopharyngeal ridges. Lateral to the tonsil is its fibrous capsule and the superior constrictor, and, outside this again, the styloglossus and stylopharyngeus muscles (Fig. 20.2) and the glossopharyngeal nerve, Lying on the superior constrictor is the tonsillar branch of the facial artery, which supplies the tonsil, but the bleeding that may occur after tonsillectomy is often from the *paratonsillar vein*, which lies on the deep surface of the tonsil. Like the rest of the mouth, the tonsil is covered by stratified squamous epithelium, which also lines a transverse cleft near the upper pole of the tonsil, called the *intratonsillar cleft*.

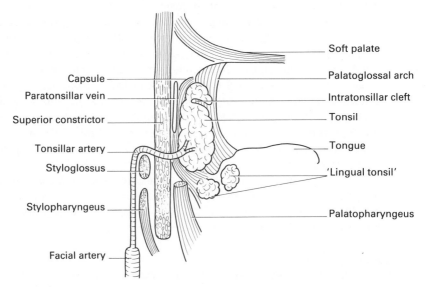

**Fig. 20.2** Relations of the tonsil as seen in a coronal section.

## THE MOUTH CAVITY AND TONGUE

The *vestibule of the mouth* is the cleft between the outer (labial) surface of the teeth and gums and the cheek. Normally it is a very narrow cleft, since the tone of the buccinator muscle keeps the cheeks in contact with the gums and can be used, along with the tongue, to remove any food that accumulates here. The main cavity of the mouth is separated from the nasal cavity by the hard and soft palates, the latter being movable. Complete separation of the mouth cavity and oropharynx from the nasopharynx occurs when the soft palate is elevated to press against the ridge of Passavant, and this is necessary for two reasons: firstly, during swallowing, to prevent food going upwards instead of downwards, and, secondly, in speech when pronouncing palatal sounds, such as 'k' and 'g'. Both these functions are disturbed in cases of cleft palate and a third function is also upset in these cases in babies. Sucking depends on isolation of the mouth cavity, after which depression of the tongue and the floor of the mouth leads to a fall in pressure so that milk is taken into the mouth.

### The tongue

The tongue is divided anatomically and developmentally into an anterior two-thirds and a posterior one-third. The two parts are separated by a V-shaped sulcus, the *sulcus terminalis*, at the apex of which is the *foramen caecum*, the site of the outgrowth of the thyroglossal duct (p. 327). In front of the sulcus there are a number of *vallate papillae* (Fig. 20.3). These are so called because each papilla, which is several millimeters in diameter, is surrounded, like a Roman fort, by a ditch, outside which is a slight elevation or wall (the 'vallum', in Roman terms). Taste buds are found in the 'ditch'.

The tongue, as you would expect, is covered by stratified squamous epithelium, which is thin on the under-surface but thick on the upper surface, where two other types of papillae are seen. You can see them best if the tongue is first dried,

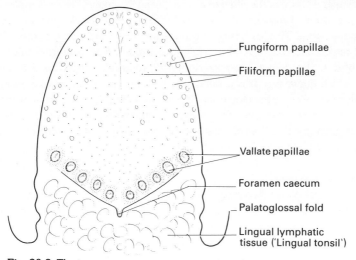

Fungiform papillae

Filiform papillae

Vallate papillae

Foramen caecum

Palatoglossal fold

Lingual lymphatic tissue ('Lingual tonsil')

**Fig. 20.3** The tongue.

when the thin, *filiform papillae* and the deep red, flat-topped, *fungiform papillae* are visible.

## Musculature of the tongue

The intrinsic muscles run in three directions, vertically, horizontally and longitudinally. The complex interlacing mass of striated muscle makes the diagnosis of 'tongue' in a histology examination very easy. It is also responsible for the astonishing way in which the tongue can change shape, becoming wide and flat, narrow and thick, or rolled up laterally to form a gutter shape. The latter shape is impossible to achieve for a small number of people and this inability is genetically determined. It is of no practical importance but try it yourself.

The extrinsic muscles of the tongue arise from bone and their fibres blend in with the intrinsic muscles. *Hyoglossus* has already been mentioned, as have *palatoglossus* and *styloglossus*. *Genioglossus* is important. Arising from the superior genial tubercle behind the front of the mandible, its fibres fan out to be inserted into almost the whole length of the tongue. It can thus be used to protrude the tongue, acting in conjunction with the intrinsic muscles.

## Nerve supply of the tongue

The muscles of the tongue are all supplied by the *hypoglossal nerve* except for palatoglossus. Since the tongue develops from a pair of *lingual swellings* which later fuse in the midline, the two sides of the tongue are independent as far as their nerve supply and blood supply are concerned so that, if, for example, the left hypoglossal nerve is damaged, the muscles on the right can protrude the right half of the tongue but those on the left are paralysed. Thus, when the patient is asked to protrude the tongue, the tongue will deviate to the left, i.e. to the same side as the lesion.

Embryology is again concerned in the sensory supply. The anterior two-thirds develops from the 1st branchial arch and is therefore supplied by a branch of the mandibular division of the 5th nerve, specifically the lingual. This is supplemented by a (pretrematic) branch of the nerve of the 2nd arch (the facial) in the form of the chorda tympani. This, as has been seen, joins the lingual nerve in the infratemporal region and is involved in carrying the sensation of taste from the anterior two-thirds of the tongue. The posterior one-third develops from the 3rd branchial arch, which grows forwards to submerge the 2nd, and this part of the tongue is therefore supplied by the glossopharyngeal nerve both for ordinary sensation and for taste. The extreme back of the tongue, in the region of the epiglottis, is of 4th arch origin and is therefore supplied by the internal laryngeal branch of the vagus, a branch of the superior laryngeal nerve.

## Subdivisions of the tongue

The above-mentioned division of the tongue into anterior two-thirds and posterior one-third is the classical description but is of course only very approximate, especially since the sulcus terminalis is so oblique. Also, although the vallate papillae are in front of the sulcus, they are nevertheless supplied by the glossopharyngeal nerve. Certainly, however, the appearance of the tongue behind the sulcus is different from that in front since it is much smoother and is nodular in appearance because of the presence of numerous lymphoid follicles (Fig. 20.3). This collection of lymphatic tissue is sometimes called the *lingual tonsil*, which, together with the palatine and pharyngeal tonsils, forms a lymphatic ring around the oropharyngeal isthmus. This is known as *Waldeyer's ring*, but the idea that all substances entering the digestive tract have to jump through a lymphatic hoop that forms a protective mechanism seems a little far-fetched.

## The teeth

The first of the deciduous teeth to appear are usually the lower central incisors, which erupt at about *6 months*. The first of the permanent teeth, the first molar, appears at about *6 years*. The deciduous (milk) teeth comprise two incisors, one canine and two milk molars in each dental arch, a total of 20 teeth. Note, however, that the milk molars occupy the position of the permanent *premolars*, so that when the first permanent molar appears at the age of 6 years it is *behind* the milk molars. Parents often, therefore, believe that it is another milk tooth.

The permanent teeth comprise two incisors, one canine, two premolars and three molars in each arch, a total of 32 teeth (see Fig. 20.1). The second permanent molar appears at about 12 years and hence was sometimes called the 'factory molar', because in the bad old days a child was deemed to be old enough to be sent out to work when it appeared. The third molars, the wisdom teeth, appear sometime after the age of 18 years, but they often remain unerupted for years or even for a lifetime.

The roots of the teeth are enclosed in the alveolar part of the mandible or maxilla. Before any teeth have erupted this part is non-existent so that the mandible and the maxilla both appear to be relatively much smaller than they do in the adult. Similarly, after the teeth have been lost, the edentulous mandible and maxilla revert to their infantile appearance as the alveolar part of the bone is resorbed. Hence the mental foramen no longer lies halfway between the upper and lower borders of the mandible, but is situated at the upper border, where it may be pressed upon by dentures. In the absence of well-fitting dentures, the lips sink in and the mouth is surrounded by vertical wrinkles.

### The nerve supply of the teeth and gums

Once again, embryology is the key. Figure 20.4 shows how the gums and their enclosed teeth develop in the lower jaw. They are formed between two semicircular furrows around the tongue, which, in its early stages, is just a cushion sitting on the floor of the primitive mouth cavity. The outer furrow becomes the vestibule of the mouth and the inner becomes the space between tongue and gums. Thus the nerve supply of the skin, the mucous membrane of the lips and the outer surface of the gums are all derived from the same nerves, i.e. the mental nerve anteriorly and the buccal nerve posteriorly. Similarly, the inner surface of the gums and the lateral surface of the tongue have the same nerve supply: the lingual nerve. The lower teeth themselves are supplied by the inferior alveolar nerve. Similarly, in the upper jaw, the skin of the inner surface of the lips and the outer surface of the gum are all supplied by the infra-orbital nerve anteriorly and the buccal nerve

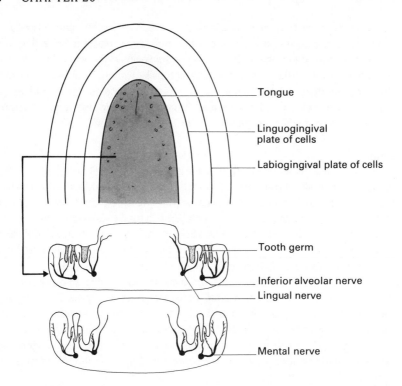

**Fig. 20.4** The upper diagram is the view you would see if you were hovering in the oral cavity of a fetus, looking down. The lower two diagrams are sections across the upper diagram to show how the plates of cells break down to form the furrows between tongue and gums and gums and lips.

posteriorly, while the inner surface of the gums and the palate are supplied by the nasopalatine and the greater palatine nerves.

## DEGLUTITION

The bolus of food, moistened by saliva, is passed back into the pharynx by the movements of the tongue, which is pressed against the palate. As the bolus passes the oropharyngeal isthmus, breathing is arrested for a moment as the soft palate is elevated and comes into contact with the ridge of Passavant. The pharyngeal muscle is relaxed at this stage but a wave of contraction then propels the bolus down towards the oesophagus. The contraction of the inner longitudinal layer of muscle elevates the larynx, which is drawn up under the epiglottis. The epiglottis itself is moved downwards, partially to cover the opening into the larynx, by the aryepiglottic muscles, to be described later. The bolus of food then passes over the back of the larynx and past the cricopharyngeus into the oesophagus. Elevation of the larynx is, for most people, essential for swallowing. If you have visited Spain, you may have tried drinking wine from a vessel that does not touch the lips. The wine is poured directly into the open mouth — very hygienic but a difficult feat since the head has to be held extended and in this position the trachea is stretched and tethers the larynx so that it cannot be elevated.

## THE LARYNX

The skeleton of the larynx is cartilaginous, although the cartilages may become calcified in the elderly so that they are visible in an X-ray. The principal cartilages are the *cricoid*, the *thyroid* and the *two arytenoids*. There are also some nodules associated with the arytenoids. The cricoid cartilage is shaped like a signet ring with the wide part posteriorly. It is, in fact, the only *complete* cartilaginous ring in the respiratory system. The thyroid cartilage has two leaves or *alae*, which meet at an acute angle in the male but a more obtuse angle in the female (like the pubic rami, Fig. 20.5). The arytenoid cartilages are perched on top of the back of the cricoid, articulating with it by means of a tiny plane synovial joint. There is another similar synovial joint between the inferior horn of the thyroid cartilage and the side of the cricoid (Fig. 20.6). The arytenoids are three-sided pyramids, with a *vocal process* directed forwards and a *muscular process* directed laterally (see Fig. 20.9). Finally, the epiglottis is a large leaf-shaped piece of elastic cartilage attached to the back of the thyroid cartilage and projecting upwards behind the hyoid bone, to which it is attached by a *hyo-epiglottic ligament*.

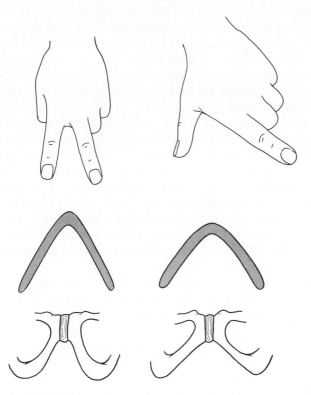

**Fig. 20.5** The angles between the two alae of the thyroid cartilage (centre) and the two pubic rami (below) are more acute in the male (left) than in the female (right).

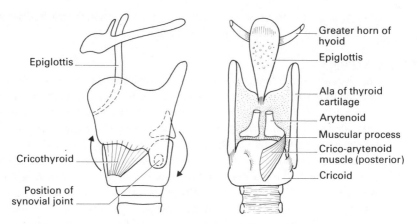

**Fig. 20.6** Left: the laryngeal cartilages and the action of cricothyroid. Right: posterior view of the laryngeal cartilages and the posterior crico-arytenoid muscle.

## The ligaments, joints and mucous membrane

Both synovial joints that have been mentioned have capsular ligaments. The joints between the inferior horn of the thyroid cartilage and the side of the cricoid permit a very small degree of rotation, so that they allow the cricoid to pivot on the thyroid in such a way that the front of the cricoid moves upwards at the same time as its posterior part moves downwards (arrows in Fig. 20.6). The recurrent laryngeal nerve enters the larynx just behind this joint. The joints between the arytenoids and the upper surface of the cricoid allow the arytenoids to pivot around a more or less vertical axis (see Fig. 20.9). They can also move very slightly laterally and medially.

The upper border of the thyroid cartilage is attached to the lower border of the hyoid bone by the *thyrohyoid ligament*, which is thickened slightly in the midline. In the midline, too, is the *median cricothyroid ligament*, but this is not as important as the main *cricothyroid ligament* or *cricovocal membrane*. This is attached to the upper border of the cricoid cartilage but passes upwards deep to the lower border of the thyroid cartilage to be attached anteriorly to the back of the thyroid cartilage on either side of the midline and posteriorly to the vocal process of the arytenoid cartilage. The upper free border of this membrane is thickened to form the *vocal ligament* (Fig. 20.7). The vocal ligament, like the inguinal ligament (p. 198), is not, therefore, a cord-like structure.

In the same plane as the cricothyroid ligament, but higher up, is a sheet of dense connective and elastic tissue called the *quadrate ligament*. Its free lower border is slightly thickened to form the *vestibular ligament*, which is about 5 mm above the vocal ligament and parallel to it. The inside of the larynx is lined by epithelium that is similar to that in the whole of the upper respiratory tract, i.e. pseudostratified columnar. Over the vocal ligaments, however, it gives way to stratified squamous epithelium and this is bound down to the ligament with no intervening submucosa so that its colour is a pearly grey rather than the pink colour of the rest of the mucosa. The ridges of mucous membrane produced by the

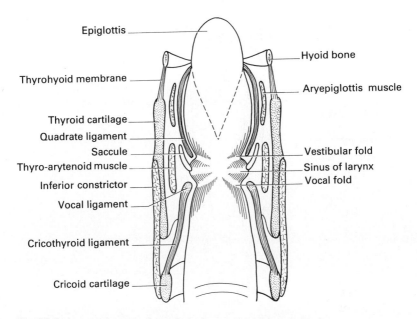

**Fig. 20.7** A coronal section through the larynx viewed from behind.

underlying vocal ligaments are the *vocal folds* (or vocal cords) and the ridges produced by the vestibular ligaments are the *vestibular folds*. The depression between the vocal and vestibular folds on each side is the *sinus of the larynx*. A tiny diverticulum of mucous membrane that passes upwards from the sinus contains mucous glands. This is the *saccule*.

The posterior edge of the thyroid cartilage projects backwards beyond the cricoid cartilage and its associated structures, so that there is a fossa medial to the alae. This is the *pyriform fossa* (Fig. 20.8) and it is important because a carcinoma here may reach a large size before producing any symptoms, in fact the first sign may be the appearance of enlarged lymph nodes in the neck. Tumours of the vocal cords, however, produce voice changes when they are still very tiny.

### The muscles and movements of the larynx

The only intrinsic muscle of the larynx that is visible from the outside is *cricothyroid* (see Fig. 18.1). If the cricothyroid joint is likened to the temporomandibular joint, cricothyroid is the masseter. That is to say, it 'shuts the mouth' so that the front of the cricoid moves upwards and the back moves downwards carrying the arytenoids with it (see Fig. 20.6). The movement is very slight indeed but the movement of the arytenoids is enough to increase the tension in the vocal ligaments since they have a very high Young's modulus. Cricothyroid is therefore a *tensor* of the vocal ligaments and its antagonist is the *thyro-arytenoid* muscle (Fig. 20.9). This is a wide sheet of muscle that passes between the back of the thyroid cartilage and the arytenoids so that it naturally moves the arytenoids forwards and thus lessens the tension in the vocal ligaments. Some of its fibres

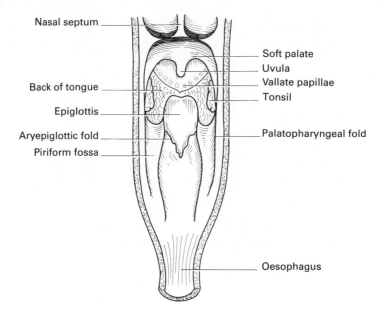

**Fig. 20.8** Posterior view of the larynx and the oral and nasal cavities.

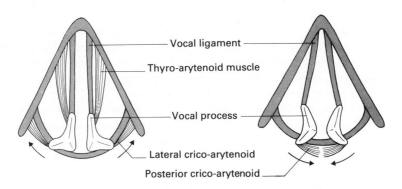

**Fig. 20.9** Abduction, adduction and relaxation of the vocal cords. Tightening of the cords is shown in Fig. 20.6.

are attached to the side of the vocal ligament itself, so that different parts of the ligament may be under different tensions. This muscle (*vocalis*) is used in whispering.

Besides tightening and slackening, the vocal ligaments may also be abducted or adducted. The *lateral crico-arytenoid* muscle is attached to the lateral side of the cricoid cartilage and is inserted into the muscular process of the arytenoid. It thus pulls the muscular process forwards so that the arytenoid pivots around a vertical axis and the vocal process moves medially, thereby adducting the vocal ligaments (Fig. 20.9). In this way the space between the vocal folds becomes reduced to a narrow slit. Adduction is aided by the *interarytenoid* muscles,

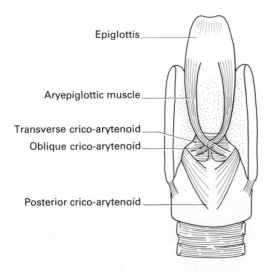

Epiglottis

Aryepiglottic muscle

Transverse crico-arytenoid
Oblique crico-arytenoid

Posterior crico-arytenoid

**Fig. 20.10** Muscles of the posterior part of the larynx.

transverse and oblique, which draw the arytenoids nearer together (Fig. 20.10). The oblique interarytenoids are continuous with a larger sheet of muscle that is inserted into the side of the epiglottis. This is *aryepiglotticus*, which, together with the interarytenoids and the epiglottis itself, acts as a sphincter, narrowing the opening into the larynx and preventing swallowed material from entering it.

The vocal ligaments are abducted by the *posterior crico-arytenoid* muscles. These arise from the cricoid on either side of the midline and are inserted into the muscular processes of the arytenoids. They pull these processes backwards, pivoting the arytenoids and producing separation of the vocal ligaments. The space between the vocal folds thus becomes diamond-shaped, since the mucous membrane covers not only the vocal ligaments but also the arytenoids themselves.

The vocal cords may be seen by indirect laryngoscopy, i.e. by means of a warmed, angled mirror placed at the back of the mouth, or, with the patient anaesthetized, by an instrument that can pull the tongue and epiglottis forwards and expose the inlet of the larynx (the glottis) directly. Only the edges of the cords can be seen because of the shadow thrown by the vestibular folds but their movements are clearly visible. If the patient is asked to say 'Ah' the cords are abducted, but a high-pitched 'Eee' causes adduction. This, of course, is why a singer can hold a high note much longer than a low one.

### Nerve supply of the larynx

The sensory nerve supply to the larynx is entirely from the vagus nerve via two of its branches: the *superior laryngeal* (the nerve of the 4th branchial arch) and the *recurrent laryngeal* (the nerve of the 6th arch). The superior laryngeal nerve arises high in the neck and passes deep to both carotid arteries (see Fig. 19.9). It then divides into *internal* and *external laryngeal branches*. The internal laryngeal is quite large and enters the larynx by passing through the thyrohyoid membrane

**Table 20.1** Motor nerve supplies

|          | All muscles supplied by:      | Except for:          | Which is supplied by:    |
|----------|-------------------------------|----------------------|--------------------------|
| Tongue   | Hypoglossal nerve             | Palatoglossus        | Pharyngeal plexus        |
| Palate   | Pharyngeal plexus (X + XI)    | Tensor veli palatini | Mandibular trigeminal    |
| Pharynx  | Pharyngeal plexus (X + XI)    | Stylopharyngeus      | Glossopharyngeal         |
| Larynx   | Recurrent laryngeal           | Cricothyroid         | External laryngeal       |

(see Fig. 18.1). It is sensory to the larynx down as far as the vocal cords and it also supplies the extreme posterior part of the tongue. The external laryngeal nerve is small. It accompanies the superior thyroid artery, lying a little behind it, and then runs on the surface of the inferior constrictor to end in the cricothyroid muscle.

The recurrent laryngeal nerve is a mixed nerve. It is sensory to the larynx below the level of the vocal cords, as well as to the trachea, and it is motor to all the intrinsic muscles of the larynx except the cricothyroid. It is closely related to the inferior thyroid artery and is sometimes damaged during thyroidectomy.

If it is completely divided on one side, that vocal cord will lie midway between adduction and abduction, in the so-called *cadaveric position*. This will not produce as much effect on voice production as you might think, since the other cord can cross the midline to become approximated to the paralysed cord. There is a problem if both recurrent laryngeal nerves are crushed but not completely divided since the nerve fibres supplying the abductors are more liable to injury than those supplying the adductors. The cords may therefore be closely approximated with consequent breathing difficulties (*Semon's law*). It has been suggested, however, that this involuntary adduction might be the result of the unopposed action of the cricothyroid, whose tensing effect on the cords also causes their adduction.

Why does the larynx need a sensory nerve supply? The answer is obvious if you have ever inhaled a particle of food or a fly. Any foreign body entering the larynx immediately sets up a cough reflex via the vagus nerve and the foreign body is expelled unless it becomes impacted. Smaller particles such as dust become trapped in the mucus and are wafted up the trachea and through the larynx by the action of the cilia.

Table 20.1 may help you to remember some of the complicated innervation of structures in this region.

# Chapter 21
# The Face and Scalp

This chapter starts with the nerve supply of the face because one of the commonest errors in examinations of any sort is a mix-up between the trigeminal and the facial nerves. The trigeminal nerve is the *sensory* nerve of the face, supplying the whole face and most of the scalp with sensory fibres, except for a small area over the parotid region that is supplied by the greater auricular nerve (p. 316). It also supplies *motor fibres* to the *muscles of mastication*, which have been described in Chapter 19. The facial nerve is the *motor supply* to the muscles of facial expression, including platysma, and also to stylohyoid and the posterior belly of digastric (Chapter 19) and to stapedius (Chapter 23). The facial expressions include astonishment, when the eyebrows are raised, so remember that the occipitofrontalis muscles are included.

## THE TRIGEMINAL NERVE AND THE SENSORY SUPPLY TO THE FACE

The intracranial portion of this nerve will be described in the next chapter but this is a convenient place to sum up its general distribution. As a major sensory nerve, it has a large ganglion (inside the cranium), in which synapses *do not take place*. The cells are pseudo-unipolar cells (p. 26), similar to those in the spinal ganglia. The three divisions of the nerve are *ophthalmic, maxillary* and *mandibular* and they supply, embryologically speaking, and respectively, the frontonasal process and its neighbouring structures, the maxillary process and the 1st branchial (pharyngeal) arch. The maxillary process is, in fact, an outgrowth from the dorsal end of the 1st arch. Thus the ophthalmic division is sensory to the nose, the region of the orbits and the frontal region of the scalp, the maxillary division supplies the upper jaw, including the teeth, and the mandibular division supplies the lower jaw and its associated structures, including the anterior two-thirds of the tongue (Chapter 19). The mandibular division also carries motor fibres to the muscles of mastication (including mylohyoid and the anterior belly of digastric) and the two tensors, tympani and veli palatini.

The sensory supply from the trigeminal nerve to the face is carried by three *major* branches. Each is derived from a different branch of the trigeminal nerve and all the foramina from which they emerge lie on a vertical line that passes between the two premolar teeth. The ophthalmic division gives rise to the *supra-orbital nerve*, which reaches the face through the supra-orbital notch or foramen, the maxillary division provides the *infra-orbital nerve*, emerging from the infra-orbital foramen, and the mandibular division gives the *mental nerve*, which emerges from the mental foramen (see Fig. 16.3).

Each of these three branches has a more extensive distribution than you might expect. The supra-orbital nerve travels right back as far as the vertex of the skull, where it meets the greater occipital nerve. The infra-orbital nerve supplies the skin of the cheek and the side of the nose and also the mucous membrane of the inside of the corresponding part of the cheek and the outer surface of the gums. The mental nerve supplies the skin of the chin, the mucous membrane of the lower lip and the outer surface of the gums. These three nerves are very large and if you dissect the face and do not find them you are not a very good dissector.

The three nerves are supplemented by a number of smaller nerves that you may be forgiven for failing to find. The most important of these are the *auriculo-temporal*, which comes from the mandibular division and supplies the temporal region and the outer ear (Chapter 19), and the *buccal nerve*, which is from the same division (Chapter 19) and supplies the skin and mucous membrane of the central part of the cheek. Do not confuse it with the buccal branch of the facial nerve, which is motor. The smaller nerves are the *supratrochlear*, *infratrochlear*, *external nasal* and *lacrimal* from the ophthalmic division, and the *zygomatico-temporal* and *zygomaticofacial* from the maxillary division. One other, which is often forgotten, is the 'Alderman's nerve', the *auricular branch of the vagus*. This supplies quite a lot of the external ear and external auditory meatus. At City banquets, when the Aldermen had eaten themselves to a standstill, they would drip a little iced water behind the ears, which would, so it is said, reflexly stimulate their gastric juices to further effort. More to the point, stimulation of the external meatus by, for example, hard wax or a surgical instrument may cause coughing.

## FACIAL MUSCULATURE

As you now know, there are two groups of muscles on the face. These are the muscles of mastication, supplied by the mandibular division of the trigeminal, and the muscles of facial expression, supplied by the facial nerve.

### The muscles of mastication

The *temporalis* muscle can be seen in the temporal region (see p. 348) covered by the tough *temporal fascia* and the side of the *epicranial aponeurosis*, to be described below. Masseter is on the side of the face and its anterior border can easily be palpated when the teeth are clenched. Both these muscles, as well as the pterygoid muscles, which lie more deeply, have been described in Chapter 19.

### The muscles of facial expression

The most important of these are shown in Fig. 21.1. They are very superficial muscles so that they can move the skin and superficial fascia in various directions. There is no need to learn their precise attachments to the bones of the face and they will be described only briefly. The two major groups of muscles are *orbicularis oris* and *orbicularis oculi*.

Orbicularis oculi has two main parts. The *palpebral part* is in the eyelids, while the much larger *orbital part* surrounds the orbit and blends in with the frontal belly of occipitofrontalis. The palpebral part closes the eyes gently as in sleep while the orbital part acts like a sphincter and closes the eye forcibly, as happens, for example, during a dust storm. Its action causes radiating skin wrinkles from the lateral corner of the eye. In bright sunlight the eye is screwed up and the sunburn

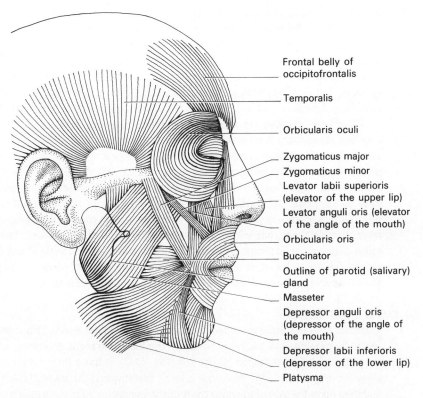

Frontal belly of occipitofrontalis

Temporalis

Orbicularis oculi

Zygomaticus major

Zygomaticus minor

Levator labii superioris (elevator of the upper lip)

Levator anguli oris (elevator of the angle of the mouth)

Orbicularis oris

Buccinator

Outline of parotid (salivary) gland

Masseter

Depressor anguli oris (depressor of the angle of the mouth)

Depressor labii inferioris (depressor of the lower lip)

Platysma

**Fig. 21.1** The muscles of the face: masseter and temporalis are muscles of mastication; the remainder are muscles of facial expression.

leaves pale radiating lines caused by the wrinkles. The antagonist to the palpebral part is *levator palpebrae superioris*, which opens the eye and is supplied by the oculomotor nerve and by sympathetic fibres (Chapter 22).

Orbicularis oris has some fibres of its own but is mainly a composite muscle formed of *levator* and *depressor anguli oris*, *levator labii superioris* and *depressor labii inferioris*, *buccinator* and *platysma* (Chapter 19). If you are a Latin scholar, the names of these muscles will tell you their actions; if not, a study of Fig. 21.1 should be almost as helpful. I cannot resist mentioning the 'grace muscle': the muscle with the longest name, which is one of the smallest muscles. A certain anatomist was dining in a Cambridge college when, to his horror, he was called upon to say grace, which is usually said in Latin. With great presence of mind, he solemnly recited, 'Levator labii superioris alaeque nasi, Amen'. It means, of course, the elevator of the upper lip and the ala of the nose.

Orbicularis oris, in gentle contraction, closes the lips by a sphincter-like action but when it contracts more powerfully can protrude the lips, as in whistling or kissing. The individual components play an important part in expressing emotions and presumably have an extensive representation in the motor cortex in actors.

*Buccinator* is the main muscle of the cheek and its most important function is to keep the cheeks in contact with the gums so that food does not accumulate in the vestibule of the mouth. The *frontal belly of occipitofrontalis* (see below) must not be forgotten, as it is responsible for elevating the eyebrows in an expression of surprise and it also contracts in looking upwards, to lift the eyebrows out of the way. Patients with an exophthalmic goitre, which causes protrusion of the eyeballs, can look upwards without contraction of this muscle, and absence of wrinkling of the forehead on looking up is one of the clinical signs of this condition.

All of these muscles are supplied by the facial nerve and the main reason for knowing about them is so that you will understand the effects of facial paralysis. This is quite a common condition as a result of lesions of the facial nerve (Bell's palsy). To test for the integrity of the nerve one can ask the patient to screw up his/her eyes, to smile or to whistle. Loss of muscle tone causes the normal skin folds to disappear on the side of the lesion. Paralysis of the palpebral part of orbicularis oculi causes the lower eyelid to fall away from the eye, with drainage of tears over the cheek, while loss of orbicularis oris can similarly cause dribbling of saliva. Paralysis of buccinator may lead to food accumulating in the vestibule of the mouth. The muscles must be supported while they are recovering or they will stretch under gravity and cause a permanent asymmetry of the face.

## BLOOD VESSELS

The main arterial supply to the face is by means of the *facial artery*. This enters the region by passing over the lower border of the mandible at the anterior border of masseter, where its pulsation can be felt (Fig. 21.2). It has a tortuous course (to allow for movement of the face, so it is said), first towards the angle of the mouth and then up at the side of the nose towards the medial angle of the eye. It gives off upper and lower *labial branches*, as well as numerous other branches to the face. Anastomoses are very free, both between the facial and other smaller arteries on the same side and across the midline. The face is therefore a very vascular region and cuts bleed freely and heal quickly. The corresponding vein is the *facial vein*, which follows a straighter path and communicates freely with deeper veins, such as those of the pterygoid venous plexus. In particular, it communicates at the medial angle of the eye with the veins of the orbit and thence with the cavernous sinus (Chapter 22). The central area of the face is therefore sometimes known as the 'danger area' as infections here can spread via the veins to the cranial cavity. This risk, however, is much less important than it was in pre-antibiotic days but it should still be borne in mind.

The other artery on the face where pulsation can be felt is the *superficial temporal* artery, a branch of the external carotid. This pulsation can be felt just in front of the tragus of the ear, and the course of the anterior branch of the artery on the forehead can be clearly seen in bald, angry men. It becomes noticeably more tortuous with increasing age. There are wide anastomoses between this and the facial artery.

## THE PAROTID GLAND

This major salivary gland is situated in a position well known to victims of mumps. The investing layer of deep cervical fascia splits to enclose it in a capsule of tough

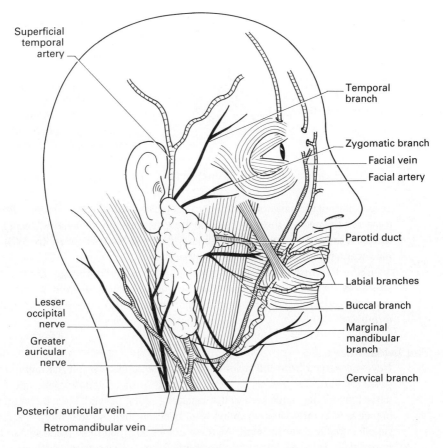

Superficial temporal artery

Temporal branch

Zygomatic branch

Facial vein

Facial artery

Parotid duct

Labial branches

Buccal branch

Marginal mandibular branch

Cervical branch

Lesser occipital nerve

Greater auricular nerve

Posterior auricular vein

Retromandibular vein

**Fig. 21.2** Some vessels and nerves of the face.

inelastic fibrous tissue, so that swelling of the gland gives rise to a rapid increase in pressure with consequent servere pain. A thickening in the capsule forms the *stylomandibular ligament*.

The gland itself is a mainly serous gland, whose duct leaves the anterior border and crosses the masseter muscle, following a line from the tragus of the ear to the middle of the moustache region. The duct then turns round the anterior border of masseter and pierces the buccinator muscle to enter the mouth opposite the second upper molar tooth. It can easily be rolled under the finger when masseter is contracted by clenching the teeth. Try your own. Now.

The gland occupies all the space available to it. In other words, it fills a space between the back of the mandible and the sternocleidomastoid muscle and is moulded around all the structures that bound this space (Fig. 21.3). Thus it overflows anteriorly on to the superficial surface of the mandible and masseter and has an accessory lobe that lies alongside its duct on the surface of masseter. It is, in fact, wrapped around the posterior border of the mandible so that it is related to the medial surface of the medial pterygoid. Posteriorly, it is similarly wrapped

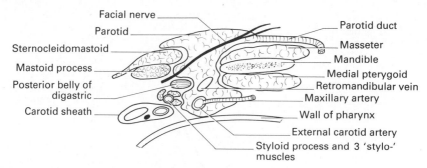

**Fig. 21.3** Horizontal section through the parotid to show its relations.

around sternocleidomastoid and the posterior belly of the digastric muscle. Above, it reaches the neck of the mandible and the external auditory meatus. Deeply, it is moulded around the styloid process and the three 'stylo-muscles' (p. 343) and may even reach the lateral wall of the pharynx.

By far the most important relationships of the parotid, though, are the three structures that actually pass through the gland. These are, from superficial to deep, the *facial nerve*, the *retromandibular vein* (p. 343) and the *external carotid artery*.

### The facial nerve

Having emerged from the stylomastoid foramen and given off the unimportant *posterior auricular branch* to the occipital belly of occipitofrontalis, the facial nerve immediately finds itself in the substance of the parotid. Here it divides into branches with intercommunicating rami between them so that there is a rather simple intraglandular network. All these nerves lie in the same plane so that they imperfectly divide the gland into a superficial and a deep part, which intercommunicate by means of glandular tissue in the interstices of the network. Finally, five branches emerge from the anterior border of the gland, although some of them may be duplicated. These are the *temporal, zygomatic, buccal, marginal mandibular* and *cervical*. These branches break up to supply the local muscles of facial expression and they should be remembered as five fingers radiating out from the palm of a hand placed over the parotid. Remember particularly that the marginal mandibular branch runs *below* the lower border of the mandible before crossing it to supply the muscles in the region of the chin. This is why incisions are not made along the lower border of the mandible but at least a finger's-breadth below it.

### The external carotid artery

Having given off the branches that have already been mentioned, beginning with the superior thyroid and ending with the posterior auricular, the external carotid divides in the substance of the parotid into *maxillary* and *superficial temporal* arteries. The former passes forwards, deep to the neck of the mandible (see Fig. 19.1) to enter the infratemporal region, where its further course has been described in Chapter 19 and will be further described in Chapter 23. The superficial temporal artery lies in front of the ear and divides into anterior and posterior branches for the region of the forehead and for the side of the scalp respectively.

## THE EYE

A full description of the eyeball will be given in the next chapter but important parts of the eye and its adnexa appear on the face and will be briefly described here.

### The eyelids

The 'skeleton' of the eyelids consists of the *tarsal plates*, particularly evident in the upper eyelid. The tarsal plates consist of dense fibrous tissue and into the superior plate is inserted a major part of levator palpebrae superioris. Partly embedded in its deep surface are large modified sebaceous glands (*Meibomian* or *tarsal glands*), which occasionally get blocked and become cystic. Their ducts open on to the edge of the eyelid in relation to the eyelashes, as do smaller *ciliary glands*. The stiffness of the upper tarsal plate makes it possible to evert the upper eyelid by 'flipping' it over, so that foreign bodies can be removed from the deep surface of the eyelid. Superficial to the tarsal plates is the thin palpebral part of orbicularis oculi and deep to them and the Meibomian glands is a layer of conjunctiva. The conjunctiva is reflected above and below on to the anterior surface of the eyeball, where it is a loose, very vascular layer over the sclera but is firmly adherent to the cornea, forming its anterior epithelium. The region where it is reflected is the *fornix*. The conjunctiva is so thin that, when a subconjunctival haemorrhage occurs as the result of an injury, the blood is fully oxygenated and bright red instead of becoming dark in colour as happens in a bruise.

The medial and lateral extremities of the eyelids are attached to the medial and lateral *palpebral ligaments*. The medial ligament is very strong and tethers the eyelids to the medial wall of the orbit.

### Nerve supply of the cornea

This is the most important paragraph in this chapter. The cornea (as you will have guessed) is supplied by the ophthalmic division of the trigeminal nerve and the reason this sensory nerve supply is so important is because, if the nerve is anaesthetized (for example, to remove a foreign body from the eye) or if it is damaged, the cornea will not react to the presence of a foreign body and may become ulcerated and eventually suffer severe damage. It is therefore important to keep the eye covered when the sensory nerve supply is interrupted in any way.

### The lacrimal apparatus

The *lacrimal gland* occupies the upper lateral corner of the orbit, lying in a shallow depression in the bone. It wraps itself around the lateral border of levator palpebrae superioris in much the same way as the submandibular gland wraps itself around the posterior border of the mylohyoid muscle. The superficial part of the gland therefore occupies the most lateral part of the upper eyelid. Its secretions pass through 9–12 ducts in the lateral part of the superior fornix of the conjunctiva. Thence the tears pass across the front of the eyeball, propelled by blinking and kept to the surface of the eyeball by the lower lid, which itself is held against the eyeball by the tone of the palpebral part of orbicularis oculi (see p. 368). The tears drain into the *lacrimal puncta*, which are two minute holes at the medial ends of the upper and lower eyelids. The puncta lead into *canaliculi*, which open into the *lacrimal sac*. This lies in a deep groove bounded by the lacrimal bone and the frontal process of the maxilla. The lacrimal sac is deep to the medial palpebral ligament and leads down into the *nasolacrimal duct*. This is probably shorter than you expected. It is only about 1.8 cm long and leads from the medial angle of the eye to the inferior meatus of the nose (p. 397).

## THE SCALP

The most obvious feature of the scalp is the *galea aponeurotica* (*epicranial aponeurosis*) and its associated muscles. The galea is an extensive and tough aponeurosis into which the occipital and frontal bellies of the occipitofrontalis muscle are inserted. Laterally it is attached to the zygomatic arch. The posterior end of occipitofrontalis is attached to the occipital bone above the superior nuchal line but the frontal belly has no bony attachments, its anterior part merging with the muscles of facial expression above the orbits. Alternate contraction of the frontal and occipital bellies, a feat which only a select band of people can do, can move the whole scalp backwards and forwards but the main function of the frontal belly is to raise the eyebrows (p. 370), thus wrinkling the forehead.

The layers of the scalp can be remembered by means of a perfect mnemonic: SCALP. They are Skin, Cutaneous fat and areolar tissue forming the superficial fascia, Aponeurosis (epicranial), Loose areolar tissue and Pericranium. Surprisingly, the most important of these layers from a clinical point of view is the layer of loose areolar tissue since this provides an easy plane of cleavage in injury and a plane in which blood from severed vessels can spread for a long distance. The first three layers of the scalp cannot easily be separated so they remain attached to each other when the scalp is avulsed, as may happen when long hair becomes caught in machinery. Bleeding beneath the aponeurosis may spread downwards as far as the zygomatic arch and into the upper eyelids.

### Blood vessels and nerves

The scalp is extremely vascular and bleeds freely when injured, especially as the arteries are held in the dense connective tissue of the second layer and cannot retract. The vessels anastomose freely and haemorrhage cannot be controlled by compressing any single vessel. Bleeding can always be stopped, however, by direct pressure on and around the wound. In infections of the scalp, pus may spread beneath the aponeurosis and it is well to remember the presence of the valveless *emissary veins*. These pass through the bone of the skull and unite the scalp veins with the intracranial venous sinuses, so they provide a potential channel for the spread of infection into the cranial cavity. Fortunately, the blood flow is usually from the inside to the outside.

The vessels and sensory nerves all enter the scalp from the periphery and since they have all been mentioned previously in this and other chapters it will suffice to

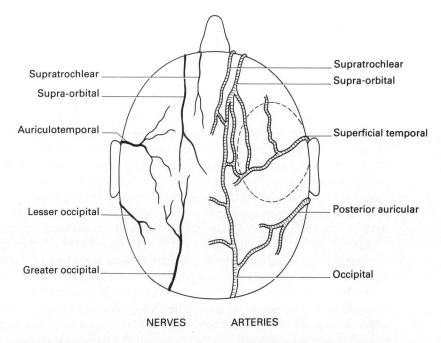

Supratrochlear

Supra-orbital

Auriculotemporal

Supratrochlear

Supra-orbital

Superficial temporal

Lesser occipital

Greater occipital

Posterior auricular

Occipital

NERVES            ARTERIES

**Fig. 21.4** Nerves and arteries of the scalp. The dotted line shows a temporal 'flap'.

show them in the form of a diagram (Fig. 21.4). Anastomoses are so plentiful that if most of the scalp is torn away it will survive as long as one of the main arteries is intact. For this reason, too, neurosurgeons' incisions are made to form flaps of scalp based on one or more of the major arteries (Fig. 21.4).

# Chapter 22
# Cranial Cavity and Orbit

It is important to understand that the cranial cavity is a rigid box containing the brain, important blood vessels and the cerebrospinal fluid. It communicates with the vertebral canal through the foramen magnum and with the tissues outside the cranium via various foramina. Hence extensive bleeding or a growing tumour inside the cranium has little scope for expansion and, while compensation for the presence of such a lesion can occur to some extent by the displacement of some cerebrospinal fluid or blood, an early rise in intracranial pressure will occur.

The concept of the cranium as a closed box also explains the structure of the intracranial blood vessels. In order to avoid the collapse of the larger veins when the pressure rises, for example during systole, they have relatively rigid walls, composed of dura mater, and are known as *venous sinuses*. The arteries have very thin walls. In arteries outside the skull the distending effect of the blood pressure is counteracted by the tissue pressure and by the tension in the vessel wall. Inside the cranium, when the vessel distends during systole, the intracranial pressure also rises so that there is less need for a thick vessel wall. The arteries are, however, prone to localized distension, *aneurysm*, wherever the vessel wall is weak and this may lead to a life-threatening haemorrhage.

Before reading the following account you should revise pp. 302–306.

## THE MENINGES

The meninges (sing. meninx) are the three layers that cover the brain — the *dura mater, arachnoid mater* and *pia mater*. The dura is often called the *pachymeninx* because it is the thickest layer (cf. pachyderm, such as an elephant), while the pia and arachnoid together form the *leptomeninges* (soft). The dura lines the skull and also forms various incomplete septa, which will be described shortly. The arachnoid covers the brain (and spinal cord) but bridges across the deep fissures and sulci. The pia, however, is closely adherent to the brain, following its contours, so that there is a space, the *subarachnoid space*, between the two leptomeninges and this contains cerebrospinal fluid. The subarachnoid space continues through the foramen magnum so that the cerebrospinal fluid in the cranium is continuous with that around the spinal cord (Chapter 14).

### Dura mater

The dura has two layers, an outer fibrous layer that is adherent to the bone and forms, in fact, the fibrous layer of the periosteum, and an inner smooth *serous layer*. The serous layer, in general, follows the fibrous layer but leaves it in places to form the venous sinuses and various septa. The dura is supplied by numerous small arteries and one large one: the *middle meningeal*. This arises from the maxillary artery and enters the skull through the foramen spinosum. It then finds itself in the middle cranial fossa, runs forwards and laterally and then begins to ascend on the inside of the squamous temporal bone before dividing into anterior and posterior branches. The anterior branch runs upwards and backwards towards the vertex while the posterior branch passes backwards, its surface marking being adequately indicated by a pencil placed behind the ear. The artery is accompanied by a large vein and the vessels make a prominent marking on the bone. They supply the bone as well as the dura and are important because, being so closely related to the bone, they are liable to be torn in injuries to the skull. In such cases access may be obtained through a small opening made in the skull. The anterior division may be found at a point 4 cm above the midpoint of the zygomatic arch.

## The falx cerebri and tentorium cerebelli

There are two major infoldings of the serous layer of dura. The largest, the falx cerebri, has the *superior sagittal sinus* in its attached margin (see below) and it forms an incomplete septum in the sagittal plane, partly subdividing the cranial cavity into two. It projects into the cleft between the two cerebral hemispheres and is sickle-shaped, like the falciform ligament (p. 227 and Fig. 22.1), from which you will deduce that falx means sickle. The point of the sickle is near the cribriform plate and its base is continuous with the *tentorium cerebelli*. The tentorium itself forms an incomplete roof over the posterior cranial fossa, i.e. over the cerebellum, and the brainstem projects through the tentorial opening. The falx and the tentorium are rather rigid sheets of dura and the combination of their sharp edges

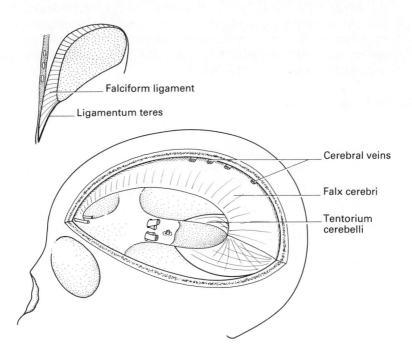

Falciform ligament

Ligamentum teres

Cerebral veins

Falx cerebri

Tentorium
cerebelli

**Fig. 22.1** The falx cerebri and the tentorium cerebelli.

and the soft brain may have potentially lethal consequences when the brain is displaced to one or other side by a space-occupying lesion.

## THE DURAL VENOUS SINUSES

The dural venous sinuses are formed by the serous layer of the dura leaving the fibrous layer so that there is an endothelium-lined channel between them. The largest sinus is the superior sagittal sinus and it is particularly important because it is associated with drainage of the cerebrospinal fluid.

### The superior sagittal sinus

This begins just in front of the cribriform plate and runs backwards, gradually increasing in size until it reaches the region of the internal occipital protuberance, where it usually turns to the right to become the *right transverse sinus*, which lies in the attached border of the tentorium cerebelli (see Fig. 22.3). The sinus then follows an S-shaped course, closely related to the back of the petrous temporal bone. This part is called the *sigmoid sinus*, which ends by passing through the right jugular foramen and becoming the internal jugular vein. A number of veins draining the cerebral hemispheres open into the superior sagittal sinus, either directly or via dilatations, called *lacunae laterales*, into each of which several veins open. Beneath the dura lies the arachnoid mater and in places projections of arachnoid pass through small openings in the serous dura and project into the superior sagittal sinus and the lacunae to form *arachnoid villi* (Fig. 22.2). The subarachnoid space is beneath the arachnoid mater and thus forms the 'lumen' of the arachnoid villi. The enclosed cerebrospinal fluid thus only has to pass through the arachnoid and one layer of endothelium to reach the bloodstream and this is the normal route for the absorption of cerebrospinal fluid. In later life the villi become clumped together to form *arachnoid granulations* and these make depressions on the inside of the skull.

### Inferior sagittal sinus

This is much smaller than the superior sagittal sinus. It runs in the free edge of the

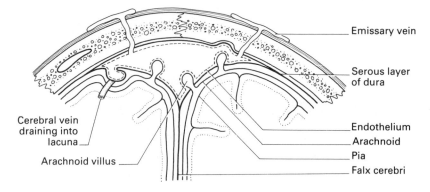

**Fig. 22.2** The arachnoid villi project into the superior sagittal sinus, seen here in coronal section.

falx until it reaches the tentorium. It then receives part of the venous drainage of the brain (*great cerebral vein* or *vein of Galen*) and becomes the *straight sinus*. This lies along the line where the falx and tentorium come together and it then usually passes to the left to form the left transverse and left sigmoid sinuses and finally the left internal jugular vein. Hence in most skulls you will find that the right jugular foramen is bigger than the left. This is variable, however, and quite often the ends of the superior sagittal and the straight sinuses join up to form the *confluence of the sinuses* near the internal occipital protuberance.

## Cavernous sinuses

These are a pair of sinuses lying on either side of the sella turcica and, therefore, of the pituitary gland. Anteriorly, the *superior* and *inferior ophthalmic veins* and posteriorly some of the minor venous sinuses open into them (Fig. 22.3). The cavernous sinuses are joined to each other across the midline by *intercavernous sinuses* that lie in front of and behind the pituitary. Veins from the lower parts of the brain also drain into the sinuses.

The cavernous sinuses are important because of the cranial nerves 3, 4, 5 and 6 that are intimately related to them, as will be seen later, and because the internal carotid passes through the middle of the sinus, being the only artery with endothelium on both its internal and its external surface (Fig. 22.4).

The smaller venous sinuses are shown in Fig. 22.3 and need not be mentioned further here except to say that the *inferior petrosal sinus* passes out through the jugular foramen before joining the internal jugular vein as its first tributary. The

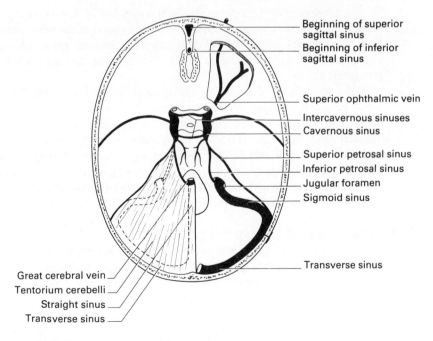

Beginning of superior sagittal sinus
Beginning of inferior sagittal sinus

Superior ophthalmic vein

Intercavernous sinuses
Cavernous sinus

Superior petrosal sinus
Inferior petrosal sinus
Jugular foramen
Sigmoid sinus

Transverse sinus

Great cerebral vein
Tentorium cerebelli
Straight sinus
Transverse sinus

**Fig. 22.3** The venous sinuses of the cranial cavity.

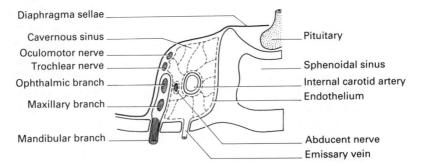

**Fig. 22.4** A section through the cavernous sinus to show the nerves related to it.

majority of the dural venous sinuses communicate with veins outside the cranium by emissary veins.

## RELATIONS OF THE BRAIN

Although this book does not cover neuro-anatomy, the relations of the brain to the structures forming the cranial cavity are obviously important. The lobes of the cerebrum are related to the bones with the corresponding names: the frontal lobe with the frontal bone, the parietal lobe with the parietal bone and so on. The temporal lobe is not only related to the temporal bone, however, since it occupies the middle cranial fossa and is therefore also related to the sphenoid. Its medial surface is close to the cavernous sinus.

The *pituitary gland* (*hypophysis*), which is attached by its stalk to the under-surface of the hypothalamus, occupies the pituitary fossa of the sella turcica, the fossa being roofed in by a fold of dura called *diaphragma sellae*. Above, and usually just in front of, the pituitary lies the *optic chiasma*, an important relationship as will be seen.

The under-surface of the frontal lobe is supported by the orbital plate of the frontal bone and the lesser wing of the sphenoid, i.e. the floor of the anterior fossa, but the under-surface of the occipital lobe is supported by the tentorium, not the occipital bone. Below the tentorium the cerebellum sits snugly in the posterior fossa with the lower end of the medulla projecting through the foramen magnum. The pons, too, is in the posterior fossa, lying on the slope (the *clivus*) formed by the occipital and sphenoid bones in front of the foramen magnum. The midbrain passes through the opening in the tentorium to reach the diencephalon.

## THE BLOOD SUPPLY OF THE BRAIN

The brain is supplied by the internal carotid arteries anteriorly and the vertebral arteries posteriorly, the actual share-out of the brain between these two arteries depending on the form of the *circulus arteriosus* or the *circle of Willis*.

### The internal carotid artery

This forms a series of steps inside the skull. These form a characteristic pattern in carotid angiograms that radiologists call the *carotid siphon*, although none of

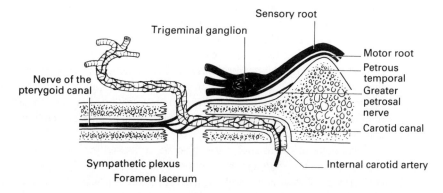

**Fig. 22.5** A section through the foramen lacerum, shown here as a canal rather than a foramen. Note that nothing goes completely through the foramen.

them seems to know why! The artery starts by entering the carotid canal just in front of the place where the internal jugular vein leaves the jugular foramen. It immediately turns forwards to run horizontally through the *carotid canal* in the petrous temporal bone. It then turns upwards to enter the cranial cavity through the *internal* opening of the foramen lacerum. This leads to some difficulty since it is usually stated that *nothing* of importance passes through the foramen lacerum. If, however, it is thought of as a canal instead of a foramen it becomes *easy* to understand (Fig. 22.5). Thus the internal carotid enters the cranial cavity via the *internal* opening of the foramen lacerum and, as will be described later, the greater petrosal nerve *leaves* the cranial cavity also via the internal opening, to reach the pterygoid canal. As the internal carotid artery reaches the end of the carotid canal, it often comes directly into contact with the trigeminal ganglion because, with increasing age, the bony roof of the canal often disappears so that the ganglion sits directly upon the internal carotid and may appear to pulsate.

In the next part of its course, the internal carotid artery turns forwards to enter the cavernous sinus (see Fig. 22.4). It finally turns upwards and backwards in the hollow behind the anterior clinoid process before breaking up into its three terminal branches — the *anterior* and *middle cerebral arteries* and the (usually) small *posterior communicating artery*. As it turns, it gives off the important *ophthalmic artery*, which enters the optic canal with the optic nerve.

Only very small branches to the hypophysis and the other local structures are given off by the artery before it breaks up into its terminal branches.

### The vertebral artery

The course of the vertebral artery prior to its entry into the cranium has already been described (p. 331). As it enters the cranium it gives off small *anterior* and *posterior spinal arteries*, and then joins with its opposite number to form the single midline *basilar artery*. Before joining, however, it usually (variations are common) gives off the *posterior inferior cerebellar artery*. This is badly named since as well as supplying the cerebellum it also provides branches to supply part of the medulla, so that occlusion of this vessel leads to widespread signs and symptoms.

The basilar artery passes along the pons in the midline, supplying branches to the brainstem, cerebellum (*anterior inferior cerebellar*) and the inner ear (via the internal auditory meatus), and it ends by dividing into the right and left *superior cerebellar* and *posterior cerebral arteries*. The oculomotor nerve passes between these two arteries. The posterior cerebral artery communicates with the middle cerebral artery by means of the posterior communicating artery, which, along with the anterior communicating artery between the two anterior cerebral arteries, completes the circle of Willis.

Textbooks of neuro-anatomy should be consulted for details of the blood supply and venous drainage of the brain, which are extremely important from a clinical point of view since cerebral vascular disease is one of the common causes of sickness and death.

## THE CRANIAL NERVES WITHIN THE CRANIUM

There are 12 cranial nerves. You should take pains to learn the numbers and names and not rely on obscene mnemonics, since neurologists refer to them constantly as in, for instance, '3rd nerve palsy'. They all leave the brainstem on its ventral surface, except for the 4th (trochlear) nerve, which arises dorsally just behind the colliculi and passes around the brainstem to reach its ventral aspect.

### 1 Olfactory

The olfactory nerves pass through the cribriform plate of the ethmoid. The cell bodies are peripheral, in the olfactory mucosa, and their axons travel upwards to reach the olfactory bulbs.

### 2 Optic

The optic nerves are really prolongations of the brain rather than true nerves. They develop from the *optic stalk*, which is an outgrowth from the embryonic forebrain, and they are surrounded by dura, pia and arachnoid. The subarachnoid space contains cerebrospinal fluid, so that this extends forwards as far as the back of the eyeball. If there is a rise in intracranial pressure, the optic nerve will therefore be affected and there will be changes in the optic disc within the eyeball. Since the nerve is really brain tissue rather than a true nerve, the nerve fibres will not regenerate if they are divided.

Having left the eyeball, the optic nerve enters the cranium through the optic canal (which also contains the ophthalmic artery) and then forms the *optic chiasma*. Here the temporal fibres continue their course on each side but the nasal fibres cross the midline after making a slight swerve into the optic nerve (Fig. 22.6). When working out the effects of damage to the chiasma, it is important to remember that light rays reaching the *nasal* side of the retina have come from the *temporal* field of vision and vice versa. Thus, if the central (crossover) part of the chiasma is affected, perhaps by a pituitary tumour, both *nasal* areas of the retina and therefore both *temporal* fields of vision will be affected (*bitemporal hemianopia*). If the left optic tract is divided, the temporal side of the left retina and the nasal side of the right retina will be affected so

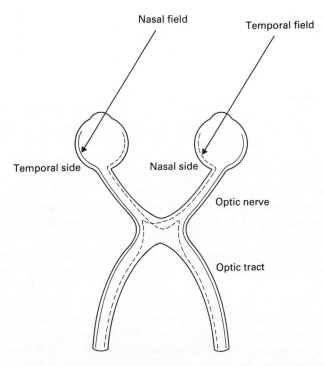

**Fig. 22.6** The optic chiasma and the path of light from the *right* half of the visual field.

that the visual field will be deficient on the right side of the body (*homonymous hemianopia*).

### 3, 4 and 6 The oculomotor, trochlear and abducent nerves

It is as well to group these nerves together since they all supply the muscles that move the eyeball. The *abducent nerve*, naturally enough, supplies the *abductor* muscle, i.e. lateral rectus. The *trochlear nerve* supplies the muscle with a pulley (cf. the pulley-like *trochlea* of the humerus), i.e. superior oblique. All the other muscles are supplied by the oculomotor nerve, which also supplies *levator palpebrae superioris*. All these nerves are found in relation to the cavernous sinus.

The 3rd nerve, having passed between the superior cerebellar and posterior cerebral arteries, pierces the dura near the posterior clinoid process and travels forwards high up on the lateral wall of the cavernous sinus (see Fig. 22.4). It gradually inclines downwards and enters the orbit through the lower part of the superior orbital fissure, where it divides into upper and lower branches (see Fig. 22.9). The 4th (very slender) nerve, having passed round the midbrain, pierces the dura behind the oculomotor nerve and lies in the lateral wall of the cavernous sinus below it. It travels straight forwards and so enters the orbit through the upper part of the superior orbital fissure to enter the upper border of the superior oblique. The 6th nerve also has a long intracranial course since it leaves the brain at the lower border of the pons and therefore in the posterior cranial fossa. It pierces the dura on the clivus (p. 304) and then has to climb to the top of the ridge formed by the petrous temporal bone before entering (literally) the cavernous sinus. Here it lies lateral to the internal carotid artery before entering the orbit through the lower part of the superior orbital fissure.

## 5 Trigeminal nerve

This emerges at the junction between the pons and the middle cerebellar peduncle and therefore starts in the posterior cranial fossa. It crosses the petrous temporal bone at the medial smooth part of the ridge and then enlarges to form its ganglion in the middle cranial fossa overlying the roof of the carotid canal (see Fig. 22.5). It does not pierce the dura in the posterior fossa but carries a diverticulum of posterior fossa dura forwards beneath the middle fossa dura of the lateral wall of the cavernous sinus (see Fig. 22.4) and its three branches finally pierce the dura in the cavernous sinus (Fig. 22.7). The diverticulum is called the *cavum trigeminale* or *Meckel's cave*.

The trigeminal ganglion, like those of other cranial nerves, is a sensory ganglion exactly comparable to a spinal (dorsal root) ganglion of a spinal nerve, i.e. it consists of a collection of pseudo-unipolar cells, each of which has an axon passing into the brainstem and a dendrite extending peripherally to a sensory nerve ending. The trigeminal motor root bypasses the ganglion and joins the nerve distal to it, just as in a spinal nerve the motor (anterior nerve) root joins the spinal nerve distal to the ganglion. The motor root joins the mandibular division, which is therefore a mixed nerve. *The other two divisions are entirely sensory.*

The trigeminal nerve has three divisions: the *mandibular, maxillary* and *ophthalmic*. The mandibular division leaves through a trapdoor in the floor, i.e. it turns downwards soon after the nerve enters the cavernous sinus and leaves the cranial cavity through the foramen ovale, usually accompanied by the lesser petrosal nerve (p. 401). Its further course has been described (pp. 350 ff.). The maxillary division travels forwards and slightly downwards, pierces the dura of the

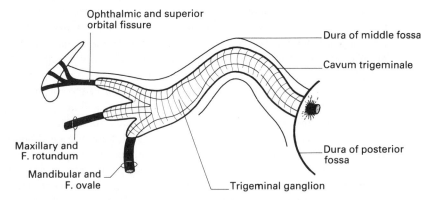

Ophthalmic and superior orbital fissure

Dura of middle fossa

Cavum trigeminale

Maxillary and F. rotundum

Mandibular and F. ovale

Dura of posterior fossa

Trigeminal ganglion

Fig. 22.7 A transverse section through the petrous temporal bone with the posterior cranial fossa on the right and the middle fossa on the left. The cavum trigeminale is a pocket of posterior fossa dura projected forwards around the trigeminal nerve into the middle fossa.

cavum trigeminale and finally leaves through the foramen rotundum. Its further course will be described in the next chapter. The ophthalmic division pierces the dura and leaves through the superior orbital fissure, where it divides into *frontal, lacrimal* and *nasociliary* divisions.

The superior orbital fissure therefore transmits three motor nerves (3, 4 and 6) and three sensory nerves (frontal, lacrimal and nasociliary), as well as a couple of ophthalmic veins (superior and inferior).

### 7 The facial nerve

The facial nerve runs, along with the 8th nerve, into the internal auditory meatus. The nerve fibres that are destined to become part of the *chorda tympani* and *greater petrosal nerve* (i.e. parasympathetic fibres) travel separately in the *nervus intermedius*, so called because it lies between the 7th and 8th nerves. The further course of the facial nerve will be described later in this chapter.

### 8 The vestibulocochlear nerve

This is the nerve that is responsible for hearing and balance, hence the two divisions. It enters the internal auditory meatus and at its far end it divides into vestibular and cochlear branches.

### 9 The glossopharyngeal nerve

Arising from the side of the medulla in front of, but in line with, the rootlets of the vagus and cranial accessory nerves, the glossopharyngeal nerve immediately traverses the jugular foramen in a small anterior subdivision of the foramen.

### 10 The vagus nerve

This nerve, too, passes through the jugular foramen, in company with the cranial part of the accessory nerve.

### 11 The accessory nerve

This nerve is in two parts; it has been described on p. 345.

### 12 The hypoglossal nerve

This is a purely motor nerve and so arises from the side of the medulla rather anteriorly, in line with the anterior nerve roots of the spinal nerve. Its numerous rootlets unite to form a single nerve that leaves through the hypoglossal (anterior condylar) canal.

## THE ORBIT

Most of the bones of the orbit have already been mentioned so only a summary need now be given. The margin of the orbit is formed by frontal, maxillary and zygomatic bones (see Figs 16.3 & 22.8). The medial wall is formed by the lacrimal and ethmoid bones and the lateral wall by the zygomatic bone. The floor is the maxilla and the roof the orbital plate of the frontal. At the back are the greater and lesser wings of the sphenoid separated by the superior orbital fissure (Fig. 22.8). The inferior orbital fissure is a gap between the maxilla and the greater

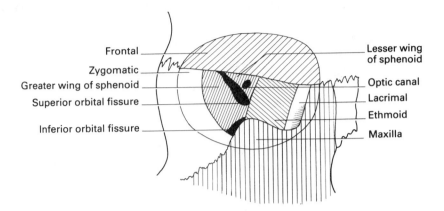

Frontal — — Lesser wing
of sphenoid

Zygomatic — — Optic canal

Greater wing of sphenoid — — Lacrimal

Superior orbital fissure — — Ethmoid

Inferior orbital fissure — — Maxilla

**Fig. 22.8** The bones of the right orbit. The orbital process of the palatine bone is too small to be worth recording.

wing, while the optic canal is in the root of the lesser wing. It is most important to understand that the medial walls of the two orbits are parallel but the lateral walls diverge so that the long axis of the orbit is at an angle to the long axis of the eyeball (see Fig. 22.10).

## The muscles of the orbit

The muscles, with one exception, originate from or near a fibrous ring that includes the lower part of the superior orbital fissure and the optic canal (Fig. 22.9). From this fairly restricted origin the muscles spread out to be attached to the eyeball around its periphery, like the muscles of the rotator cuff around the head of the humerus (p. 47). Since the muscles are each enclosed in a fascial sheath that is continuous as a single layer between the muscles (cf. the investing layer of deep fascia of the neck as seen in Fig. 17.3), the whole forms a

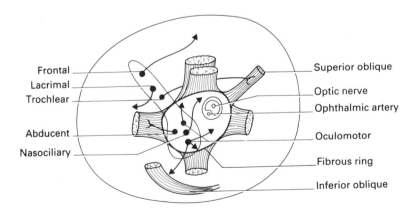

Frontal — — Superior oblique

Lacrimal — — Optic nerve

Trochlear — — Ophthalmic artery

Abducent — — Oculomotor

Nasociliary — — Fibrous ring

— Inferior oblique

**Fig. 22.9** Muscles of the orbit are arranged around the two major foramina, except for inferior oblique.

fibromuscular cone whose blunt apex is at the back of the orbit. The interior of the cone is filled with fat and the optic nerve runs through it. Just outside the cone, above it, is *levator palpebrae superioris*, which is attached not to the eyeball but to the eyelid (p. 373).

### Levator palpebrae superioris

This muscle arises from the back of the orbit just above the fibrous ring and lies immediately beneath the roof of the orbit. Its actions have already been described but it is worth reminding you again that it is supplied both by the oculomotor nerve and by sympathetic nerves so that loss of either of these nerves can lead to drooping of the eyelid (*ptosis*).

### Lateral rectus

This lies along the lateral side of the orbit (Fig. 22.10) and so is an 'abductor' of the eyeball and is supplied by the abducent nerve.

### Medial rectus

This lies along the medial wall of the orbit and therefore turns the eyeball so that the pupil points medially. The left and right medial rectus muscles contracting simultaneously cause convergence of the gaze, which is necessary when both eyes focus on a near object such as this book.

### Superior rectus

This is where you need to study Fig. 22.10 carefully, noting that the long axes of the orbits are divergent. The line of pull of the muscle relative to the eyeball is therefore oblique and, as well as turning the eye upwards, it also turns it medially.

### Inferior rectus

This is also obliquely aligned relative to the eyeball so that it turns the eyeball upwards and medially.

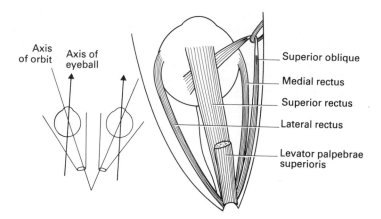

**Fig. 22.10** The actions of the muscles that move the eyeball (see text).

### Superior oblique

This is more complicated still. The muscle passes along the medial wall of the orbit to reach its *trochlea*. Here it loops backwards through a fascial sling before being attached to the eyeball. Figure 22.10 shows that it will turn the eyeball downwards and outwards. It is supplied by the trochlear nerve.

### Inferior oblique

This is the exception mentioned above. It arises from the floor of the orbit and travels laterally below the eyeball like a sling and is finally attached to the lateral side. It turns the eyeball to look upwards and outwards.

### Muscle actions

The actions of medial and lateral rectus muscles are straightforward. Superior rectus, however, turns the eye upwards and medially so that to turn the gaze directly upwards involves simultaneous contraction of superior rectus (up and in) and inferior oblique (up and out). Similarly in order to look at one's feet, it is necessary to use inferior rectus (down and in) and superior oblique (down and out). All this sounds fairly simple but in fact the muscle actions are extremely complex when one considers the two eyes together. Thus to look to the left, the left lateral rectus (abducent) and the right medial rectus (oculomotor) contract together but, if the object is near, there has to be appropriate convergence as well. If the object is travelling from left to right and also upwards and at the same time away from the observer, the nerve and muscle interactions have to be very finely co-ordinated so that it is not surprising that this co-ordination may easily be disturbed. An excessive dose of alcohol, for example, is well known to cause double vision.

## The blood vessels

The ophthalmic artery (do not forget the 'h' after 'p' as well as after 't') is derived from the internal carotid artery and enters the orbit through the optic canal just lateral to and below the optic nerve. Its most important branch is the *central artery of the retina*, which is an end artery (p. 20). It enters the optic nerve and travels near its centre towards the eyeball. Within the eye it divides up into *nasal* and *temporal branches*, which are accompanied by corresponding veins. These can be seen very clearly with the aid of an ophthalmoscope. The eye is the only place where direct observation of the blood vessels is possible and ophthalmoscopy plays an important part in the diagnosis of diseases of the cardiovascular system.

Once within the orbit, the ophthalmic artery (Fig. 22.11) crosses to the medial side of the orbit, usually above the optic nerve. It then travels forwards along the medial wall towards the front of the orbit and ends by dividing into the *dorsal nasal* and *supratrochlear arteries,* which emerge on to the face. The first large branch in the orbit is the *lacrimal artery,* which follows the lateral wall of the orbit, supplies the lacrimal gland and ends on the face by supplying the eyelids. Next come the *long* and *short posterior ciliary* arteries. These enter the eyeball to supply the choroid, the short arteries (many) supplying the back of the eyeball and the long (two of them) entering the back of the eyeball but travelling within it to reach the

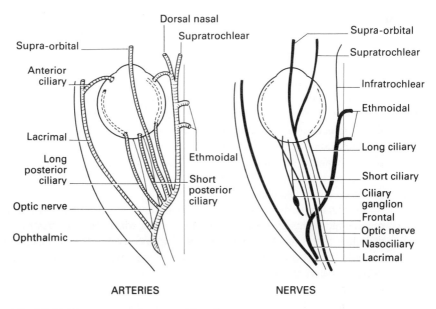

**Fig. 22.11** The arteries and nerves of the orbit.

region of the corneoscleral junction before breaking up into branches. Some of the branches mentioned above give *anterior ciliary arteries* to the front of the eyeball. The supra-orbital artery runs below the roof of the orbit and supplies the scalp (p. 367). The *anterior* and *posterior ethmoidal arteries* enter canals in the ethmoid bone to supply the ethmoidal air cells and the posterior (and larger) ethmoidal artery finally enters the nasal cavity.

The veins accompany the arteries but drain into superior and inferior ophthalmic veins to communicate respectively with the cavernous sinus and the pterygoid plexus of veins. Like the arteries, the veins communicate with those on the face (p. 370).

## The nerves

Two of the three motor nerves follow a very simple course (see Fig. 22.9). The trochlear nerve enters the orbit above the fibrous ring and therefore outside the muscular cone, and immediately crosses to the medial side and enters the upper border of superior oblique. The abducent enters within the ring and immediately turns laterally to enter lateral rectus.

The oculomotor nerve divides into two branches that lie above and below the nasociliary nerve within the muscular cone. The superior branch travels upwards to reach superior rectus and levator palpebrae superioris. The inferior branch breaks up to supply medial rectus, inferior rectus and inferior oblique. Just as inferior oblique is an exception to the other muscles, so its nerve is an exception to the other motor nerves in that it carries parasympathetic fibres from the *Edinger–Westphal nucleus* in the midbrain. These travel in a branch to the *ciliary ganglion*, one of the four parasympathetic ganglia in the head and neck. The fibres synapse in the ganglion and then travel in the short ciliary nerves to the back of the eyeball. They are destined to supply the *ciliaris muscle* and the *sphincter pupillae*.

The sensory nerves are all branches of the ophthalmic division of the trigeminal nerve. The frontal and lacrimal nerves enter the orbit above the fibrous ring and the nasociliary within it. The frontal nerve is directly below the roof of the orbit, lying on levator palpebrae superioris. It divides into *supra-orbital* and *supratrochlear nerves*, which leave the orbit at its upper border and supply the scalp, where their further course has been described on p. 367. The lacrimal nerve travels along the lateral wall of the orbit to supply the skin of the eyelids and it also carries parasympathetic secretomotor fibres to the lacrimal gland. Since these fibres are associated mainly with the maxillary division of the trigeminal, they will be described in the next chapter.

Finally, the nasociliary nerve has a course and branches similar to those of the ophthalmic artery (Fig. 22.11), except that the *short ciliary nerves* are derived from the ciliary ganglion, not the nasociliary nerve, and the lacrimal nerve is a separate branch of the ophthalmic division. The nasociliary nerve thus gives off two long ciliary nerves, which enter the back of the eyeball and are important because they carry sympathetic fibres to dilator pupillae and sensory fibres to the cornea (p. 373). The *anterior* and *posterior ethmoidal nerves* have a similar course to the arteries except that, for some reason, the nasociliary nerve is said to end by turning medially to *become* the anterior ethmoidal nerve, giving off as it does so a small *infratrochlear nerve*.

## THE EYEBALL

A section through the eyeball that shows the main features, but not the histology, is given in Fig. 22.12. There are three coats, the outer two of which are developmentally comparable to a continuation of the layers of the meninges. Thus the

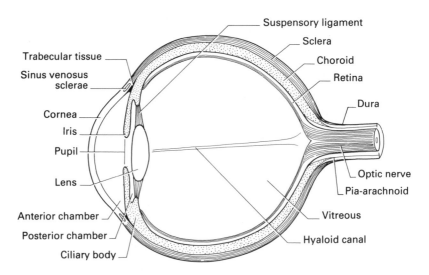

**Fig. 22.12** A section through the eyeball.

tough dura is represented by the even tougher *sclera*. Anteriorly, however, this is replaced by the transparent *cornea*. The vascular pia-arachnoid is represented by the vascular coat, the *choroid*, which anteriorly forms the *ciliary body* and the *iris*. The vascular layer as a whole is known as the *uveal tract*. The innermost layer of the eyeball represents the brain itself and is the retina.

## The sclera and cornea

The sclera is composed of tough fibrous tissue and forms the 'white' of the eye. The cornea has a more complex biochemical structure to make it transparent but, whereas the sclera contains blood vessels, the cornea is avascular and has no lymphatics and can therefore be transplanted with comparative ease. Its most superficial layer is the conjunctiva, already described in Chapter 21. At the corneoscleral junction an important venous structure is the *sinus venosus sclerae* (*canal of Schlemm*). Between this and the anterior chamber is only a meshwork of trabeculated tissue and it is through this that the aqueous humour drains from the anterior chamber into the venous system. Any form of obstruction to this drainage may lead to a rise in intra-ocular pressure (*glaucoma*).

## The uveal tract

The choroid is an extremely vascular layer, fed by the long and short posterior ciliary arteries and the anterior ciliary arteries. The venous drainage is into four large venous confluences on the outer surface of the choroid called *vorticose veins* and thence into the ophthalmic veins. The choroid is responsible for the nutrition of the outer layers of the retina. An enlargement of the uveal tract anteriorly forms the circularly arranged *ciliary body*, from which project the radial ciliary processes. As well as blood vessels, the ciliary body contains smooth muscle, *ciliaris*, whose fibres are aligned both circularly like a sphincter and radially. The ciliary processes secrete the *aqueous humour*, which flows through the pupillary aperture into the anterior chamber and thence into the canal of Schlemm. The ciliary body is continuous with the iris, which also contains smooth muscle. This, like the ciliaris muscle, is arranged both circularly (*sphincter pupillae*) and radially (*dilator pupillae*). The former is supplied by parasympathetic nerves and the latter by sympathetic. Thus an oculomotor nerve lesion may cause dilatation of the pupil (*mydriasis*). The pupil constricts as a reaction to increased light falling on the eye and also when the eye is focused on a near object (reaction to accommodation). The black appearance of the pupil is due to the large amount of pigment in the retina and the choroid.

## The retina

The retina is in two layers, an inner *nervous* and an outer *pigmented* layer. These represent the fused two layers of the embryonic optic cup and they sometimes separate again in adult life to cause the condition erroneously known as a *detached retina*. The most superficial (innermost) layer of the retina consists of a layer of nerve cells, whose axons pass backwards to form the optic nerve. The nervous layer of the retina is replaced anteriorly by a simple layer of pigment cells (the *pars caeca* or blind part of the retina). The main blood supply to the retina is from the branches of the central artery, described above. The main cavity of the eyeball is filled with the gelatinous *vitreous humour* (or vitreous body), which is therefore in direct contact with the retina. A *hyaloid canal*, representing the embryonic hyaloid artery, runs from the back of the eyeball to the lens.

## The lens

The lens is composed of elongated cells with an anterior epithelium. The cells are so long that they are called *lens fibres* and they are arranged in a complex fashion. The lens is rich in glycosaminoglycans and normally remains transparent but with increasing age and in certain diseases it may become opaque (*cataract*). The lens is highly elastic and is kept in a state of tension by the *lens capsule*, which in turn is attached to the ciliary processes, by a *suspensory ligament.*

For close vision, the ciliaris muscle pulls the ciliary processes centripetally (by means of its circular fibres) and forwards (by its radial fibres). The lens capsule thus becomes slack and the elasticity of the lens causes it to become fatter and therefore to shorten its focal length. Remember this, because it is a common error to suppose that the ciliary muscle pulls on the lens capsule to compress the lens.

Quite a good way to remember these complications is to realize that the oculomotor nerve is the *nerve of close vision*. It supplies the *ciliary muscle*, which shortens the focal length of the lens; it supplies *sphincter pupillae*, which causes the reaction of the pupil to accommodation; it supplies *medial rectus*, for convergence; it supplies levator palpebrae superioris, to open the eye. You should also remember that when day-dreaming during a lecture, your muscles are *relaxed* and your eye is focused on infinity.

The region in front of the lens and behind the iris is the *posterior chamber* and that in front of the iris is the *anterior chamber*. The two chambers communicate via the pupil and this is the route taken by the aqueous towards its drainage channel.

# Chapter 23
# ENT Territory

Ear, nose and throat (ENT) surgeons are officially known as otorhinolaryngologists. The larynx has been dealt with in Chapter 20 and the remainder of the ENT territory will now be described. A key region that is not well understood by students (justifiably) is the *pterygopalatine fossa*, a remote and secret region of the skull that is a sort of distribution centre since it communicates downwards with the mouth, medially with the nasal cavity, forwards with the orbit and face, backwards with the cranial cavity and laterally with the infratemporal region.

## THE PTERYGOPALATINE FOSSA

Note how the pterygoid plates support the back of the maxilla like a flying buttress, leaving a gap, the *pterygomaxillary fissure*, between the two bones. This leads directly into the pterygopalatine fossa but in the intact skull very few of the features of the fossa can be seen. Before you can understand its boundaries it is first necessary to examine the palatine bone, whose *horizontal plate* has already been described as forming the posterior part of the hard palate (see Fig. 20.1). Its *perpendicular plate* ends in two processes, an *orbital process*, which makes an insignificant contribution to the back of the orbit, and a *sphenoidal process*, which articulates with the sphenoid. The important thing is the gap between the processes, which, together with the body of the sphenoid, forms a foramen called, naturally, the *sphenopalatine foramen*. Figure 23.1 shows how the perpendicular plate articulates with the medial surface of the maxilla and projects backwards beyond the maxilla. The projecting part articulates with the medial pterygoid plate, forming a flush joint so that if you look at the lateral wall of the nasal cavity from behind it is very difficult to see where one ends and the other begins. The pterygopalatine fossa, then, is the funnel-shaped space, tapering downwards, bounded by the rounded surface of the back of the maxilla in front, the projecting part of the perpendicular plate of the palatine medially, and the vertical plate of bone that joins the two pterygoid plates behind. Laterally the pterygomaxillary fissure leads into the fossa from the infratemporal region.

The three walls of the fossa may be represented in a flat diagram if they are opened out as shown in Fig. 23.2. In the centre is the medial wall, with the *sphenopalatine foramen* leading into the nasal cavity at the top. The anterior wall is to the left, with the *inferior orbital fissure* leading into the orbit at the top, and the posterior wall is on the right, with the *foramen rotundum* leading back into the cranial cavity at the top. There is also a small opening, the *pterygoid canal*, near by. The articulation between the maxilla and the perpendicular plate of the palatine leaves a tunnel between the two bones, which is the *greater palatine canal*. This is shown at the bottom of the diagram; it opens on to the hard palate (see Fig. 20.1). A few smaller canals, the *lesser palatine canals*, lead from it. You

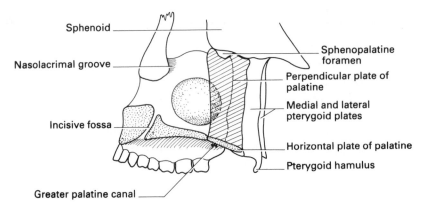

Sphenoid

Nasolacrimal groove

Incisive fossa

Greater palatine canal

Sphenopalatine foramen

Perpendicular plate of palatine

Medial and lateral pterygoid plates

Horizontal plate of palatine

Pterygoid hamulus

**Fig. 23.1** The medial surface of the maxilla and the palatine bone. Note the huge hole in the maxilla, opening into the sinus. The perpendicular plate of the palatine is cross-hatched.

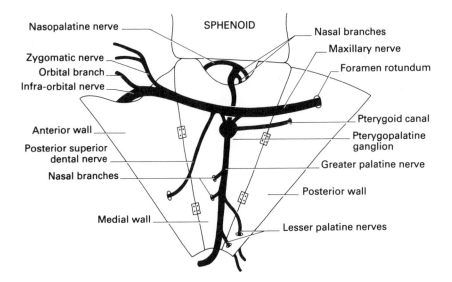

Nasopalatine nerve

Zygomatic nerve

Orbital branch

Infra-orbital nerve

Anterior wall

Posterior superior dental nerve

Nasal branches

Medial wall

SPHENOID

Nasal branches

Maxillary nerve

Foramen rotundum

Pterygoid canal

Pterygopalatine ganglion

Greater palatine nerve

Posterior wall

Lesser palatine nerves

**Fig. 23.2** The pterygopalatine fossa opened out to show its three walls and its content of nerves.

will need a skull and a short piece of wire to understand all this properly. The main contents of the fossa are the maxillary nerve and artery and another of the four parasympathetic ganglia. This is the *pterygopalatine (sphenopalatine) ganglion*, the other three being the submandibular, otic and ciliary.

### The maxillary nerve

Having left the anterior end of the cavernous sinus (see Fig. 22.7), the maxillary nerve passes through the foramen rotundum and enters the pterygopalatine fossa.

It leaves through the inferior orbital fissure, entering a bony canal in the floor of the orbit that emerges on the face as the *infra-orbital foramen* (see Fig. 16.3). Hanging down from the nerve in the fossa is the *pterygopalatine ganglion* and this is joined posteriorly by the *nerve of the pterygoid canal*. This nerve is formed within the foramen lacerum by the joining together of the *greater petrosal nerve* (p. 401) and sympathetic fibres from the plexus around the internal carotid artery (see Fig. 22.5). The pterygoid canal runs from the anterior wall of the foramen lacerum to the pterygopalatine fossa. Although there seem to be many nerves coming from the spenopalatine ganglion, they are only using the ganglion as a convenient route. *The only neurons that synapse in the ganglion are the parasympathetic secretomotor nerves for the lacrimal gland.*

The branches of the maxillary nerve, some of which pass through the ganglion without synapsing, are as follows. The *greater palatine nerve* runs downwards through the greater palatine canal to reach the hard palate (see Fig. 20.1). It gives off *nasal branches*, which pierce the perpendicular plate of the palatine, and also *lesser palatine nerves*, which pass backwards to supply the soft palate. The *nasopalatine nerve* passes medially through the sphenopalatine foramen to enter the nasal cavity; some nasal branches remain on the lateral wall of the nasal cavity to supply it with sensory fibres, while the nasopalatine nerve itself crosses the roof of the nose to reach the septum. It then runs downwards and forwards on the septum and ends by passing through a foramen in the incisive fossa and supplying the anterior part of the palate (see Fig. 20.1). The *posterior superior dental nerve* runs down on the back of the maxilla and enters a small foramen to supply the molar teeth.

The maxillary nerve itself, while in the canal in the floor of the orbit, gives off *orbital branches*, which supply some smooth muscle in the orbit (the *orbitalis* muscle) and then form two small cutaneous nerves, the *zygomaticotemporal* and *zygomaticofacial*. Branches are also given off to the teeth. The largest of these is the *anterior superior dental*. This runs down in the anterior wall of the maxillary sinus (remember that the floor of the orbit is the roof of the sinus, see Fig. 16.9) to reach the anterior teeth while another branch may be large enough to be given a name of its own, the *middle superior dental*. The maxillary nerve finally forms the *infra-orbital* nerve (see Chapter 21).

The nerve supply to the lacrimal gland is complicated. A *lacrimatory nucleus* in the pons gives axons that travel in the nervus intermedius before joining the facial nerve. They leave the facial nerve while it is inside the petrous temporal bone as the *greater petrosal nerve*, which emerges through a small hole in the anterior surface of the petrous temporal in the middle cranial fossa. It runs towards the foramen lacerum, deep to the trigeminal ganglion, and is joined by sympathetic fibres from the internal carotid artery to form the *nerve of the pterygoid canal*. This emerges into the pterygopalatine fossa; the fibres synapse in the ganglion and then join the maxillary nerve. They leave this in one of the *orbital* branches and reach the *lacrimal* nerve, probably, but not for certain, via the *zygomaticotemporal nerve* and a communicating loop between it and the *lacrimal nerve*. The fibres finally enter the lacrimal gland.

The maxillary artery is easy. After giving off various branches in the infratemporal region (Chapter 19), it enters the fossa through the pterygo-maxillary fissure and breaks up into the branches that correspond to and are distributed along with the branches of the maxillary nerve.

# THE PARANASAL SINUSES

Various reasons have been given for the presence of these sinuses, such as to make the skull lighter or to help the resonance of the voice, but it is still not at all clear why they are there. They develop after birth as diverticula from the main nasal cavity but they do not become fully developed until puberty. As you would expect, they are lined by a similar ciliated columnar mucous epithelium to that of the nasal cavity itself.

### The frontal sinuses

These are situated in the frontal bone above the orbits. They are very variable in size and usually asymmetrical. They open into the middle meatus of the nose.

### The maxillary sinus (or antrum)

This occupies the greater part of the maxillary bone and is the largest of the sinuses. Its relations are important and are shown in Fig. 16.9. The floor of the sinus is the upper surface of part of the hard palate, and the roots of the teeth, especially the molars, make elevations in the floor. The roof of the sinus is the floor of the orbit and contains the maxillary/infra-orbital nerve. Medially situated is the nasal cavity.

The sinus opens into the middle meatus but unfortunately the opening is at the top of the sinus (see Fig. 16.9) so that when the sinus contains mucus or pus it cannot escape into the nose until the sinus fills to the level of the opening. It can be drained by postural drainage, i.e. by holding the head back, or surgically, by making an opening into the lower part of the nasal cavity or through the alveolar part of the maxilla into the mouth.

### The ethmoidal sinuses or air cells

These consist of a mass of thin-walled cells of bone lined by mucous membrane. They open into the middle meatus and the most posterior of them into the superior meatus.

### The sphenoidal sinus

Within the body of the sphenoid are a pair of sinuses separated by a thin bony septum. They are below the floor of the pituitary fossa and behind the extreme upper and posterior part of the nasal cavity. They open into this region, the *spheno-ethmoidal recess*.

# THE NASAL CAVITY

The nasal cavities, right and left, are separated by the nasal septum. This is formed by the *vomer* and the *perpendicular plate of the ethmoid* (Chapter 16) and is completed by a plate of cartilage. The lateral wall is more complicated. Its basis is the medial surface of the maxilla (Fig. 23.3), which has a surprisingly large hole in

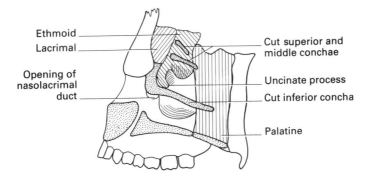

Ethmoid

Lacrimal

Opening of
nasolacrimal
duct

Cut superior and
middle conchae

Uncinate process

Cut inferior concha

Palatine

**Fig. 23.3** The bones that help to close off the huge hole shown in Fig. 23.1.

it that opens widely into the maxillary sinus. This opening is closed off almost completely by a number of other bones and by mucous membrane. The *palatine bone* was mentioned at the beginning of this chapter. Anteriorly, the fragile *lacrimal bone* contributes, as does the *inferior concha*, which, you will remember, is a separate bone, unlike the superior and middle conchae, which are part of the ethmoid (see Fig. 16.9). A curved prolongation of the ethmoid, the uncinate process, meets the inferior concha. The remainder of the opening is closed off by the covering of mucous membrane, except for the opening into the maxillary sinus.

The features of the intact lateral wall are shown in Fig. 23.4. The three *conchae* overhang three *meatuses*. Into the inferior meatus opens the *nasolacrimal duct*. This is a gap between the lacrimal bone, together with the inferior concha, and the medial surface of the maxilla, which leads from the medial angle of the orbit to the inferior meatus. It has been described in Chapter 21. The middle meatus in Fig. 23.4 has been exposed by removing the middle concha. The curved uncinate process of the ethmoid raises a ridge of mucous membrane that forms the lower boundary of a curved groove, the *hiatus semilunaris*. The frontal

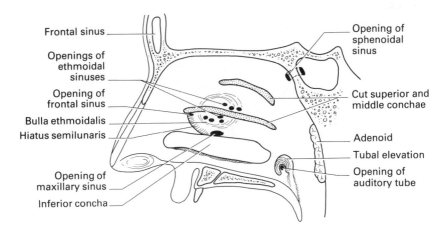

Frontal sinus

Openings of
ethmoidal
sinuses

Opening of
frontal sinus

Bulla ethmoidalis

Hiatus semilunaris

Opening of
maxillary sinus

Inferior concha

Opening of
sphenoidal
sinus

Cut superior and
middle conchae

Adenoid

Tubal elevation

Opening of
auditory tube

**Fig. 23.4** The lateral wall of the nasal cavity.

sinus opens into the upper end of this. Above the hiatus semilunaris is the *bulla ethmoidalis,* a bulge caused by the underlying ethmoidal air cells, and just under cover of the bulla is the opening of the maxillary sinus. The ethmoidal air cells open on to the surface of the bulla and into the superior meatus. The opening of the sphenoidal sinus is above the superior concha into the *spheno-ethmoidal recess.*

The floor of the nasal cavity is made up of the upper surface of the hard palate, i.e. the maxilla and palatine bones. The roof consists of the sphenoid and the ethmoid (cribriform plate). The uppermost part of the nasal cavity is lined by olfactory mucosa. The axons of its contained nerve cells form the olfactory nerves, which climb through the holes in the cribriform plate.

## THE AUDITORY TUBE

Also known as the *Eustachian* or *pharyngotympanic tube,* part of this is made up of a triangular plate of folded cartilage (Fig. 23.5). It extends from the anterior wall of the tympanic cavity to the nasopharynx, where the cartilage produces a semilunar bulge. The muscles attached to it are levator palati, tensor veli palatini and salpingopharyngeus, which have all been described in earlier chapters. The function of the tube is to equalize the pressures on either side of the tympanic membrane. On ascending a mountain, for example, as the atmospheric pressure decreases the tympanic membrane bulges outwards until the action of tensor veli palatini during swallowing opens the tube and allows air to escape from the middle ear. The membrane resumes its original shape with an audible click. On coming

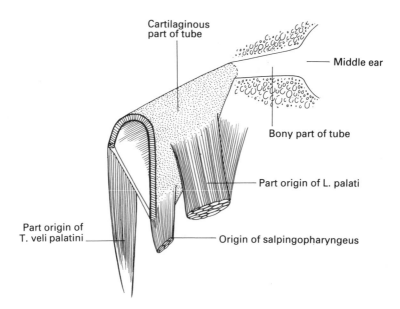

**Fig. 23.5** The auditory tube and the three muscles attached to it (cf. the styloid process, p. 343). The narrowest part of the tube is at the junction of the bony and cartilaginous parts.

down again, air passes up the tube. Unfortunately, the slit-like opening into the pharynx makes it easier for air to escape than to pass up the tube so that, if you fly with a cold, which narrows the tube because of swelling of the mucous membrane, it is easy to ascend but descent may be painful and rupture of the tympanic membrane can occur.

## THE MIDDLE EAR

This is a narrow cleft in the petrous temporal bone and has four walls, a roof and a floor. It may therefore be represented in a flat diagram (Fig. 23.6) by an opened-out box. It is separated from the external auditory meatus by the *tympanic membrane* (ear-drum) and medial to it lies the inner ear.

### The lateral wall

The major feature of the lateral wall is the *tympanic membrane*. It has three layers and the *handle of the malleus* is embedded in its middle layer, which is dense connective tissue. The outer layer of stratified squamous epithelium is continuous with the skin of the external auditory meatus. The inner layer of (partially) ciliated columnar or cubical epithelium is continuous with the epithelium of the middle ear. The handle of the malleus is directed downwards and posteriorly and has a *lateral process*, which produces a prominent elevation on the outer surface of the membrane. Above it is the *flaccid part* of the membrane between *anterior* and *posterior malleolar folds*. On inspection with an auriscope, the lateral process of the malleus can clearly be seen through the pearly-grey membrane, and the concavity of the membrane causes a reflection of the auriscope light, which spreads out downwards and forwards from the end of the handle. It is known as the 'cone of light'.

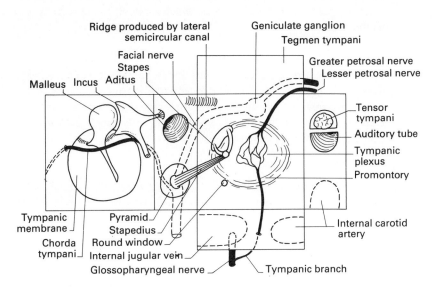

**Fig. 23.6** The middle ear cavity, opened out like a box.

The rounded *head of the malleus* is in that part of the tympanic cavity that is above the level of the membrane and this region is called the *epitympanic recess* or *attic*. The head articulates with the body of the *incus* by means of a minute synovial joint. This incus has a *short process* and a *long process*. The former is attached to a small fossa on the posterior wall of the canal by a tiny ligament and this forms a pivot round which the three ossicles move. The long process bends medially at its tip and articulates with the *stapes*.

## The posterior wall

The main features of the posterior wall are the *aditus* to the *mastoid antrum* and the *pyramid*. The aditus is an opening leading into a cavity in the petromastoid part of the temporal bone, which, in turn, leads downwards into a mass of mastoid air cells in the mastoid process. They can easily be seen in X-rays. The mastoid antrum lies directly deep to the *suprameatal (MacEwen's) triangle*, a depression that can be felt above and behind the auricle. The pyramid is a small volcano-like elevation below the aditus. From its tip emerges the *stapedius muscle* on its way to be attached to the stapes.

## The medial wall

Beyond the medial wall is the inner ear, and two components of this project so far laterally that they produce bulges in the medial wall. The largest of these is the first turn of the cochlea, which produces the *promontory*. High up in the angle between medial and posterior walls is a small ridge produced by the *lateral semicircular canal*.

Two openings in the medial wall lead into the inner ear. Above and behind the promontory is the *fenestra vestibuli* or *oval window*. This leads into the vestibule and is closed by the footpiece of the stapes. Below the promontory is the *fenestra cochleae* or *round window*, which is closed by the secondary tympanic membrane.

## Anterior wall

This has two openings, one above the other, separated by a thin shelf of bone. The upper opening is that of the *canal for the tensor tympani muscle* and the lower is the *auditory tube*. The muscle turns through a right angle after leaving its canal and is inserted into the neck of the malleus. It is supplied by the mandibular division of the trigeminal nerve.

## Roof

The roof is formed by the *tegmen tympani*, a thin plate of bone that separates the tympanic cavity from the middle cranial fossa and the temporal lobe of the brain.

## Floor

Beneath the floor, the bone is hollowed out to accommodate the *jugular bulb*, the expansion at the upper end of the internal jugular vein. In front of this, the first bend in the *internal carotid* is related to the floor and anterior wall.

## Nerves

The *facial nerve* approaches the tympanic cavity from the internal auditory meatus and therefore first encounters the medial wall. Here it bends abruptly backwards, forming the *genu*, which is decorated by the *geniculate ganglion*. The nerve then turns downwards in the posterior wall, usually in a bony canal but sometimes just under the mucous membrane. It finally emerges from the *stylo-mastoid foramen*.

The geniculate ganglion is a sensory ganglion and so does not contain synapses. It is made up of the cell bodies of pseudo-unipolar cells, like those of the spinal or trigeminal ganglion, and their peripheral processes convey taste sensation from the anterior two-thirds of the tongue via the chorda tympani.

The facial nerve has three branches in this region. The first is the *greater petrosal nerve*, which leaves the region of the genu to emerge from a foramen in the petrous temporal bone in the middle cranial fossa. Its course has been described earlier in this chapter. The second is the *nerve to stapedius*. The third is the *chorda tympani*, whose name tells you something about its course. It leaves the facial nerve in the posterior wall, ascends towards the posterior part of the tympanic membrane and then inserts itself between the layers of the membrane medial to the handle of the malleus. It enters the bone again at the front of the membrane and then descends to emerge through a minute foramen medial to the spine of the sphenoid.

The *glossopharyngeal nerve* also has a contribution to make. Its *tympanic branch* leaves the nerve just below the jugular foramen and enters a small canal between the jugular foramen and the carotid canal. This branch breaks up beneath the mucous membrane of the promontory to form the *tympanic plexus* and from this arises the *lesser petrosal nerve*. This emerges from the front of the petrous temporal lateral to the greater petrosal (Lesser is Lateral), passes (usually) through the foramen ovale and provides the secretomotor nerve supply to the parotid.

## THE INNER EAR

This will not be described in detail but some common errors will be pointed out. The first common misapprehension is that the *osseous labyrinth* can be illustrated in three dimensions. The osseous labyrinth is, in fact, a complicated *space* in the petrous temporal bone so that, just as one cannot draw the London Underground system, one can only illustrate a *cast* of the labyrinth, not the labyrinth itself. It comprises the *semicircular canals*, the *vestibule* and the *cochlea*. It contains the *membranous labyrinth*, comprising the *semicircular ducts*, the *utricle and saccule* (within the vestibule) and the *duct of the cochlea* (within the cochlea). The osseous labyrinth contains *perilymph* and the membranous labyrinth *endolymph*. The *vestibulocochlear nerve* divides into a *vestibular* division, supplying the utricle and saccule and the semicircular ducts, and the *cochlear nerve*, which runs up the central canal of the cochlea and supplies the organ of Corti in the cochlear duct.

The function of the semicircular ducts is to convey a sense of motion in any direction, and the utricle and saccule convey a sense of position in space. In other words, the ducts tell you where you are going and the utricle and saccule tell you where you are.

The semicircular canals and their contained ducts are placed in three planes, approximately at right angles to each other. The *superior* and *posterior canals* are vertically placed, the former being at right angles to the long axis of the petrous temporal and the latter along its long axis. The superior canal makes a bulge, the *arcuate eminence,* on the upper surface of the petrous temporal and the lateral canal, which is horizontal, makes a bulge on the medial wall of the middle ear. Since the long axes of the two petrous temporal bones are approximately at right angles to each other, the superior canal of one side is parallel to the posterior canal of the other. (Draw a diagram to convince yourself of this.)

In a cross-section through one of the turns of the cochlea, the membranous labyrinth (duct of the cochlea) can be seen to be triangular in shape. Above and below it are parts of the osseous labyrinth, the *scala vestibuli* and the *scala tympani*, which communicate with each other at the apex of the cochlea through an aperture called the *helicotrema* (Fig. 23.7).

When sound waves set up vibrations in the tympanic membrane, the malleus and incus rotate together about the axis of the short process of the incus. The stapes is therefore pushed backwards and forwards in the oval window and sets up vibrations in the perilymph. These travel up the scala vestibuli, through the helicotrema and down through the scala tympani, ending up at the secondary tympanic membrane. This sets up vibrations in the *basilar membrane* and, by a complex process, these are transformed into nerve impulses.

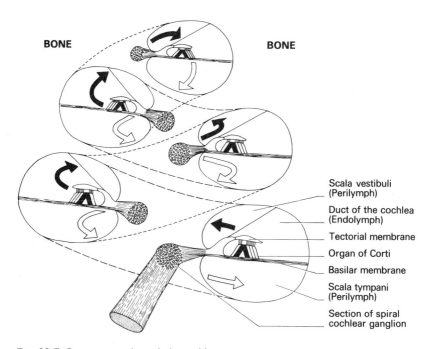

**Fig. 23.7** Cross-section through the cochlea.

# LYMPHATIC DRAINAGE OF THE HEAD AND NECK

This is not exactly ENT territory but it is convenient to deal with it in the last chapter of the head and neck section. There are many small groups of lymph nodes, each with its own name, in the head and neck region but it is pointless to break down the complex system of nodes too far since the groups are not discrete collections of nodes. A simplified system of the most important nodes will therefore be followed here.

There are two major groups into which all the other lymph nodes eventually drain. These are the *upper* and *lower deep cervical groups* of nodes and they are shown in the inset in Fig. 23.8. The upper deep cervical group is situated in the angle between the lower border of the mandible and sternomastoid, and also deep to sternomastoid itself. These are related to the internal jugular vein and the posterior belly of digastric and are therefore sometimes called the *jugulodigastric nodes*. One of them may be tender and enlarged in infections of the tonsil and is therefore often called the *tonsillar gland*. This group drains the head and upper part of the neck, either directly or via one of the outlying groups (Fig. 23.8). The second large group, the lower deep cervical group of nodes, is found in the posterior triangle in the angle between sternomastoid and the clavicle, lying partly deep to sternomastoid. They are often called the *jugulo-omohyoid* or *supraclavicular nodes*. They drain the lower part of the neck and also receive lymph from

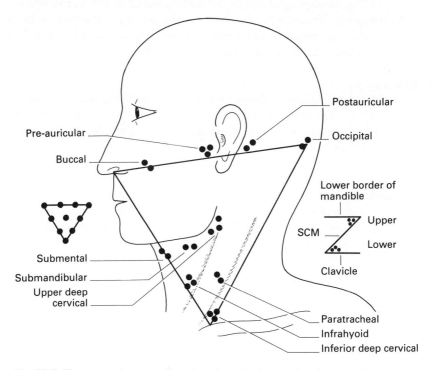

**Fig. 23.8** The principal groups of lymph nodes in the head and neck arranged as a triangle (inset, left). The inset (right) shows the two biggest groups into which the others drain. SCM = sternocleidomastoid.

the upper deep cervical nodes, from the breast and even from structures in the thorax and abdomen. One node on the left may be enlarged as a result of spread from a carcinoma of the stomach (Virchow's gland). The efferents from this group form the *jugular trunk*, which opens into the thoracic duct or the right lymph duct.

The lymphatic drainage of the head and neck is shown very diagrammatically in Fig. 23.8 and may be remembered as an inverted triangle of nodes in horizontal groups of four, three, two and one.

The uppermost group of four consists of the *occipital* (around the occipital vessels), the *postauricular* (or retro-auricular), the *parotid* (on surface of, and in the substance of, the parotid gland) and the *buccal*, on the surface of buccinator.

The second group of three comprises the *submental* (under the chin), the *submandibular* (embedded in the salivary gland of the same name) and the *upper deep cervical nodes* (already described). The submental and submandibular nodes receive lymph from the face and deeper structures, such as the tongue, and from the outlying groups, as shown in the diagram. They drain into the upper deep cervical nodes.

The third group of two comprises the *infrahyoid nodes*, lying in relation to the larynx on the thyrohyoid and cricothyroid membranes, and the *paratracheal*, lying between the trachea and oesophagus. They drain the structures in the middle of the neck and their lymph passes to the upper or lower deep cervical nodes.

The final group of one is the lower deep cervical group of nodes, already described. These nodes are situated appropriately at the apex of the inverted triangle (cf. the apical group of axillary nodes, p. 64) and all the lymph from the head and neck eventually passes through these nodes.

### The tongue

The tip of the tongue drains directly to the submental nodes. Further back, the anterior two-thirds of the tongue drains into the submandibular nodes, the lymphatics near the midline passing to both sides and directly to the upper deep cervical nodes. The posterior one-third drains directly into the upper deep cervical nodes.

### The larynx

The upper part, above the vocal cords, drains to the infrahyoid nodes and to the upper deep cervical nodes. The lower part drains to the paratracheal and inferior deep cervical nodes.

Fortunately, you do not have to remember details of the lymphatic drainage of the brain since there are no lymphatics inside the cranial cavity.

# Index